THOUSAND OAKS LIBRARY
01024 9806

D0743190
CARD

THE NEW ENCYCLOPEDIA *of the* Saltwater Aquarium

THE NEW ENCYCLOPEDIA

of the

Saltwater Aquarium

Greg Jennings

FIREFLY BOOKS

Published by Firefly Books Ltd., 2007

First printing

Publisher Cataloging-in-Publication Data (U.S.)
Jennings, Greg.
The new encyclopedia of the saltwater aquarium / Greg Jennings.
[304] p. : illus., photos. (chiefly col.) ; cm.
Includes bibliographical references and index.
Summary: An illustrated encyclopedia covering all aspects of the setup, maintenance and stocking of a saltwater aquarium for the home, including profiles of suitable fish and invertebrates.
ISBN-13: 978-1-55407-182-1
ISBN-10: 1-55407-182-8
1. Marine aquariums. 2. Marine aquarium fishes. 3. Marine invertebrates as pets. I. Title.
639.34/2 dc22 SF457.1.J466 2007

Library and Archives Canada Cataloguing in Publication
Jennings, Greg, 1955-
The new encyclopedia of the saltwater aquarium / Greg Jennings.

Includes bibliographical references and index.
ISBN-13: 978-1-55407-182-1
ISBN-10: 1-55407-182-8
1. Marine aquariums. . Marine aquarium fishes. 3. Marine invertebrates as pets. I. Title.

SF457.1.J45 2007 639.34'2 C2006-902790-0

Published in the United States by
Firefly Books (U.S.) Inc.
P.O. Box 1338, Ellicott Station,
Buffalo, New York 14205

Published in Canada by
Firefly Books Ltd.
66 Leek Crescent,
Richmond Hill, Ontario L4B 1H1

The Brown Reference Group
(incorporating Andromeda Oxford Limited)
8 Chapel Place, Rivington Street,
London EC2A 3DQ
www.brownreference.com

For The Brown Reference Group plc:
Editorial Director: Lindsey Lowe
Project Editor: Graham Bateman
Editor: Virginia Carter
Design: Steve McCurdy

Printed in China

Photographs
All photographs © Hippocampus Bildarchiv/www.Hippocampus-Bildarchiv.com
Except:
© Photomax Specialist Aquarium Picture Library: 27, 54, 57*t*, 74, 148, 151, 182, 187, 254*t*, 254*b*, 267, 271, 272, 273, 274, 275, 276, 277, 278l, 278r, 281, 282, 283*t*, 283*b*, 287, 288, 289.
Courtesy Interpet 34, 37, 40, 42*b*, 43*t*, *b*, 51*tl*.

Pages 1: Cherubfish (*Centropyge argi*).
Pages 2–3: Shoal of pajama cardinalfish (*Sphaeramia nematoptera*).
Opposite: Yellow-fin goatfish (*Mulloidichthys vanicolensis*) on a Red Sea reef.

Contents

INTRODUCTION

Interest in saltwater aquaria has grown greatly over recent years for a number of reasons. Firstly, there is the obvious fact that these fish are interesting to keep and, even if they are not especially colorful in all cases, they are frequently bizarre in appearance.

There is also tremendous scope within the hobby. You may decide to opt for a species-only setup, perhaps containing just an individual fish, which you can tame to the extent that it will come to recognize you and take food from your fingers, or a group that will live together in harmony. Alternatively, you may prefer to build a mixed community aquarium, where a number of different species can be housed together.

The other possibility is to keep not just fish but also a range of invertebrates in what is often described as a reef tank. This can be a microcosm of reef life and can have a similar structure, since the tank is decorated with natural rock. Although this type of aquarium places more demands on the fishkeeper, the end result can be very striking. Colorful crustaceans and fish create movement here in the company of live corals, sea anemones, starfish, and sponges, not overlooking other reef invertebrates such as clams.

As nature programs on television have brought the beauty and fascination of the reef directly into people's homes, it is not very surprising that there has been a growth of interest in recreating this environment in miniature. It is perhaps also not entirely unexpected that some of the keenest saltwater aquarium enthusiasts are also enthusiastic divers who have experienced the world's reefs at first hand.

▶ *Lined surgeonfish (*Acanthurus lineatus*) with accompanying common cleaner wrasse (*Labroides dimidiatus*).*

The Natural History of Saltwater Fish

Fish occupy every watery niche on the planet from freshwater through brackish water to saltwater. Marine fish live in temperate seas, tropical seas, and arctic seas. Some roam open oceans, while others spend all their lives in restricted lagoons and on coral reefs. Unlike their freshwater relatives, they live in a concentrated medium that contains many dissolved substances, the commonest of which by far is salt, or sodium chloride. The presence of so much salt makes it difficult for any organism to live, but fish have managed to adapt in order to survive.

▶ *A typical reef aquarium with a variety of fish.*

What is a Saltwater Fish?

There are approximately 22,000 recognized species of fish. It is not difficult to identify fish because they can usually be distinguished by their fins or their scales. They are vertebrates, and yet they generally need to live in water in order to breathe. They rely on gills to extract oxygen from the water. Fish depend on their fins to swim. In spite of considerable variations in body shape—from the elongated and rounded body profile of eels to the vertical form of seahorses—the basic design of the fins tends to be similar. There are individual variations—for example, some fins are enlarged for display purposes.

When swimming, a fish's major propulsive thrust tends to originate from the tail, or caudal fin, although a fish's speed depends more on its body shape. Those with a rectangular body shape, such as boxfish, are much slower at swimming through the water than damselfish, for example, simply because they lack a streamlined body shape.

Fish with highly elaborate fins may also have difficulty in swimming, but they may use their fins to help disguise their presence, turning them into ambush, rather than active, predators. A number of fish have powerful teeth in their mouths, which can be used for rasping algae, breaking open the hard shells of mollusks, or seizing live prey.

The Senses

Reef fish tend to have good eyesight, and the water in this part of the ocean is generally clear. It is thought that they can see colors even more vividly than we can, which may explain why they are often brightly colored. The position of the eyes on a fish's head gives a clue to its behavior. Predatory species such as lionfish have eyes that are located relatively centrally, so they can see ahead and target their prey. Other fish (in fact, the majority of fish) have eyes positioned on the sides of the head, giving them a wider field of vision and enabling them to spot predators approaching from the side.

Sound travels through water in the form of pressure waves, and fish have evolved a particular way of detecting these waves. They rely on jelly-filled canals known as lateral lines, which extend down each side of the body in the midline. Tiny pits on the lateral lines serve as detectors. By sensing pressure waves that reverberate off any solid object, they provide early warning of danger as well as allowing the fish to swim in the dark without colliding with rocks or other

◄ ▼ *The eyes of fish may be centrally placed, as in predatory fish such as Russell's Lionfish (Pterois russelli)—left—or on the sides, as in the African Pygmy Angelfish (Centropyge acanthops)—below.*

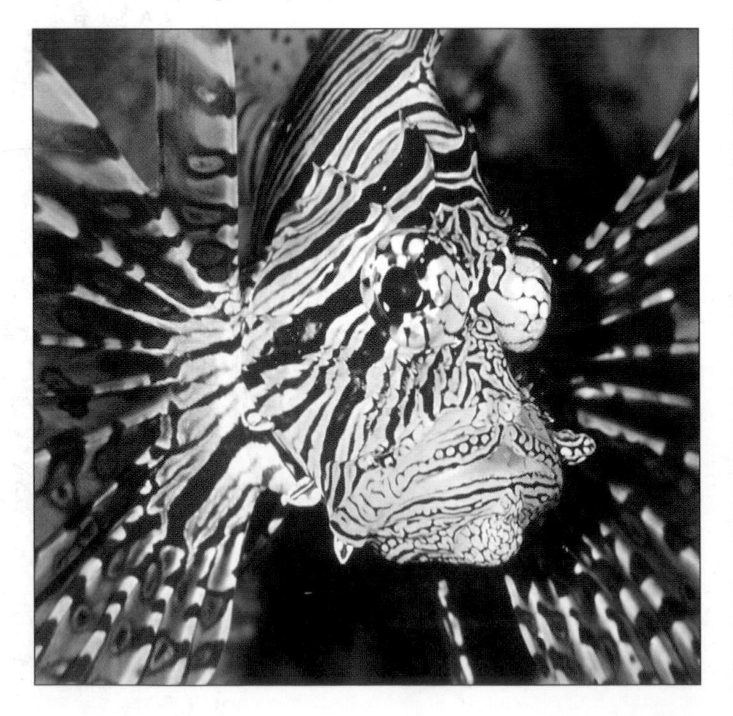

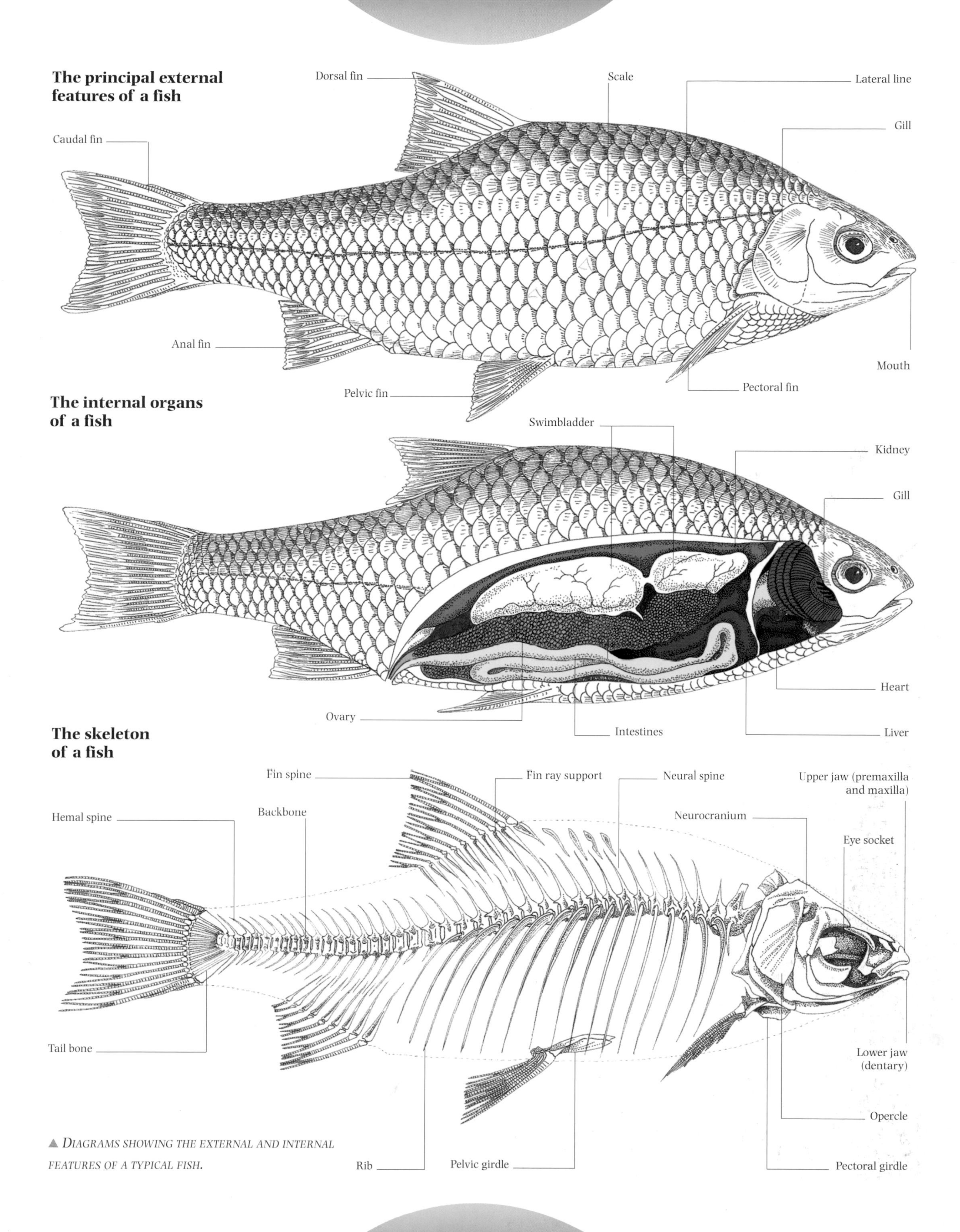

▲ *DIAGRAMS SHOWING THE EXTERNAL AND INTERNAL FEATURES OF A TYPICAL FISH.*

objects. Water is a denser medium than air, and sound travels more quickly and farther through the marine environment. Fish can communicate with each other using sound as well as more evident visual displays.

Fish possess a much more acute sense of smell than humans, which enables them to home in on food. They can use this sense to detect chemical messengers called pheromones that are released into the water by other fish to indicate when spawning is imminent.

Respiration

Fish need oxygen to survive. Saltwater contains only 4 percent of the amount of oxygen present in air, so a marine fish's respiration system has to process large amounts of relatively oxygen-poor water in order to take up enough oxygen. The fish sucks water in through its mouth and it passes over the gas-absorbing surfaces of specialized organs called gills. Oxygen is absorbed into the bloodstream through the gills, and carbon dioxide passes out from the gills at the same time, by a process known as diffusion. The water leaves the body through a flap—the operculum—that covers the gills.

Air is contained within the fish's body in an organ known as the swimbladder. The fish adjusts its position in the water by moving air in and out of this organ as required, thereby regulating its buoyancy. If a fish tends to hang at an abnormal angle in the water or appears to have difficulty swimming, this is usually indicative of a swimbladder disorder.

Osmoregulation

Although in many respects freshwater and marine fish are similar, there is one significant way in which they differ physiologically, relating directly to the different environments in which they are found. Whereas a saltwater fish has to conserve freshwater in its body, a freshwater fish needs to conserve salts.

Osmosis is the process whereby water moves from a dilute solution to balance out a more concentrated solution, through a semipermeable membrane, while salts move in the other direction. A fish living in the sea, surrounded by a higher concentration of salt than exists in its body, is under constant threat of losing water from its body by osmosis and becoming dehydrated.

Unwanted salts can enter its body through the gills and in the saltwater it swallows, as well as in its food. Saltwater fish have special cells in the gills that actively remove salt. They can also selectively absorb freshwater alone from the intestinal tract, and they produce a highly concentrated urinary output to conserve water.

▼ *Osmoregulation in marine fish. The skin acts as a biologically active, semipermeable membrane, transporting molecules from inside the body to the denser saltwater surrounding it.*

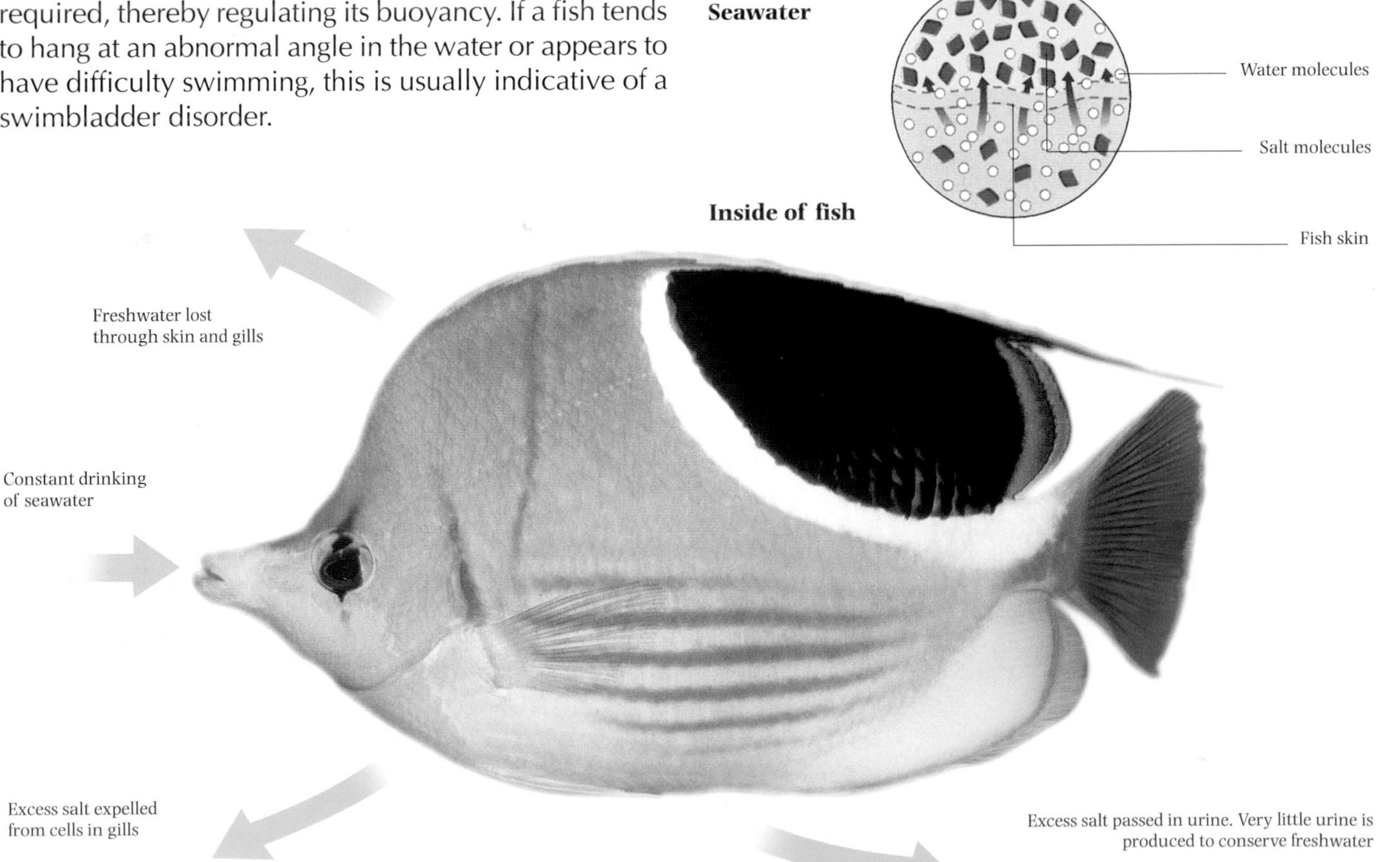

NAMING FISH

The way in which fish, marine invertebrates, and all other organisms are divided into groups forms the basis of the science known as taxonomy. In the past this exercise relied heavily on morphological similarities in appearance. Increasingly, relationships between organisms are being established on the basis of similarities in their DNA—the genetic material present in every living cell of an organism's body.

The classification process uses a series of so-called ranks. All fish are vertebrates and, at the most general level, they can be divided into two groups. Firstly, there is the older lineage of cartilaginous fish, which includes sharks and rays. The majority of fish, however, are contained in the second group—known as bony fish, or teleosts, based on the structure of their skeleton.

HOW FISH ARE CLASSIFIED

Taxonomy has been likened to a tree with a series of branches, becoming progressively more specific. It is the so-called lower ranks that are generally of most significance to aquarists. The fish in the following section are arranged initially by family groups.

Families themselves may sometimes be split into subfamilies or other subgroups. Beneath this level in the taxonomic tree is the genus, a group of closely related species—the next level of classification. The lowest order in the taxonomic tree is the subspecies, a subdivison of a species. One species can have two or more subspecies, but subspecies are unusual in saltwater fish. At this level the differences in appearance between individuals can be almost indistinguishable.

TAXONOMY IN ACTION

The tomato clownfish shows how the classification system works in practice. The group of bony fish is divided into about 60 different orders. The tomato clownfish belongs to the very large order Perciformes. (In fact, the vast majority of saltwater fish kept by amateur aquarists belong to this order.)

Orders are then divided into a number of families. Family names can always be recognized because they end in -idae. All clownfish and relatives belong to the family Pomacentridae. In some instances families are divided into subfamilies, in which case the letters -inae are used at the end.

Very closely related fish are placed in the same genus, in this case the genus *Amphiprion*. Finally, a species is given its unique name, which is a combination of the genus title and a species epithet. The tomato clownfish is *Amphiprion frenatus*. From the level of genus downward, the names are italicized and the genus name always begins with a capital letter.

A summary of the classification of the tomato clownfish is given below.

Order Perciformes—Perchlike Fish

Family Pomacentridae—Anemonefish and Damselfish

Genus *Amphiprion*—Clownfish

Species *Amphiprion frenatus*—Tomato Clownfish

▶ *TOMATO CLOWNFISH (AMPHIPRION FRENATUS).*

VALID NAMES AND SYNOMYMS

Whenever a new species is discovered, a detailed account of the so-called type specimen—in other words, the first example to be described—is published in a scientific journal and it is accorded a recognized scientific name. Although a species of saltwater fish or an invertebrate is often known under several common names, there will be only one valid scientific name.

It is possible, however, for fish to be reassigned to a different species or genus on occasion. This may be the result of further study after the initial report. Alternatively, as often happens, it could be that the same fish has been described several times before it was recognized as a single species. In the accounts that follow in this book, other familiar scientific names (synonyms) are given. For example, the tomato clownfish has in the past been known under the synonyms *Amphiprion macrostoma* and *A. polylepis*.

CORAL REEFS

Coral reefs are found in the oceans around the world, occurring in tropical waters on each side of the equator. Some, such as the Great Barrier Reef off Australia's east coast, have become very well known and are proving to be major tourist attractions. They draw visitors from all over the world, keen to see the unique life forms ranging from tangs to turtles that can be encountered in this type of environment. Today it is not only divers who can appreciate this aquatic landscape—glass-bottomed boats and underwater viewing observatories give other people the opportunity to behold the beauty of the reef with great clarity and absolutely no need to get wet.

CORAL REEFS AND THE SALTWATER AQUARIUM

Since the 1970s when marine fishkeeping began in earnest, a limited trade in marine creatures—not just fish, but invertebrates too—has developed from reefs in some areas of the world, notably the Philippines and Indonesia as well as from the vicinity of the Red Sea and around Hawaii. This development has enabled the fascinating ecosystem of the reef to be appreciated at first hand in home aquaria by a wider audience.

The impact of this trade on the reef has been damaging in some areas in the past, largely because of the

use of unsuitable and environmentally damaging methods of collecting fish. Chemicals, specifically sodium or potassium cyanide, were used to drug all the fish in an area so that they could be collected easily.

This type of collecting is totally counterproductive because it causes widespread mortality. Those fish that survive the acute phase of poisoning often suffer permanent harm and their life span is dramatically shortened. The chemicals can affect the swimbladder, causing buoyancy problems and limiting the fish's ability to swim properly. Thankfully, catching fish in this way has largely ceased as a result of pressure from the trade, and emphasis is now being placed on reef management systems that are sustainable and that contribute to the long-term health of the reef and its inhabitants.

Fish for marine aquaria are now caught with nets, along with invertebrates. For many communities in far-flung areas of the world, this trade provides a vital influx of foreign capital and creates employment for local people, ranging from the fisherman themselves to those who acclimatize and care for the fish on land and organize their shipping around the world.

There are areas of the reef where collecting is outlawed to prevent overfishing. A feature of most marine fish is their vast reproductive capacity, which means that it is quite possible to catch young fish without causing any long-term harm to the population of the species or the reef itself.

Coral reefs also offer divers the opportunity to see and experience one of the most diverse and striking natural environments on the planet. This so-called ecotourism also brings in money, which effectively adds to the value of the reef, and this growing interest should encourage its future protection. It is very important, however, that associated developments catering for visitors to the reef are carefully controlled.

◀ *Red Sea coral reef with various corals and Jewel Lyre-tailed Anthias, or Sea Goldies (Pseudanthias squamipinnis).*

▲ *A shoal of Dory Snappers (Lutjanus fulviflamma) patrolling the surface of an Indian Ocean coral reef.*

Reef Structure

It is misleading to think of reefs as just shallow areas of water. At the edge of some reefs is an area described as the "wall"—a sudden near vertical falloff resembling a mountainside. One of the most spectacular examples can be found off the coast of Grand Turk, one of the Turks and Caicos Islands in the Caribbean. The reef here lies under a depth of about 50 feet (15 m) of water before ending abruptly. The "wall" reaches down to the seabed 10,000 feet (3,000 m) below.

Reefs are often divided into different types based on their structure and the way in which they formed. The most conspicuous, the ones that are closest to the shoreline, are described as fringing reefs. They effectively extend out from the shore over a period of time. The active area of the reef, where new coral growth is occurring, gradually spreads away from the shore over many years. The presence of the reef creates a calmer area of water, known as a lagoon, between this area of coral and the shore behind.

On the seaward side of the reef, where the depth of the water increases dramatically, there is another pronounced wall. The reef can extend farther from the shore only if debris falls down and provides a shallower base which the coral can colonize. Otherwise, there would not be sufficient light penetrating down into the depths of the ocean to support the growth of coral. In favorable areas however, in suitably shallow water, fringing reefs can extend for a distance of 1,640 feet (500 m) from the shoreline.

CLASSIFICATION OF REEFS

The basic classification of different reef types was first proposed by the scientist Charles Darwin in 1842 and remains valid today. Aside from fringing reefs, he described barrier reefs, although the way in which they form is more controversial. Changes in sea level are believed to be a factor in their development. Such reefs probably started to develop millions of years ago, originally forming when the sea was much lower than it is today. The growth of the coral has kept pace with the rising sea level, and it has clearly risen up in height, creating a protective barrier in front of the coastline.

Changes relating to Earth's geological movements may have been involved in some cases. A barrier reef may develop from what was once a fringing reef if the land sinks into the ocean, creating a greater distance between the shoreline and the reef. The reef then divides into two—the barrier reef is located in deeper water, while the fringing reef lies closer inshore.

Ocean currents, which are linked to Earth's rotation, have also played a part in the location and formation of barrier reefs. For example, the sea off Australia's east coast is warmer than the oceans around the west coast. This has favored the development of the Great Barrier Reef on the east side of the continent.

One of the most characteristic types of reef in the Pacific is the coral atoll, represented on a map by places such as Bikini Atoll near the Marshall Islands. Coral atolls typically occur in deepwater areas and have a circular shape, having developed around what may have been a small island at the outset. Alterations in sea level and land height resulting from geological movements can help shape these structures.

The Maldives in the central Indian Ocean are a chain of coral atolls that have formed around the tops of a ridge of extinct volcanoes. They are among the many coral atolls under threat as huge volumes of water are being unleashed from the polar ice cap into the oceans, raising the sea level.

Since Darwin's time, other subdivisions of reef structure have been proposed. There are patch reefs, for example. They occur in sandy lagoons in shallow water and look like small hills. Larger reefs with a hill-like structure are known as platform reefs, while at the other extreme, small outcrops are described as coral knolls, (also known as bommies in Australia).

The actual structure of the reef can be divided into distinct areas, too. This pattern of "zonation" has a significant impact on where particular types of reef inhabitant are to be found. A characteristic feature of the coral atoll's structure is the reef face that surrounds it and slopes down into deep water. In areas where the current is strong there may be little sign of life, but on the calmer lees side coral growth on the reef face is likely to be evident, particularly near the surface.

Probably the best-known example of zonation, however, is the central lagoon, which is a calm area of water in the center of a coral atoll. It is surrounded by sand that has been deposited there, creating dry areas of land. They are still close to sea level, which means that any human settlements on these islands are now prone to being overrun by the sea as the result of rising sea levels.

THREATS TO REEF LIFE

Having suffered previously from tourism and over-collection for the aquarium trade, these unique habitats are now facing the additional human-made threats of pollution and global warming. The signs are that both phenomena are already starting to damage the very structure of reefs by turning the sea into a more acidic environment. This is largely as a direct consequence of the increased output of carbon dioxide. A greater volume of the gas is now entering the oceans and raising the level of carbonic acid.

There are real fears among scientists that this process will cause a mass extinction of marine life, especially of corals and other creatures, such as mollusks, that protect themselves using calcium carbonate. It has been estimated recently that within a century the pH of the water could fall from its current value of approximately 8.2 to as low as 7.7. This increasing acidity will attack the calcium carbonate, breaking it down more quickly than it can be replaced.

There are also natural threats to the reef, such as the crown-of-thorns starfish (*Acanthaster planci*). It feeds on the delicate coral polyps and can multiply at a prodigious rate. The starfish can dramatically reduce coral cover, resulting in a serious disturbance to the entire ecosystem. Its devastating effects can be seen across large areas of Australia's Great Barrier Reef.

A coral reef requires very stable environmental conditions if it is to thrive, in terms of water temperature, sunlight, and water clarity. If the conditions needed for coral growth cannot be sustained, then the entire reef and its inhabitants will die.

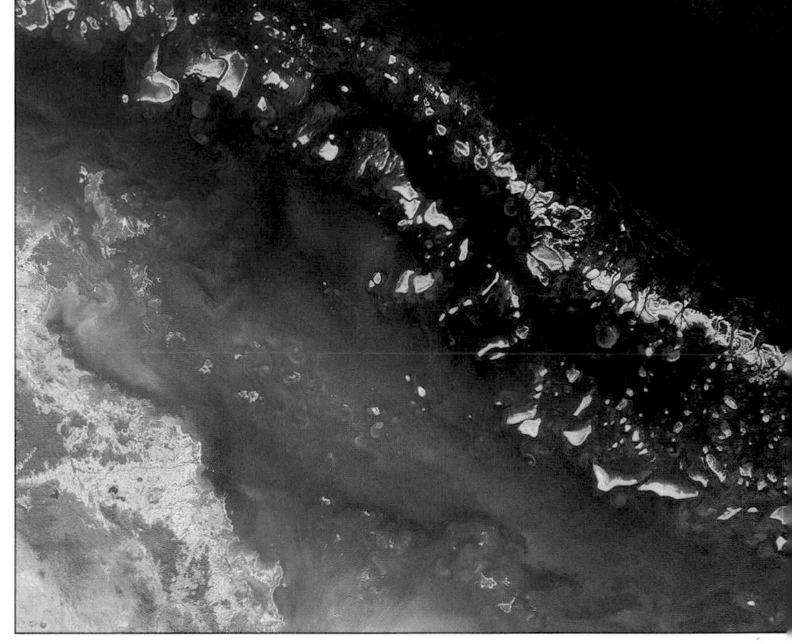

▲ *Image from space of part of the Great Barrier Reef off the coast of Queensland, Australia.*

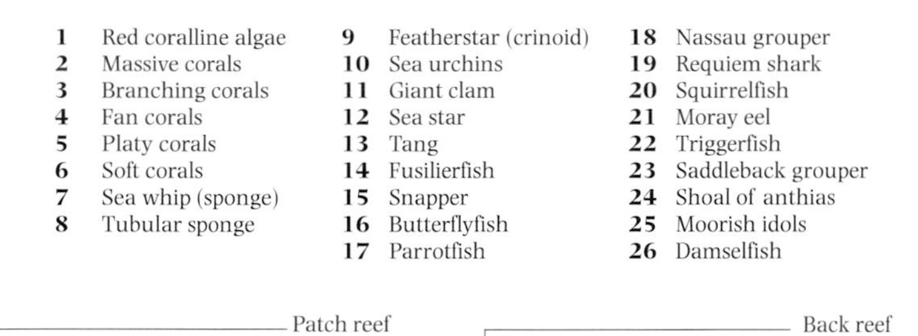

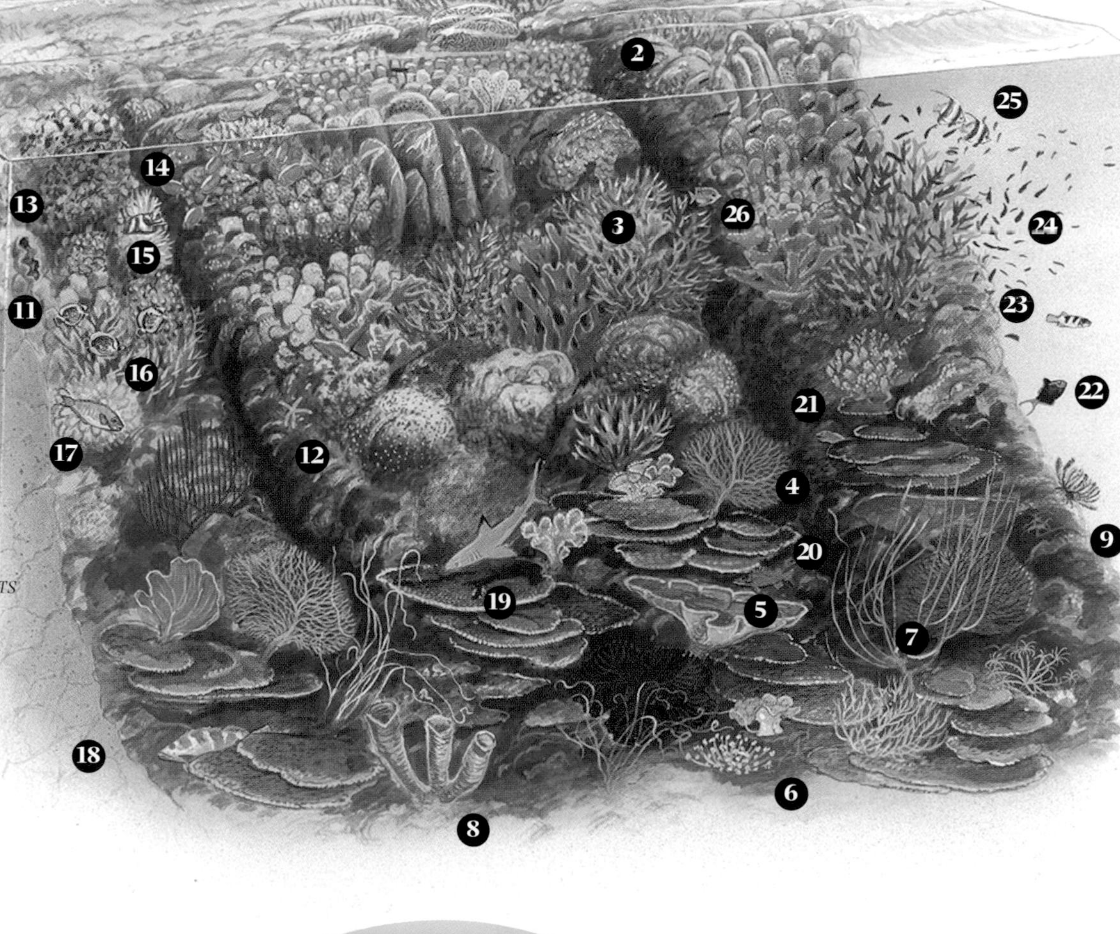

▲ *Diagram showing the formation and components of a typical coral reef. A coral reef is a complex habitat comprising hard and soft corals, sea fans, echinoderms, sponges, polychaete worms, and a kaleidoscope of different fish species.*

Saltwater Aquarium Setup and Maintenance

Recreating a marine environment in aquaria would not be possible without the technology to replicate what is generally regarded as the most stable ecosystem on earth. Coral reefs are not subject to significant shifts in temperature or to water changes that beset many freshwater fish in their natural habitats. This means that reef-dwelling fish and invertebrates need to be kept under tightly prescribed conditions. They cannot adapt well outside these parameters, and their care is potentially more demanding as a consequence.

Many people who take up the saltwater aquarium hobby have experience in keeping freshwater fish. This is not a prerequisite for maintaining a saltwater aquarium successfully today, because technology has simplified the task greatly. What is essential, however, is the commitment to monitoring conditions on a regular basis in order to avoid any catastrophic changes in, for example, water quality, which must be avoided at all costs.

The most challenging time in terms of keeping saltwater fish and invertebrates is the initial period—which lasts for two months or so—before the system has stabilized. Particular attention to detail at this stage is vital to prevent problems from arising.

▲ *Reef aquarium with corals, sea anemones, and fish.*

Early Beginnings and Modern Technology

Saltwater fishkeeping is a relatively new hobby that began in the mid-1800s. It started essentially with the public aquaria that opened and were visited by members of the public at many British seaside towns. Visitors could marvel at seeing native sea life for the first time at close quarters—fish were housed in tanks filled with water that was pumped in from the sea outside these buildings.

The risk of pollution and the threat of introducing illness to a tank mean that marine aquarists today do not rely on seawater to create the conditions prevailing on the reef. They use specially formulated sea salts added to ordinary freshwater in the correct proportions.

▲ *The technical equipment required to run a saltwater aquarium efficiently in order to maintain the correct water conditions is exemplified by this view of the kit below an aquarium.*

No matter how far away you are from the sea, it is possible to replicate the water requirements of the reef.

Interest in saltwater fishkeeping is centered largely on fish originating from coral reefs rather than fish from more temperate waters, although there is no reason why such species cannot be maintained successfully, along with invertebrates. It is worth bearing in mind, however, that many such species are likely to outgrow the typical aquarium.

Water movement on the reef ensures that biological pollutants from the fish's waste do not accumulate in the water—unlike in the confines of an aquarium where they will be detrimental to the health of the occupants unless they can be removed efficiently from the water. This has now been made possible by a range of highly effective filtration systems and other types of equipment, such as ozonizers, that help maintain high water quality by killing off pathogens that could harm the tank occupants.

Accurate heaters, combined with thermostats, create what are known as "thermostat" units to maintain water temperature, while thermometers reveal the precise figure. Ways of monitoring the environment of the saltwater aquarium accurately have contributed to the growth of the hobby, since they ensure that conditions remain favorable.

Externally, tank design has also played a part. In the early days of the hobby there were simply iron-framed tanks held together by putty. The frame itself was at risk of corrosion by the salty water. Today, however, there are frameless glass tanks or acrylic designs to choose from, both of which provide excellent visibility of all the tank occupants.

Lighting is important, not only in an aesthetic sense, helping highlight the aquarium occupants, but in many cases also for their well-being. It can encourage the growth of marine algae in the aquarium, a vital component in the diet of many fish.

You can now set up a saltwater aquarium that corresponds to the reef habitat, confident that reliable equipment is available to maintain it adequately. What is more, an ever-increasing range of fish foods has simplified the task of caring for the different types of fish kept in these surroundings. The diets of invertebrates can also be easily catered to.

Starting Out

The first decision you need to make is what type of setup interests you and whether you want to keep only fish or invertebrates as well. You will then be able to design an aquarium that matches their needs. Some saltwater species are much more secretive than others, for example, and need plenty of retreats in order to feel secure in their surroundings. The decor within the aquarium is therefore not just for appearance, but should first and foremost reflect the needs of the aquarium occupants.

A poorly furnished tank will place the occupants under stress, which can be manifested in various ways. Nervous species may refuse to eat. As a result, they will lose condition, which in turn is likely to lead rapidly to signs of illness. Weaker individuals are likely to be subjected to bullying, which can prove fatal as well.

There may be aspects of the behavior of certain fish that are of particular interest to you. Although aquarium breeding of saltwater fish is in its infancy compared with the freshwater hobby, increasing numbers of some species are now being bred successfully in aquaria. Notable examples are clownfish—more than 30 different species have been reared successfully in aquaria—as well as seahorses, although the latter are much harder to maintain and are not recommended for your first saltwater tank.

Another important consideration when starting out is the likely adult size of a particular fish. Species can vary in size from just a few inches to more than 18 inches (45 cm) in length. Although in general fish will not grow as large in aquarium surroundings as they do in the wild, this remains an important consideration, firstly in terms of the size of tank you need to buy and secondly in terms of possible companions. You may find, for example, that if one fish rapidly outgrows others that share its accommodation, it will end up eating some of its companions unless they are separated before this stage is reached.

▼ *The Tomato Clownfish (Amphiprion frenatus) is one of the few examples of saltwater fish successfully bred in aquaria.*

Tank Size

It is usually false economy to buy a small saltwater aquarium. Clearly, if the fish are likely to outgrow their surroundings, you will soon need to replace the aquarium with a larger one. In addition, you will require a new lighting and filtration system, which will add to the expense. If you buy a bigger tank to start with, this additional expenditure will be unnecessary.

Giving the aquarium occupants extra space will be beneficial too, especially during the critical initial period when they are settling into their new surroundings. It is, in fact, much easier to maintain a larger aquarium than a small setup, because the greater volume of water in the tank means that any deterioration in water quality is less likely to have an immediate serious effect on the fish and other marine life.

As a minimum, aim for an aquarium that measures 36 inches long by 12 inches wide and 12 inches deep (91 x 30 x 30 cm), for which standard-sized hoods and lighting equipment are readily available. A large surface area is important, because that is where oxygen diffuses into the water and carbon dioxide leaves. As a rule, allow 1 inch (2.5 cm) of fish length per 2.4 U. S. gallons (2 Imp. gal/9 l) of water. When choosing fish for the aquarium, it is important that you have some idea of how large they are likely to grow. Bear in mind that most saltwater aquarium fish are of a similar size when young, and that juveniles predominate in the trade. The range of sizes that can develop subsequently, even among related species, is unlikely to be apparent at the time of purchase.

◀ *An ultramodern plastic saltwater tank.*

Choosing the Tank

There is no difference today between saltwater and freshwater tanks, and they can be used interchangeably. Glass aquaria remain popular, and their panes are held together by a special silicone sealant. It has a slight elastic quality that expands to absorb the water pressure when the aquarium is full, preventing the joints from leaking (provided they are adequately covered with the sealant). Glass tanks offer good visibility but they have the disadvantage of being heavy and cumbersome to move. They can be easily damaged, particularly at the corners in the case of frameless designs that have no plastic molding around the edges. Never be tempted to stand a glass tank on its end, since tipping the weight of the tank back to lift it up is likely to squeeze the glass at the corners, causing what are effectively compression fractures. Always lay the tank down horizontally, and be sure to lift it up carefully from beneath.

Glass tanks can be constructed in a range of sizes, but larger designs are likely to require supporting struts running across the top from front to back. They are also made of heavier sheets of glass. The majority of glass tanks are rectangular in shape, although triangular designs can be obtained to fit into the corner of a room.

Acrylic aquaria offer much greater flexibility in terms of design compared with those made of glass. The early acrylic aquaria gained something of a bad press, because they could be scratched easily and tended to discolor over time, taking on a yellowish hue. Recent advances in manufacturing techniques, however, have helped reduce these problems significantly, and acrylic tanks are now available in a wide range of shapes, although it may not be easy to obtain suitable lighting equipment for all shapes. Acrylic tanks have a significant advantage over their glass counterparts because they are much lighter and therefore more easily and safely handled when necessary.

Where to Site the Aquarium

Where you place the aquarium in your home is an important consideration from the practical standpoint of setting up the tank and maintaining it, as well as enjoying it. The best location will obviously be a room in which you spend long periods of time, so that you can enjoy watching the fish swimming around and observe the behavior of any invertebrates present. Avoid siting the tank in a hallway, partly for this reason, but also because it is more likely to be subjected to temperature variations that will affect the occupants.

Within the room select a spot near a power point to minimize the need to trail electrical cables around the room. Do not place the aquarium next to a television or a music system, because fish are very sensitive to

▲ A SALTWATER FISH-ONLY TANK WITH THE APPEARANCE OF A REEF SETUP. BUT THE CORAL IS CURED AND DEAD AND THERE ARE NO INVERTEBRATES.

vibration, and the output from such sources could disturb them. It is also vital not to position the aquarium in front of a window, exposed to direct sunlight. On hot days the sun's rays will be magnified by the glass and are likely to cause the water temperature in the aquarium to rise significantly, with potentially fatal results for the occupants. Increased light exposure is also likely to encourage algal growth, which may start to take over the tank, detracting from its appearance. Although a degree of algal growth can be beneficial, it is better to regulate it by means of artificial lighting.

The height of the tank in the room is also significant. Position it so that you can see the occupants easily when you are sitting down. This is also likely to be a convenient height at which to service the tank easily, a task that will need to be carried out regularly. You may have a suitable chest or similar piece of furniture that can be used as a base on which to stand the aquarium. Alternatively, you can buy a simple stand or a cabinet designed to match the dimensions of the aquarium.

Cabinets are available in a wide range of styles, so it should not be difficult to find one that blends well with the decor in your room. One of the major advantages of a cabinet over a stand is that essential equipment—even a filtration system—can be kept out of sight beneath the tank.

SECOND-USE AQUARIA

Established saltwater aquaria are sometimes offered for sale, especially in local newspaper advertisements or on the Internet. A second-hand system can provide a convenient shortcut to setting up your own tank, but it is vital for the welfare of the marine life that you follow some basic instructions. You will need sealable buckets and water-carrying containers, because you will not want to throw the water away. It is very important, particularly if you are living in a temperate area, to be able to strip down the tank and move the fish with minimal delay, so they do not become chilled.

Start by siphoning the aquarium water out into the buckets. Special aquarium siphons are available for this purpose—the water flows through the tubing into the

▲ *A glass-constructed reef tank.*

buckets positioned at a lower level. Once you have partially filled the buckets, catch and transfer the fish into them. It is also important to keep living rock submerged, along with any invertebrates. The remainder of the water can then be siphoned off into the water containers. Once the aquarium is empty of water and rockwork, it should be possible to transport it without having to remove the substrate and disturb the filtration bed, although this may be necessary if the tank is still too heavy to lift easily.

When it is ready to move, rest the tank horizontally in the vehicle, wedging it in place with blankets and towels so there is no risk of it sliding around on the journey. Seal the buckets and place them in cardboard boxes in footwells of the vehicle, ensuring they will not tip over. Transport the rest of the water in the water containers, but don't forget to check that the lids are screwed on tightly before moving them.

Once you arrive home, set up the tank straightaway. You will need to tip in the water with care, onto a plate placed on top of the substrate to avoid disturbing the base. Once the tank is about half full, add the living rock—it is important that it is left out of water only for a very short period of time. Ensure that rocks and other tank decor are firmly in place, including electrical equipment such as a heaterstat and power filter if you are using these items, but do not connect them to the power supply at this stage. Finally, top up the tank with the remainder of the water, pouring it slowly out of the containers. Once everything is complete, transfer the fish to the tank, floating them in gently from their traveling containers.

Only then should you switch on the heater and filter. The water temperature will almost inevitably have dropped a little, but hopefully the fish will adjust to this slight dip without any problem—assuming they have been well fed and cared for beforehand. The stress experienced by the fish and other marine life will be significantly lower if you allow their surroundings to warm up again gradually, rather than transferring them directly into warmer water.

It is at this stage that you should switch on the lighting in the aquarium if you have coral or similar invertebrates with zooxanthellae, so that they can start to photosynthesize, but delay switching on the lights for a few hours in the case of a fish-only setup. This will allow the fish to settle down with less stress after their journey and find their way around in the aquarium—they are likely to become nervous if suddenly exposed to bright light. Even if you do not switch the lights on, be sure to cover the aquarium, because a number of species can jump well and are more likely to display this ability if they are uncertain of their surroundings and feel threatened. They could end up on the floor as a consequence.

Transferring an established aquarium in this way often proves less problematic than starting out with a new one, simply because the entire ecosystem—from the bacteria that aid the filtration process through to the fish themselves—has become established. While there may be a slight fall-off in the efficiency of the natural filtration process following a move, it will not take as long for its efficiency to be restored as it would to establish an entirely new biological filter bed.

▼ *An acrylic tank with preformed reef background.*

Maintaining Water Quality

Marine fish and invertebrates are very sensitive animals. The utmost care and attention to water quality is required to ensure that waste materials are removed, that there is a correct balance of dissolved salts, and that the water is disease-free.

Filtration Methods

There are three different types of filtration processes that work to break down and neutralize the waste produced by fish in aquarium surroundings. The first is mechanical filtration. It involves removing large particles of debris, effectively by sieving them out of the water. This is achieved in the core of a cartridge at the heart of a power filter. The second type is chemical filtration, in which the filter has a charcoal lining, which appears black in color. It strips out harmful molecules that are adsorbed (held) by the carbon.

The third method is biological filtration, which is at the heart of filtration in nature. The fish's waste and other material, such as uneaten food, is broken down by a series of reactions brought about by bacteria. In order to function efficiently, bacteria require oxygen and a large surface area on which to live, which will allow them to multiply rapidly.

Undergravel Filtration

If you use a biological type of filter, it is therefore essential that you incorporate an undergravel filter plate that extends over the entire area of the base of the aquarium. The filter plate should have holes cut in it. Water passes through the substrate, where the beneficial aerobic bacteria occur, and travels through the filter. This results in a constant flow of water passing over the bacteria, bringing the waste with it. Usually the water flow is directed down through the filter bed, but with a reverse-flow system it passes upward instead. The advantage of reverse-flow systems is that the water can be constantly filtered for detritus and debris that would otherwise be drawn into the coral sand, packing it down and clogging it.

One of the most important aspects of an undergravel filter is therefore to ensure a good flow rate. This in turn depends partly on the type of substrate being

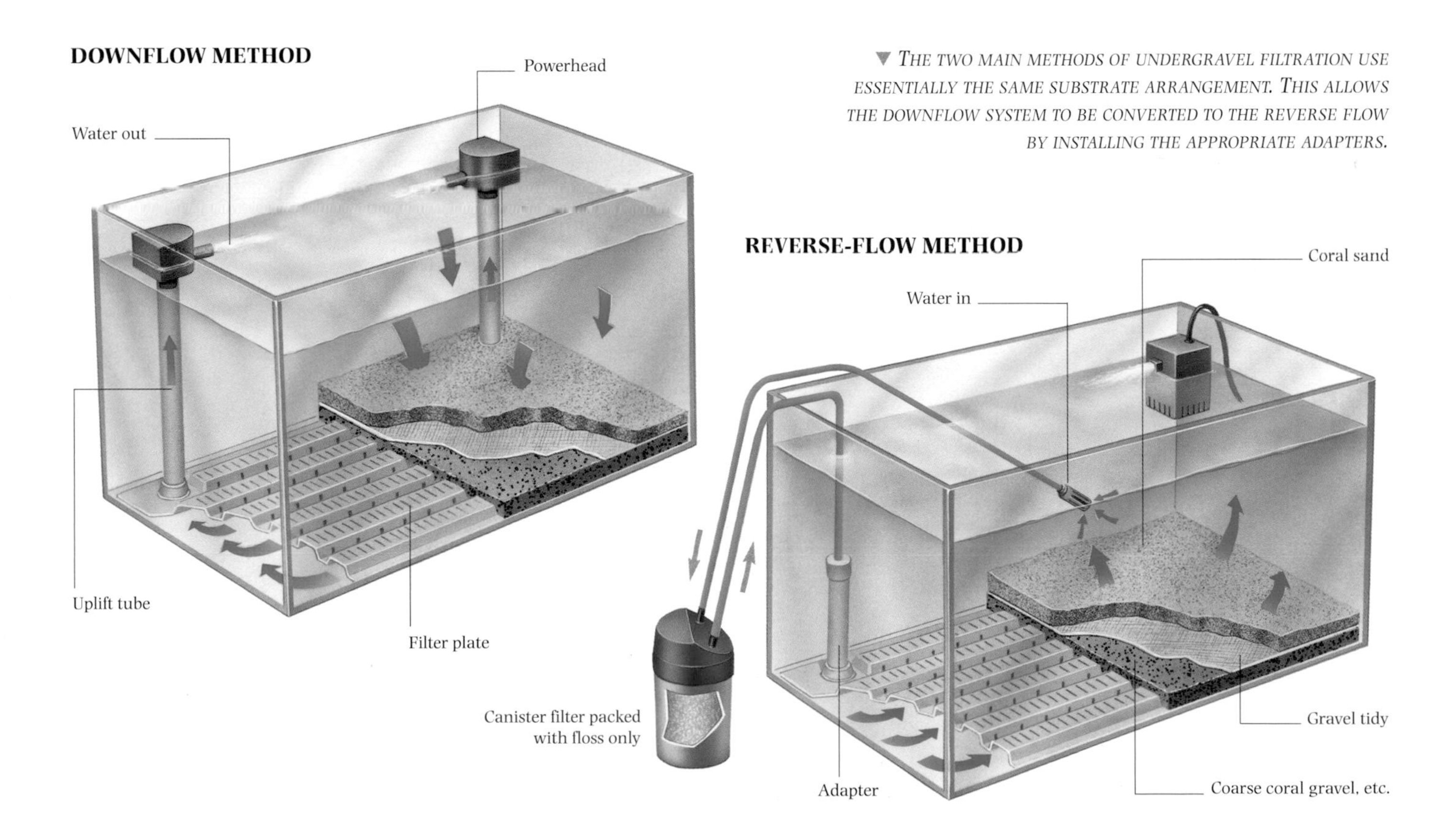

▼ *The two main methods of undergravel filtration use essentially the same substrate arrangement. This allows the downflow system to be converted to the reverse flow by installing the appropriate adapters.*

used. If the particles are small and easily compacted, the flow will be compromised and the filter will not be able to work efficiently. A build-up of nitrogenous compounds in solution will result, endangering the health of the aquarium occupants. The depth of the filter bed is also significant, because a very shallow layer of substrate over the filter will not provide an adequate medium for sufficient bacteria.

The powerheads should draw the total water volume of the tank through the filter bed at least three times every hour. If several powerheads are used, they must all be of the same make and power rating to provide maximum effeciency. The substrate will require raking through on a regular basis to prevent compaction, which reduces the through flow. Any mulm and detritus can be siphoned off at the same time.

Trickle Filters

Human waste has been processed through trickle filters since the turn of the 20th century, but it is only recently that the techniques have been adapted for use in domestic aquaria. Water containing various nitrogenous compounds (such as ammonia and nitrites) is trickled or sprayed over a medium that is ideally suited to bacterial colonization. As the water soaks through the medium, aerobic (oxygen-using) bacteria utilize the dissolved compounds as food and convert them into less toxic substances. *Nitrosomonas* species convert ammonia into nitrites and *Nitrobacter* species convert the nitrites into nitrates; yet another bacterium can be

STRENGTHS OF TRICKLE FILTERS

1. Most equipment can be concealed
2. Increased surface area promotes bacteria
3. Ease of maintenance
4. Detritus is prefiltered out
5. No water or space is lost to filter media within the tank
6. High levels of dissolved oxygen are produced as water is filtered
7. Medications are more effective since they are not interfered with by a calcareous media
8. Convenient for automatic or semiautomatic water change systems
9. A vast range of associated equipment can be installed with the filter or at later stages

WEAKNESSES OF TRICKLE FILTERS

1. Expense
2. Limited range of models
3. Some are difficult to assemble and install
4. Backup service can be poor
5. Can be noisy in operation
6. Some tanks require drilling or a proper failsafe overflow mechanism
7. Risk of leaking joints
8. Prefilters generally need daily attention
9. Livestock can be drawn into the overflow
10. KH, pH, and calcium buffers will probably be required
11. Stocking levels cannot be determined as accurately as with undergravel filters

▼ *Although more expensive to set up, a below-the-tank trickle filter sysem gives more flexibility and makes for a more natural and successful display.*

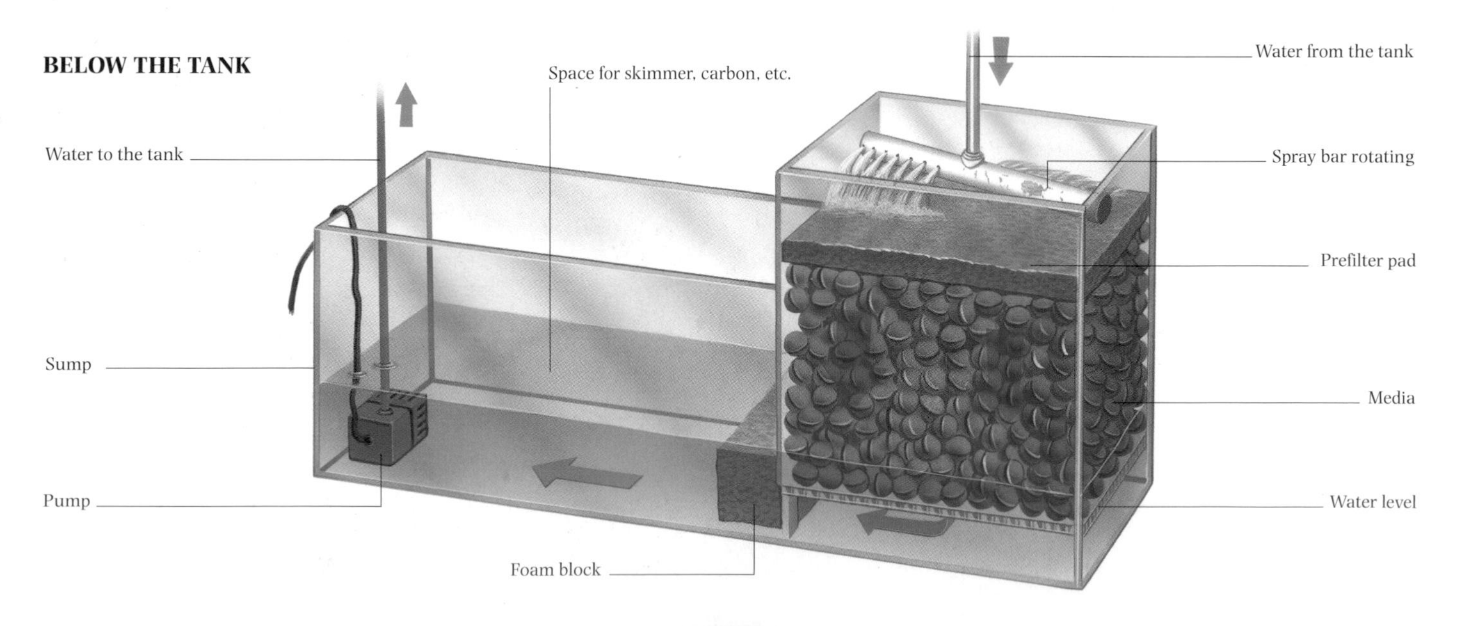

called upon to convert nitrates into free nitrogen gas but this is only achieved under anaerobic (oxygen-free) conditions and is not usually encompassed by trickle filtration systems.

Trickle filters are far more efficient than traditional undergravel filters. Again, the reasoning is surprisingly simple. The *Nitrosomonas* and *Nitrobacter* bacteria require access to large amounts of oxygen to perform their tasks efficiently. With normal undergravel systems the friendly bacteria can only call upon the available oxygen dissolved in the water. If the oxygen level is low, the bacteria count will be equally low, making the filter increasingly inefficient. Trickle filters, on the other hand, have access to unlimited supplies of free oxygen. Not only can the bacteria perform their functions at peak efficiency, they also have a favorable environment in which to multiply rapidly.

Despite this, undergravel filtration can still contribute to a successful tank setup. Some aquarists use a combination of undergravel and trickle filters on the same tank with exceptional results.

▼ *ABOVE-THE-TANK TRICKLE FILTERS ARE EASY TO INSTALL, AND NEED MINIMAL MAINTENANCE. THEY ARE PRIMARILY BIOLOGICAL IN APPLICATION AND CAN BE USED TOGETHER WITH UNDERGRAVEL FILTERS OR ON THEIR OWN.*

ABOVE THE TANK

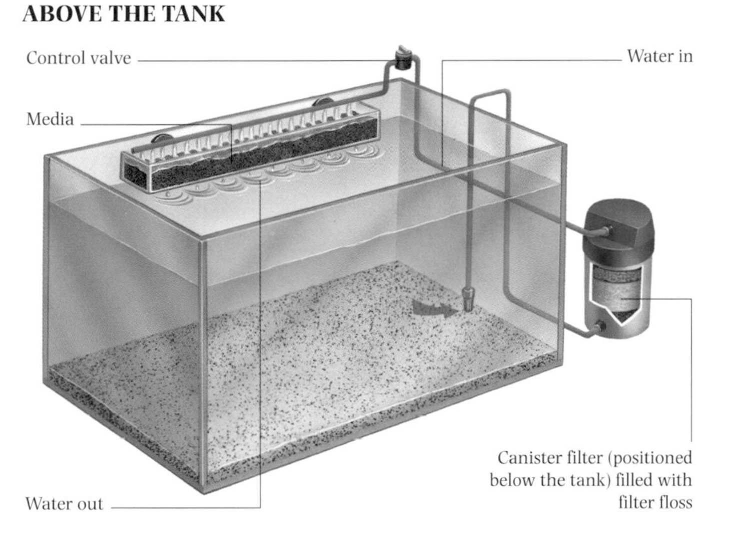

▼ *INTERNAL TRICKLE FILTERS ARE PREFERRED TO BIOLOGICAL-ONLY DESIGNS, SINCE THEY CAN BE ADAPTED TO CONTAIN A PROTEIN SKIMMER, A CARBON FILTER, AND A DENITRIFICATION UNIT.*

INTERNAL

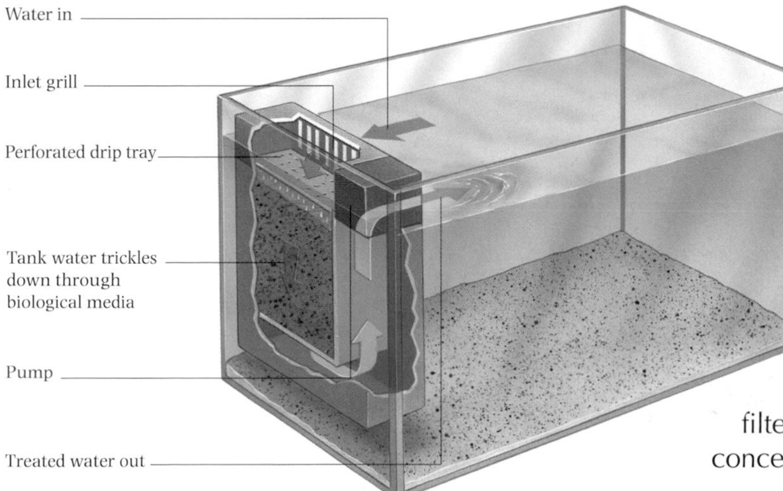

Canister Filters

The design of the aquarium is another important factor in determining the type of filter that can be used. In the case of a reef tank, rockwork covers a relatively high percentage of the area of the base of the tank. This means that there is less opportunity for water to be drawn down through the filter bed, so the efficiency of the filtration system will be reduced accordingly.

In order to maintain the water quality and the stable conditions needed for reef aquaria, therefore, it is important not to rely solely on undergravel filters but to introduce a different type of filtration system. Such systems include external canister-type filters, which incorporate a pump in their design, to draw water out of the tank. They can include a range of different filter media, and the water may pass through several stages of filtration. It often begins by flowing through a foam filter which acts as a mechanical filter, taking larger particles out of solution, and also provides a medium for bacteria to colonize. The main area of biological filtration incorporates beads or granules, which have an open structure. Their large surface area provides an ideal medium to support huge numbers of beneficial bacteria. The final stages of filtration may have a combination of activated carbon and a fine polymer wool, which traps any minute particles, helping ensure that water returning to the tank is clean. Clean water can then be pumped back through a spray bar at the top of the aquarium. The spray allows the water to have good contact with the air and to become well oxygenated as a result.

Because this type of filter is bulky and unsightly, it is a good idea to obtain a cabinet in which to hide it. This applies particularly if you decide to opt for a sump filter, which effectively creates a second aquarium concealed beneath the first. All the equipment—not

STRENGTHS OF UNDERGRAVEL FILTRATION

1. Universally understood and very widely available
2. Easy to operate—a bonus for beginners
3. Relatively cheap for smaller tanks
4. A natural environment for substrate dwellers
5. Quiet in operation
6. Calcium levels are higher and more stable due to calcium-rich substrate
7. pH is buffered by alkalinity of the substrate
8. There is a large natural surface upon which bacterial colonies can thrive
9. Since undergravel filtration has been in use for many years, the correct stocking levels can be calculated very accurately.

WEAKNESSES OF UNDERGRAVEL FILTRATION

1. Displaces large amounts of water
2. Filter bed looks unnatural
3. Tends to clog with detritus
4. Bacteria consumes oxygen from the tank (but extra turbulence from the pumps may compensate)
5. Coral sand requires gradual renewal after two years
6. Calcium carbonates in substrate may interfere with effectiveness of medication
7. Collecting coral sand and gravel for the aquarium trade is less environmentally acceptable
8. Contamination of the tank by fragments of rusting metal collected from the seabed with coral sand and gravel.

just the filtration unit—can be located here, including the heaterstat. This is beneficial, since it also ensures that the aquarium occupants cannot burn themselves by coming into direct contact with the heater.

Water is pumped out of the sump into the tank. As the water level rises, water flows back out of the aquarium into the sump, maintaining the circulation through the system. A sump system is usually used in conjunction with an acrylic main aquarium, because of the need to drill holes to connect the outflow tubing at the top of the tank.

▼ *This modular system of clip-together units shows three ways in which canister filters can be used.*

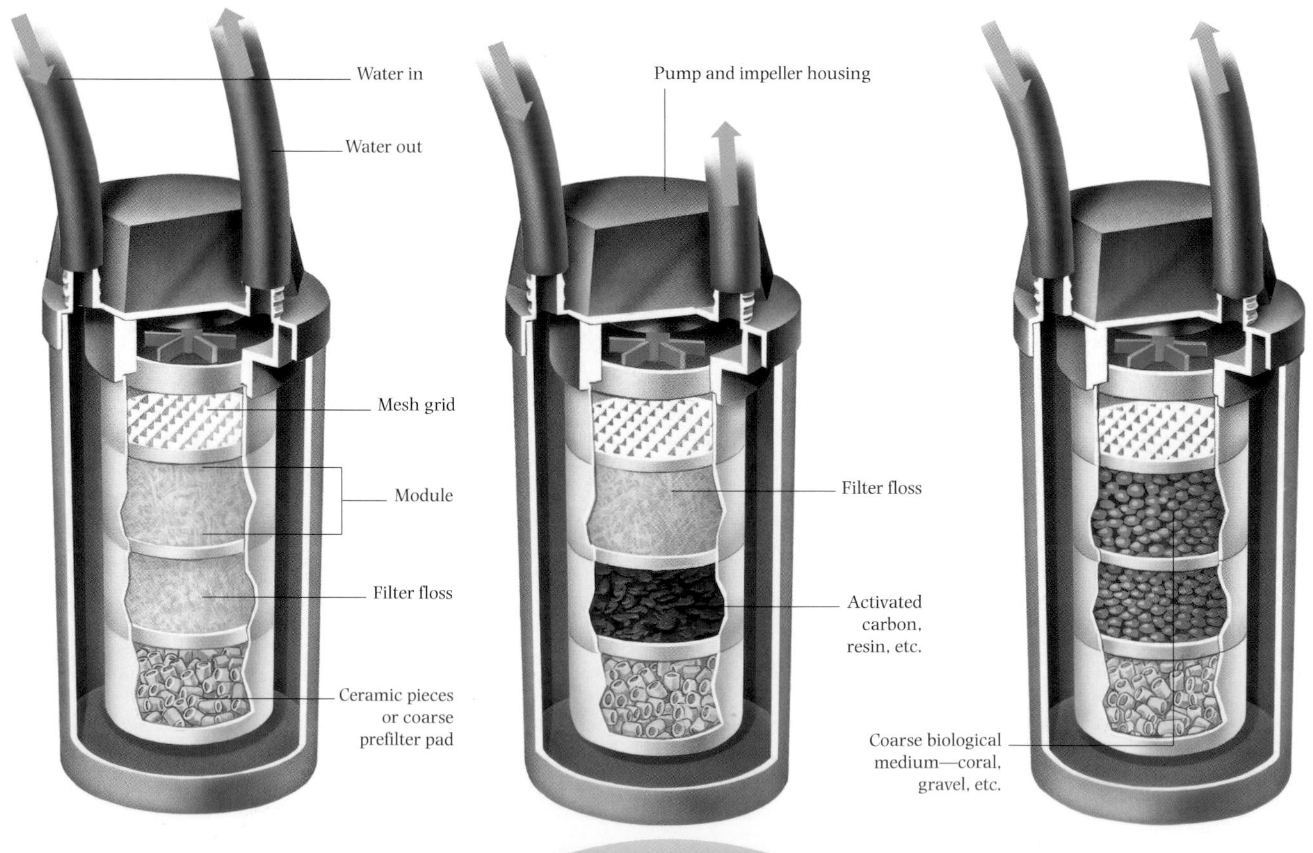

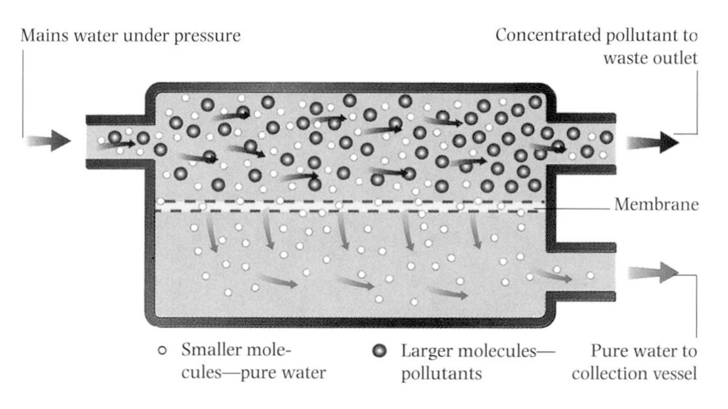

▲ *The filter in a reverse osmosis unit is so fine that only water molecules pass through. Pressure from a mains system is needed to force the water through the membrane, but if the pressure is low a pump is needed.*

A variation on the external canister filter is what has become known as the fluidized-bed biological filter. In this case, the filter medium of fine-grained sand is kept in permanent suspension so that, unlike the situation with heavy static media, there is no opportunity for the surface to become clogged with fine debris, which would reduce the filter's efficiency. Beneficial bacteria thrive under these conditions, making this a very efficient type of filter.

Reverse Osmosis

Household water may appear clean but it is full of impurities, some of which—notably chlorine or the more long-lasting alternative known as chloramine—are likely to be lethal to fish. Although these chemicals can be neutralized with a suitable aquarium product intended for this purpose, there are a number of other harmful components in household water, including nitrates, that cannot be dealt with in this way. As a result, reverse osmosis (often abbreviated to R. O.) units are now routinely used by marine aquarists to create safer and better water conditions. Reverse osmosis removes these chemicals directly. The cost of R. O. units has fallen dramatically in recent years.

The unit connects to the mains water supply and incorporates a partially permeable membrane at its core. It allows the water molecules to pass through the unit, while keeping all the contaminants back. The water in the chamber contains the impurities and will need to be discarded. In order to create ideal water conditions in the aquarium, you simply need to add an appropriate amount of salt mix to the pure water.

Protein Skimmers

Protein skimmers were developed originally for the treatment of sewage. They work by skimming off waste material in the aquarium, such as uneaten food, before it can start to deteriorate and affect the quality of the water. This is achieved by creating a stream of bubbles that pass up the length of the skimmer. The debris attaches itself by surface tension to the bubbles. Once the bubbles reach the cup of the unit, the bond is broken, and the unwanted material is left there and can be discarded. Contact between the air bubbles and the water is therefore vital for the success of this system,

▼ *Operation of an advanced counter-current protein skimmer. They are normally made of clear acrylic so the aquarist can monitor the density of bubbles and the correct water level.*

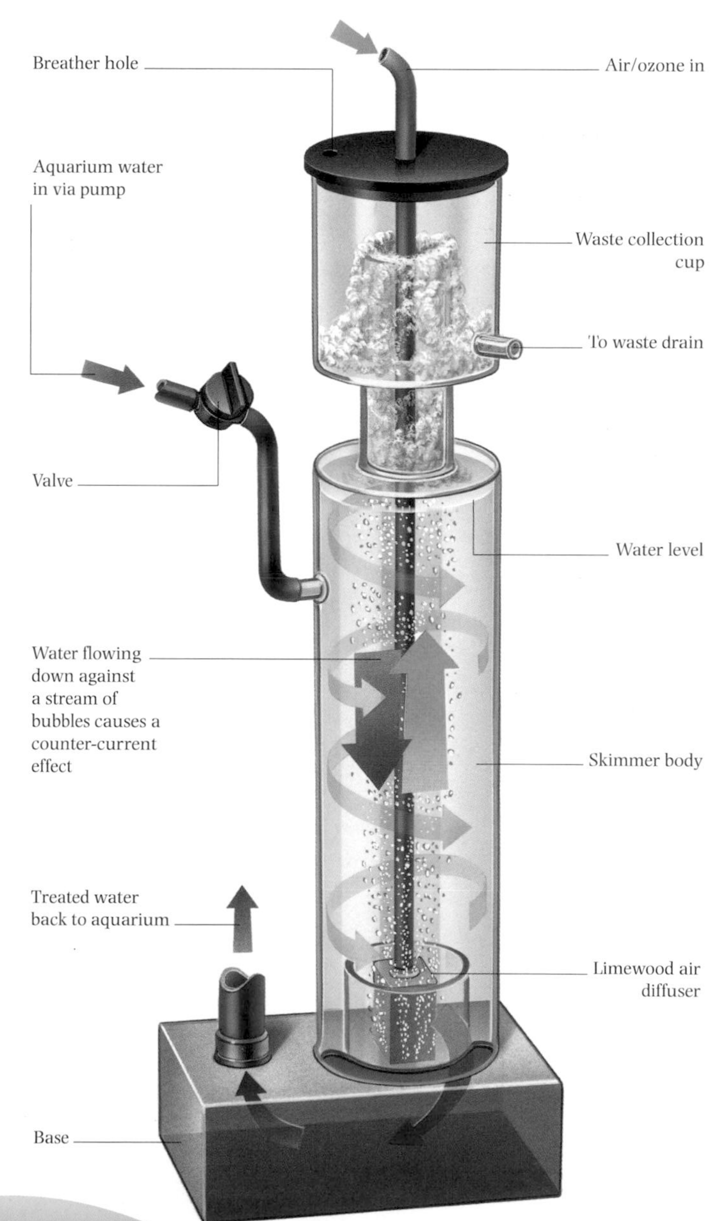

▲ ▶ *Above: Three types of standard venturi upright protein skimmers suitable for smaller aquaria. Right: More advanced skimmer with a triple-pass system that removes large amounts of protein, suitable for use in bigger tanks.*

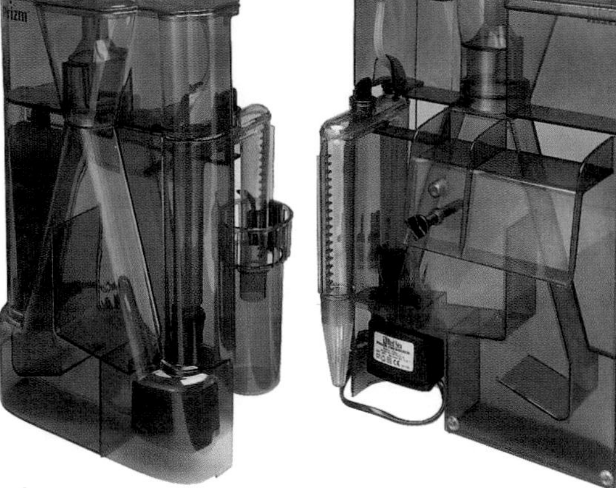

and the basic design has been improved from a single tube to the more refined system, known as the "Berlin triple-pass," which has trebled the contact time between air and water. Skimmers that use a particular type of valve (called a venturi valve) to deliver the water that is to be treated and to produce the billions of microscopic bubbles needed are called venturi skimmers.

The design of protein skimmer that you need will be influenced by the size of the aquarium. Just as with power filters, there are smaller designs that fit within the aquarium. They tend to be less efficient, however, than the larger venturi skimmers, especially those with the Berlin triple-pass technology incorporated in their design. They can be suspended from the aquarium, although it is better to run a unit of this type in a sump where it will be hidden from view.

Ozone

A further refinement that can add to the efficiency of a venturi protein skimmer is to use ozone in place of air. Ozone is a form of oxygen to which another atom has been added and it is inherently unstable. It will not only help neutralize potentially deadly ammonia, but will also kill bacteria and other harmful microbes in the water. Equally, though, used incorrectly, ozone can kill both fish and invertebrates.

The gas is produced by a piece of equipment known as an ozonizer, which is connected directly to the protein skimmer via an airline that incorporates a valve to prevent any backflow of water into the unit. A probe suspended in the tank acts as a check on the system, regulating the output of ozone. Since this gas is potentially toxic, there is a cup containing activated carbon fitted on top of the collecting chamber to adsorb any remaining ozone and prevent it from diffusing into the air. You will be able to see the efficiency of the ozone by the color of the water in the aquarium. There will be no trace of any yellowing, which is often a feature of marine setups.

The latest designs of protein skimmer have to rely on air, however, and cannot be used with an ozonizer. That is because the corrosive power of ozone will destroy the mechanism, since the air is introduced directly into the pump itself rather than through the opening (often described as the venturi cone) at the base of the unit.

Nevertheless, the so-called turbo skimmers have proved significantly more effective. This is a reflection of the design

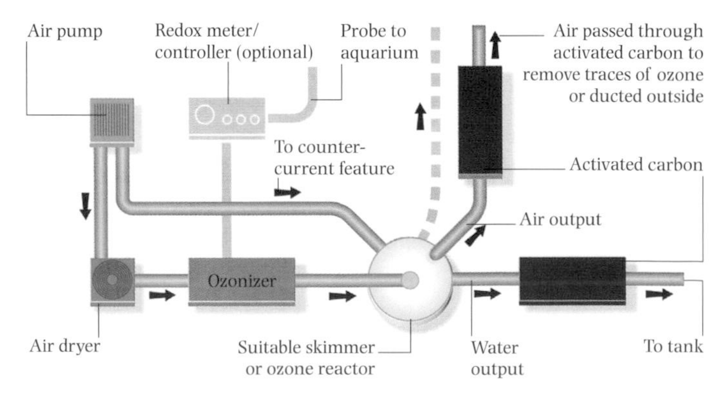

▲ *It is critical to install an ozonizer safely in order to avoid the potential hazards of using this dangerous gas.*

of the blades of the pump, which ensures that the air and water are brought into better contact with each other. Another modern option is the convergent–divergent flow skimmer. It is similar to the turbo skimmer, although the technology is slightly different, and it is ideal for smaller saltwater tanks.

ULTRAVIOLET (UV) STERILIZATION

While protein skimmers bring some benefits in terms of disease control, many saltwater enthusiasts rely on an ultraviolet (UV) sterilizer for this purpose. A number of factors determine the efficiency of these units. Firstly, it depends on how clean the water is to start with, because light will not penetrate well through murky water. The flow rate through the tube is also significant. If water passes through too quickly, the ultraviolet light it produces will not have time to kill the pathogens in it. The ultraviolet lamp itself will not be visible because it is encased in a quartz sleeve when it is operating, but you will be able to see from the connectors when it is functioning. The light must be protected in this way when the tube is switched on or it will damage your eyesight. The output of ultraviolet rays from the light source also declines over time, which can result in it becoming less efficient. You will need to change the light at regular intervals as recommended by the manufacturer, therefore, in order to ensure that it continues functioning at maximum efficiency.

▼ *1. Schematic diagram of a UV sterilizer system. A canister filter is essential to remove particles that may obstruct the UV rays in the sterilizer. 2. Detail of the internal structure of a UV sterilizer.*

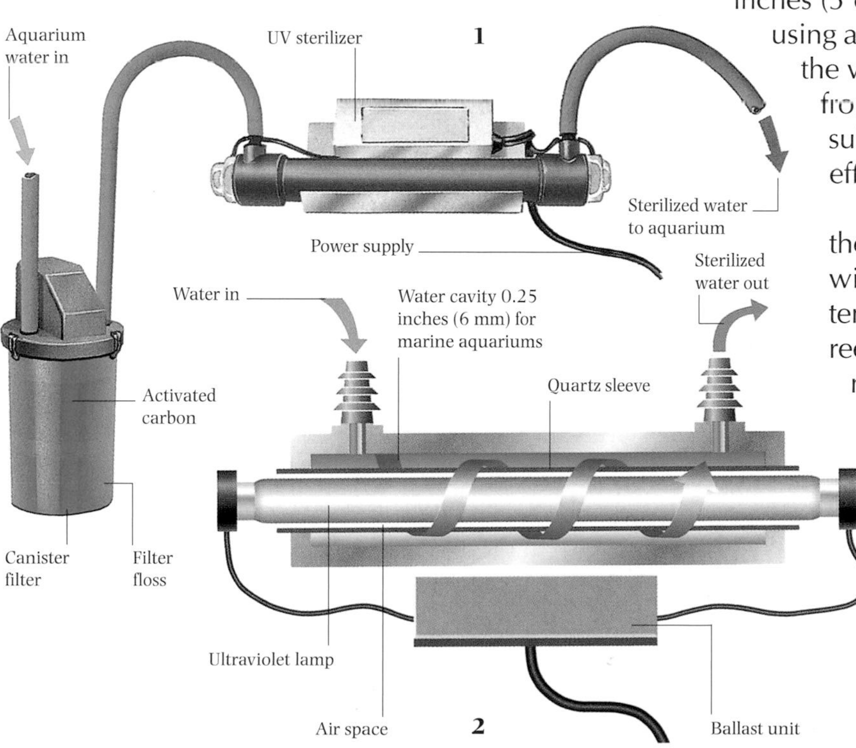

UV SPECIFICATIONS FOR MARINE TANKS

Lamp	Aquarium Size	Flow Rate per hour
8 watt	up to 48 U. S. gal (40 Imp. gal/180 l)	about 190 U. S. gal (160 Imp. gal/725 l)
15 watt	up to 96 U. S. galls (80 Imp. gal/365 l)	about 360 U. S. gal (300 Imp. gal/1,360 l)
25 watt	up to 120 U. S. gal (100 Imp. gal/455 l)	about 480 U. S. gal (400 Imp. gal/1,820 l)
30 watt	up to 180 U. S. gal (150 Imp. gal/680 l)	about 540 U. S. gal (450 Imp. gal/2,045 l)
50 watt	up to 240 U. S. gal (200 Imp. gal/910 1)	about 600 U. S. gal (500 Imp. gal/2,275 l)

THE SUBSTRATE

If you are using an undergravel filter in the aquarium, this will have a direct impact on the floor covering in the tank. You will need to incorporate coral gravel or a similar calcium-rich source, augmented with a layer of finer coral sand on top. This needs to be about 2 inches (5 cm) in total depth. Separate the two layers using a sheet called a gravel tidy. It will not restrict the water flow but it will prevent the finer sand from blocking off the spaces in the gravel substrate, which would compromise the efficiency of the undergravel filter.

In other cases the substrate can be vital to the health of the aquarium. Aragonite sand is widely used as a component of natural systems in which the aim is to set up a miniature reef. Aragonite dissolves slowly and helps maintain the high pH reading of the water. Freed from the need to have a set depth of substrate, you can opt for a shallow covering, but the aquarium occupants may influence your decision. Certain fish, including some blennies, burrow into the substrate, so they need a relatively thick covering—4 inches (10 cm) or so in depth. Other fish will regularly dig in the substrate and they too require a relatively good depth of covering.

AQUARIUM LIGHTING

In addition to enhancing the appearance of the occupants, lighting in a marine aquarium can be critical to their well-being. The type of lighting is of less significance in a fish-only setup, although it will be important if you are hoping to culture algae to supplement their diet. In a reef tank, however, correct lighting is critical to the survival of many invertebrates, including various species of corals and anemones, that have a symbiotic (mutually beneficial) relationship with living algae in their bodies.

Water on the reef is relatively clear, and so sunlight penetrates well. Since they are plants, algae must be kept under well-illuminated conditions in the aquarium in order for them to be able to photosynthesize. They produce energy from sunlight and carbon dioxide and they release oxygen into the water as a by-product of this process. The development of lighting that emits wavelengths corresponding to those of sunlight has ensured that these invertebrates can be successfully maintained in an aquarium.

Sunlight consists of a series of different colors that make up the spectrum. Light from the sun penetrates water to a variable extent depending on its color. Red light penetrates only to depths of about 33 feet (10 m), whereas blue light remains visible down to depths of 820 feet (250 m). This means that fish that may be very brightly colored to our eyes in aquarium surroundings are largely invisible in their natural habitat and are evident simply as dark shapes below this depth on the reef.

FLUORESCENT TUBES

Fluorescent tubes were the first form of lighting used above a marine aquarium. They are usually combined with a polished aluminum reflector behind, which acts like a mirror, reflecting significantly more rays of light down into the tank, thereby increasing the level of illumination. There are now several types of fluorescent tubes on the market, with different properties.

A white triphosphor tube is recommended. Not only does it create the impression of natural light, its spectral range corresponds to that required for an invertebrate tank or a fish-only setup in which macroalgal growth is to be encouraged. In many other aquarium fluorescent tubes, the spectral output gradually declines, with deleterious consequences for corals and other invertebrates that depend on lighting for their well-being, because their associated algae can no longer photosynthesize effectively. The light itself will probably appear just as bright, and so it is standard practice to change the tube before the end of its life span every 10 months or so, in accordance with the manufacturer's instructions. This fall-off in output does not occur with triphosphor tubes. In some cases, particularly if invertebrates are present, you could include a blue actinic fluorescent tube in the lighting system. The spectral range of this type of light is geared to a light output in the blue area of the spectrum.

Choose one of the 03 tubes, suitable for use above a reef tank. They have a peak in output in the actinic part of the spectrum, at 420 mm, which is the key part of the spectrum for the photosynthetic activity of zooxanthellae. Their color output also tends to be in this blue area,

◀ *FLUORESCENT TUBES, SUCH AS THESE BANKS OF TUBES HANGING OVER A REEF AQUARIUM, HAVE THE ADVANTAGE OF PRODUCING LARGE AMOUNTS OF THE CORRECT TYPE OF LIGHT BUT VERY LITTLE HEAT.*

▲ *Lighting built into the construction of an aquarium produces a stunning impact without the disadvantage of an unsightly unit.*

and they are particularly valuable for use with fish that tend to be active after dark, such as soldierfish.

Fluorescent tubes emit very little heat, which can be a significant advantage, especially above a small marine aquarium. Even so, they may not be powerful enough to give the best results.

Metal Halide Lights

Metal halide lights have become popular in invertebrate aquaria because of their "color temperature," measured in degrees Kelvin. Whereas a white triphosphor fluorescent tube's color temperature output is around 9,500 degrees K, that of a metal halide lamp is in the range of 10,000 to 13,000 degrees K. They produce a more natural white light, the wavelengths of which also penetrate better into the water, bringing substantial health benefits to the zooxanthellae.

On the downside, however, metal halide lighting produces a considerable amount of heat, so the lamps themselves have to be suspended 12 inches (30 cm) or more above the aquarium to avoid adversely affecting the water temperature. They are also quite expensive to buy and costly to operate. The bulbs usually have a power rating of 150 to 250 watts, but for deep aquaria even more powerful bulbs are likely to be needed.

▼ *Separate lighting starter units specially made for the aquarium have the advantages of ease of use and portability.*

Tank Decor

While the decor will enhance the appeal of the aquarium, it is more important to design a tank that meets the needs of the fish and other creatures you are planning to keep. Some fish, for example, require plenty of swimming space, whereas others are much shyer by nature and should have enough retreats in the aquarium in order to give them a sense of security. In the case of more delicate species it is important to encourage the growth of macroalgae, which can serve as a natural source of food, in addition to including live rock, which may provide edible items to supplement their diet. This type of rockwork is a key component of reef tanks, because it provides a substrate for sessile invertebrates and a backdrop for those that live on the reef, such as crustaceans.

Live Rock

Live rock (often also called living rock) is part of the reef that has broken off into smaller pieces as a result of the action of the sea. It is therefore likely to be covered in the smaller life forms that exist there—good and bad as far as the aquarium is concerned. It is worth bearing in mind that it is possible to introduce unwanted creatures as well as desirable ones to the aquarium by using this material. It is therefore important to check for obvious problematical species—such as hair algae, so-called because of their filamentous nature—because they can

▼ *Rockwork is a key element of reef aquaria and can be built into a wide variety of individual shapes and forms.*

▲ Varieties of natural wood and rockwork that can be bought from aquarium stockists: 1. Petrified wood; 2. Bora; 3. Dead coral; 4. Limestone; 5. Lava rock.

easily take over a reef tank very rapidly, growing over the coral and eventually smothering it.

There are a number of sources of live rock. Its origins will determine what creatures are found in association with it. There are some environmental concerns attached to the collection of such material, particularly if the reef is damaged as a consequence. Its collection has therefore been banned in some areas—around the coast of Florida, for example. As an alternative, you can opt for cultured live rock. As its name suggests, it is created by taking pieces of limestone that have been dug on land and immersing them in special tiers in the sea. Over a period of years they will be colonized by marine life and can then be harvested for the aquarium trade.

The price of live rock varies widely, depending to some extent on its origins, but also on whether or not it has been cured. Even with the best handling facilities en route from the sea to the retailer, some of the more delicate life forms on the rock are likely to die. They can affect the water quality as they start to decompose, presenting a threat to the reef inhabitants.

Cured rock has been treated to make it safer, but it is more costly. Since it is sensible to repeat the treatment, especially for a reef tank, there is probably little benefit to be gained by paying so much more for cured live rock. It should be treated as soon as you arrive back home. Examine the pieces of rock very carefully for invertebrates that may reveal themselves and that are not required for the aquarium. Keep spraying the rock with water to keep it from drying out.

Live rock is sold by weight and it is expensive, partly because of the cost of air freight. As a guide, the amount of live rock required for a reef tank is about 1 pound (2.2 kg) per 1.2 U. S. gallons (1 Imp. gal/4.5 l), based on the tank's volume. There is, however, the possibility of using so-called base rock as a cheaper alternative. As its name suggests, it can be used to form the base of an artificial reef, with living rock being placed above, the aim being to suggest that the whole reef is constructed from living rock. There are some drawbacks to this approach, however, that will not be immediately apparent. For example, macroalgae present on live rock will utilize nitrate in the water, whereas if base rock predominates, the nitrate will encourage hair algae or similar troublesome forms, whose growth can be difficult to curb.

When handling live rock, it is best to wear rubber gloves, because there may be creatures lurking there that can give you a painful bite, such as the mantis shrimp (*Odontodactylus* species)—more appropriately referred to as the thumb-splitter, because it can inflict a very nasty nip. Such animals will need to be removed from the aquarium using a special trap. Bristleworms (*Hermodice carunculata*) can also be a problem. They can cause skin inflammation if you brush against their bristlelike spines, although they are far less of a threat to other aquarium occupants than mantis shrimps.

Watch out for any small clear-looking anemones that emerge in the early days on your reef. They are likely to be glass anemones (*Aiptasia* species), which will replicate at a very fast rate, killing off other sessile invertebrates that they encounter. Unfortunately, getting rid of them is very difficult, partly because of their small size, and also because even if you can crush them, they will simply replicate again from the tissue attaching to the rock. Biological control is the best option, using the peppermint shrimp (*Lysmata wurdemanni*), which feeds on the offending pest.

PREPARING ROCKWORK

Using a clean brush, remove any creatures that appear dead. Place the rock in an acrylic tank. The water should be kept at a suitable temperature, and use a powerhead to ensure good water movement. You should then test the water regularly. At first there will be a rise in ammonia, but gradually this should decline and nitrite levels will rise over the course of several weeks. Once the nitrite is replaced by nitrate, the process is essentially complete. It will then be a matter of transferring the rock to the tank and arranging it so that it is secure. This lengthy process means that you cannot be in a rush to set up a reef aquarium.

ROCKWORK

It is important to plan the decor, especially in the case of a reef tank. If you have a particular design in mind, you need to choose rockwork accordingly. You may, for example, want to have two islands of rock—one at each end of the aquarium—rising up to create underwater peaks that will provide space for fish to swim between, or you may prefer to have a central rocky area as a focal point. Using rockwork to divide the aquarium into different areas is also recommended in cases where fish tend to be territorial, and it will reduce the risk of conflict.

If you want to pile up the rocks, they must be properly supported to prevent accidents. This can be achieved in a number of ways, the simplest being to hold rocks tightly together using plastic cable ties. Try to use pieces of rock that fit naturally together, because this will give them more natural stability and make it easier to bind them. Bear in mind that live rock needs to be correctly oriented, so that the side that has the greatest growth is directed upward as it would have been on the reef, where it was more exposed to sunlight.

▲ *SOME ARTIFICIAL ROCKS ARE PRODUCED TO MIMIC THE SHAPES OF CORAL AND COME IN A VARIETY OF COLORS, OFTEN QUITE UNNATURAL, AS SEEN HERE.*

You should be able to drill carefully through the rock using a masonry drill. This may be necessary if you want to create an overhang, or simply to give greater stability. Use cable ties to hold the pieces of rock together. The ties should prove very durable, even in saltwater, and will soon be disguised by algal growth.

Silicone sealant, as used in the manufacture of glass tanks, can be used to stabilize aquarium decor, but it is not recommended in the case of live rock, because the sealant must be applied to dry surfaces only and needs to be kept dry in order to cure (set) properly. If you use this material, be sure to use specialized aquarium sealants. They will be free from the fungicides added to household products of this type, which are likely to be toxic to the fish.

Rockwork can also be important in aquaria that house fish, and stability is a particular issue where eels are concerned. Eels will naturally seek out retreats within rockwork. They are large and powerful fish and can cause serious disruption among the tank decor. It may help if you provide several separate lairs where an eel can hide away, perhaps using a clean clay flowerpot or tubing for the purpose, incorporated in the design of the rockwork.

◀ *NATURAL SLATE FORMED INTO SHAPES WITH GAPES GIVES SPACES IN WHICH FISH AND OTHER SALTWATER LIFE CAN HIDE.*

SETTING UP

It really pays to be patient when setting up a saltwater aquarium, particularly in the case of a reef tank. Allow time for live rock to become established. In a tank that is to house herbivorous fish, especially those of a more delicate nature, allowing time for the growth of macroalgae will make it much easier to establish these fish successfully.

First wash out the aquarium to remove any dust or dirt, and place it on its base. All-glass frameless aquaria should be stood on a bed of polystyrene in order to absorb any unevenness in the surface, which can otherwise place undue pressure on the joints and result in a leak. You can color the front edge of the polystyrene with a marker pen or paint to disguise its presence.

If you are using an undertank heater, it should be slid in place so that it is directly in contact with the underside of the aquarium. It needs to extend over the entire surface and be controlled by a separate thermostat. An external unit is essential in tanks housing fish such as boxfish, which may attack the cabling and bite through it, electrocuting themselves in the process.

Insert the tank background next. Backgrounds are available in a range of styles to fit standard-sized tanks. Choosing a suitable marine backdrop will create the impression of the reef stretching out behind the tank and will mask wall coverings that would otherwise detract from the appearance of the aquarium. Extending the background around the sides of the aquarium will cut out unwanted light and will also reduce the risk of troublesome algal growth there, which would otherwise need to be cleaned off.

Next, put in the undergravel filter if you are using one. It should extend over the entire base of the tank, and the substrate should be placed on top. Other equipment, such as airstones or internal power filters, can also be put in place at this stage, followed by any inert decor. A wide range of molded seafans and corals can be used in tanks alongside fish that would otherwise destroy their living counterparts. Although they often look too bright at first, they will blend in well in due course as they become colonized by algae. This type of decor is much safer to use than items of marine origin, such as washed-up seashells. The latter can be a source of disease or may affect the water quality if they contain the decaying remains of their former occupants.

When planning the decor, try to use these items around the sides and back of the tank, in order to leave

▼ *A large purpose-built aquarium under construction as part of the decor of a living room. The first stage is to build the rockwork.*

▲ *Appearances can deceive. This fish is not swimming among natural rocks and invertebrates but in front of a plastic image placed at the back of an aquarium.*

the fish with adequate swimming space, bearing in mind the positioning of any rockwork in the aquarium.

Even if you are not including live rock, it will still be worthwhile to incorporate calcium-rich rocks such as tufa. They will dissolve gradually and will help maintain the pH of the water in the desired alkaline range by acting as buffers. Do not place rockwork or other decor in such a way that it blocks the corners of the tank, otherwise it is likely to create areas where detritus can accumulate, which will soon start to affect the water quality.

◄ ▼ *Salt mixtures, bought in solid form, contain the right balance of natural salts for a saltwater aquarium. A hydrometer is essential to ensure that the correct concentration of salts is used and maintained.*

Water and Temperature

Seawater is not recommended for the aquarium, even if available, because of the risk of pollutants entering the tank. You will need to mix up a saltwater solution. It is advisable to aerate water that has been prepared by reverse osmosis, since this will make it more alkaline. Use a suitable product to remove chlorine or chloramine from the tapwater. Make up the salt solution in a clean plastic bucket that you can then keep for this purpose. Add a known volume of water and measure out the required amount of sea salt, stirring it in well with a wooden spoon. Leftover salt can be kept dry and used again for partial water changes in the future.

If you have an inert substrate in the aquarium, you can place a saucer on the substrate and start pouring in the water on the saucer, which will avoid disturbing the base. Do not add anything live such as living rock at this stage, though, because you will need to heat the water with the heaterstat, partly to check that it is functioning properly. Most heaterstats are preset to a temperature of 77°F (25°C), which should be suitable for a saltwater tank. Allow about 1 watt for every 0.3 U. S. gallons (0.2 Imp. gal/1 l) on average, although this may need to be higher if the room is relatively cool.

To work effectively, the heaterstat should be placed at an angle so that the lower part of the unit, which incorporates the heater, is close to the substrate. Warmer water will rise up through the tank, but because the unit is angled, it will not directly impinge on the thermostat above, causing it to switch off the unit prematurely. In a large aquarium you may need two separate heaterstats, which should be positioned well away from each other. Always fit the heaterstat with a heater guard to keep the aquarium occupants from burning themselves on it.

Attach an LCD thermometer near the bottom of the tank, above the substrate on the outside of the aquarium at the front, so that you can read it easily. If you use one of the older-style alcohol thermometers, however, it will need to be placed inside the tank. In either case choose a location away from the heaterstat to give you the best indication of the water temperature.

It is possible to mix the salt solution in the aquarium itself, but there are various drawbacks to this method, even though it may appear to be a simpler option at the outset. Not least is the fact that you cannot place the living rock in freshwater, so it will be impossible to mix all

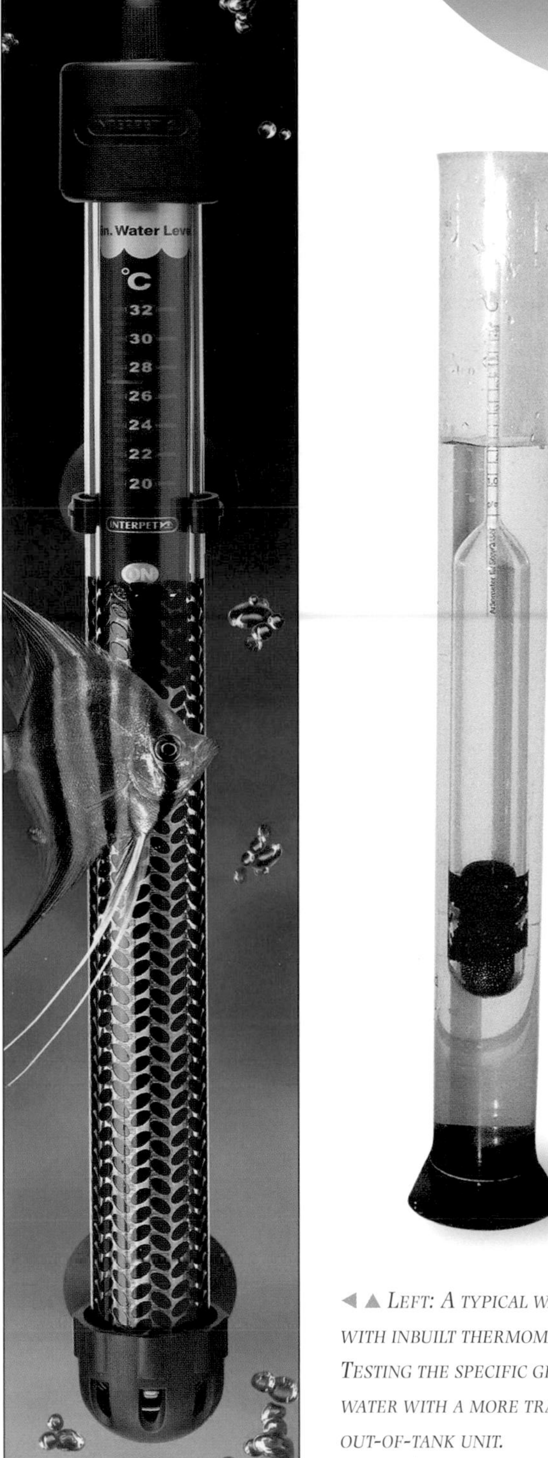

◀ ▲ *LEFT: A TYPICAL WATER HEATER WITH INBUILT THERMOMETER. ABOVE: TESTING THE SPECIFIC GRAVITY OF WATER WITH A MORE TRADITIONAL OUT-OF-TANK UNIT.*

the water beforehand, because you cannot be sure how much water the rock will displace. Using a bucket, you can measure out the water exactly rather than guess the volume needed, and add the correct amount of salt.

ADJUSTING THE SG

Once you have filled the tank and it has stabilized at the right temperature, you can then take a specific gravity (SG) reading and carry out any necessary adjustments. The concentration of the salt solution is critical to the well-being of the tank occupants and is measured on the specific gravity, or SG, scale. This requires an instrument called a hydrometer. The easiest one to use is the swing-needle type. A sample of tank water is placed in the chamber with a pipette and the result is read on the accompanying scale. The SG reading should typically be between 1.022 and 1.025, with the upper figure being recommended for a reef tank.

The SG must be kept very stable, because even tiny shifts can be harmful to the aquarium occupants. If the reading is too high, you will need to reduce the concentration of salt by removing some of the tank water and substituting dechlorinated freshwater. Should the SG reading be too low, you will have to add a little more salt directly to the aquarium, stirring it in with a wooden spoon. Do not discard the salt itself.

You will need to switch off the heaterstat once the water is at the required temperature. Always do this for safety reasons whenever servicing the tank. It will prevent you from receiving an electrical shock, which could occur if you dislodge rockwork which then falls onto the heater, for example, breaking the glass. The live rock can be put in place at this stage, once it has been cured. With the water temperatures and SG readings at the appropriate levels, there will be minimal stress on the life forms present on the rock.

TESTING WATER QUALITY

It is important to monitor the chemical changes taking place in the aquarium, using appropriate test kits, and to keep a log of the results. These kits are very easy to use. You simply need to mix a small sample of aquarium water with the reagent, follow the instructions, and then compare the results with the color band chart to read off the concentration.

▼ *THE CORRECT BALANCE OF INDIVIDUAL CHEMICALS AND pH IS CRITICAL. TESTING KITS ARE WIDELY AVAILABLE.*

TESTING WATER QUALITY

Ammonia (NH_3/NH_4^+)
Optimum level: zero at all times

Ammonia is the primary enemy of invertebrates and fish, capable of causing death in very low concentrations. Causes of ammonia include: an immature filter, overfeeding, overstocking, and dead or dying stock. By vigilance and regular testing, the presence of ammonia can be avoided.

Nitrite (NO_2)
Optimum level: zero at all times

Even trace levels of nitrite can destroy a well-presented invertebrate aquarium and can cause fish much distress. All comments regarding ammonia apply equally here.

Nitrate (NO_3)
Optimum levels: below 10 parts per million (ppm) total NO_3—preferably zero

Some fish may tolerate well in excess of 25 ppm. A reasonably harmless substance where many fish are concerned, but a good overall indicator of general water quality and one that should be kept extremely low if invertebrates are to thrive. Constantly high nitrate levels usually reflect high fish stocking ratios. This must be monitored and the aquarium destocked if necessary.

Phosphates (PO_4)
Optimum level: zero

Invertebrates do not prosper when levels of phosphate get too high. Phosphates arrive in the aquarium through unfiltered mains water (used in the mixing of fresh or saltwater changes), poor quality carbon, and marine salts, but mostly through the waste products of fish. Nuisance algae thrive where phosphate levels are high and destocking, high-quality water changes in the correct proportion, or phosphate-removing resins can all help alleviate the problem.

Temperature
Optimum level: 77°F (25°F)

A stable temperature is essential to the well-being of invertebrates and fish. Hot weather may force the temperature up and a cooler may have to be installed if valuable livestock is not to be lost. Always use an accurate thermometer.

pH
Optimum level: 8.1–8.3

pH is a measure of the alkalinity or the acidity of aquarium water. Invertebrates are sensitive to wide variations, although some natural changes should be expected during the day. Dissolved oxygen assists in the increase of pH. This builds up through the activity of photosynthesis by both micro- and macroalgae and can be detected by test meters or kits. Aquarium water could drop to as low as 7.9 at the end of the night and peak at around 8.4 just before lights out. These natural pH cycles are gradual and tend not to stress livestock to any great degree. Owing to their ingredients, pH buffers can also increase KH values to dangerously high levels. Regular water changes are essential.

Alkalinity
Optimum level: approx 600 microequivalents

Sodium carbonate is an important constituent of seawater because it helps prevent the dangerous lowering of pH by buffering it to optimum levels. As it is depleted, the buffering capacity of the water is reduced and the pH becomes unstable. Alkalinity test kits can now warn of low buffering levels and potential problems.

(K)arbonate Hardness (KH)
Optimum level: natural Seawater (NSW) is 7dKH

KH is a measurement of various carbonates and bicarbonates of calcium and magnesium, and borates within freshwater and seawater. A stable KH will prevent rapid declines in alkalinity and subsequent drops in pH. Boosting the KH of aquarium seawater to between 12 and 18dKH using a proprietary generator has been recommended. However, left to their own devices, most aquaria settle naturally to around 7dKH and there appears to be no advantage in constantly increasing dKH to unnatural levels. Indeed, it could prove harmful, since pH levels might be adversely affected.

Salinity
Optimum level:
between 1.021 and 1.024 (SG)

Salinity measures the total amount of dissolved solids in seawater. It is usually recorded as specific gravity (SG) but can also be referred to as parts per thousand (ppt) or 0/00 (for example, 35 0/00 is 1.026). Constant evaporation of freshwater from the aquarium causes the salts to become more concentrated and the salinity to rise. In order to maintain stability, automatic dosing systems, called osmo-regulators, or osmolators, are often used. These systems use conductivity meters to a very accurate level. They take their readings in micro Siemens (pS) and may be set to replenish freshwater as it evaporates.

Calcium
Optimum level: 350–400 ppm

Calcium is a vital element in any marine aquarium. A host of invertebrates draw it from the surrounding water in copious amounts, and calcium reserves need to be replenished on a regular basis. Regular water changes may achieve this but a well-stocked invertebrate tank may require the addition of biologically available calcium to keep levels optimum.

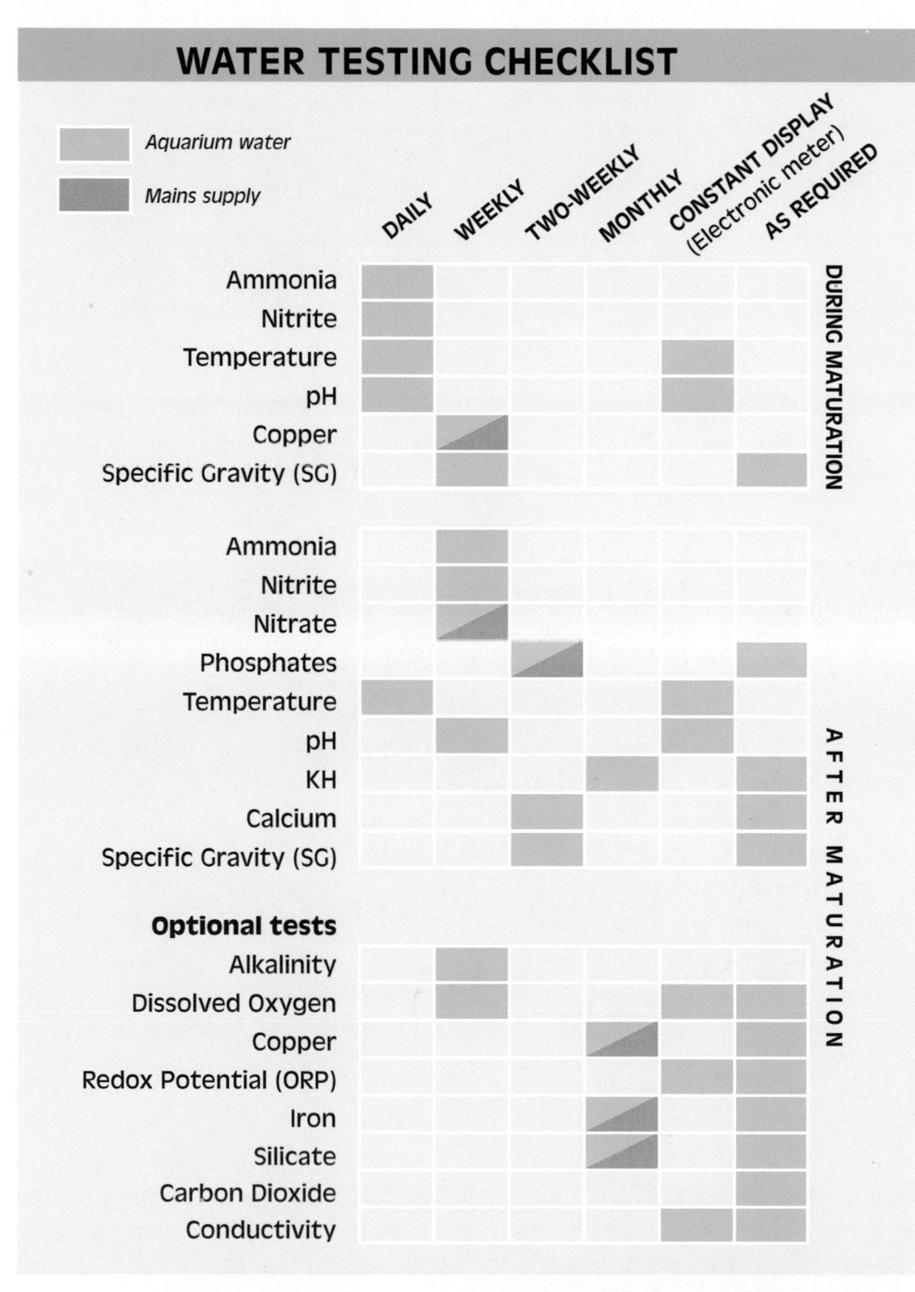

Dissolved Oxygen (O_2)
Optimum level: 6–7 ppm

Both fish and invertebrates benefit from high levels of dissolved oxygen. Good water circulation is the key, as oxygen is drawn mainly from the interface between air and water. Dissolved oxygen also affects pH. (*See* pH).

Copper
Optimum levels: zero in the invertebrate aquarium; variable in the fish-only tank

Copper-based medications have proved very reliable in the treatment of various fish diseases such as white spot and *Oodinium*. It is, however, highly toxic to invertebrates and should never be used in aquaria housing these animals. Accurate measurement of copper is essential because it can even prove lethal to fish at certain levels. Copper can even be introduced to the marine aquarium by way of the domestic water supply, and this should be tested from time to time.

Redox Potential (ORP)
Optimum level: approx 350 millivolts

Oxygen Reduction Potential is, broadly speaking, a measurement of the water's ability to cleanse itself. Highly efficient filtration and the use of ozone will help boost values. ORP can only be measured using an electronic meter with a high-quality probe. As with many "advanced" tests, ORP is not absolutely essential and the readings may be difficult to interpret without a full understanding of the multiple parameters.

Covering the Tank

You will need to fit a cover glass, known as a condensation tray, over the top of the aquarium. It will help prevent water loss from the tank by evaporation, which will have the effect of raising the SG. It will also ensure that the light casings are not splashed. (If that were to happen, salt would crystallize on the casings.) In addition, it will prevent fish from jumping out if the hood has been removed or if you are using pendant lights above the aquarium. You may need to cut holes around the edges to accommodate cabling going into the aquarium, so mark out these areas carefully to avoid cutting away more of the plastic than you have to.

The hood itself will be designed to accommodate the fluorescent tubes, which are held in place by clips. Waterproof connectors should be fitted to the end of each tube as a safety measure. The so-called starter gear that operates the lights is also contained in the hood, and obviously the entire unit needs to be kept dry.

Maturation

It will not be possible to stock the tank to capacity straightaway because of the maturation program that takes place in the aquarium. Initially there will be a buildup of ammonia in the water. The ammonia will subsequently be converted to nitrite by a process of oxidation carried out by bacteria. It takes time for the bacteria to increase in number, and so ammonia levels continue to rise at first. Later, ammonia gives way to nitrite, which is less toxic to fish and other marine life. Other bacteria then start to oxidize the nitrite to nitrate, which is relatively harmless, and may be used as a fertilizer by marine algae.

Traditionally, saltwater fishkeepers have tended to add a relatively hardy species, such as a damselfish, to the tank at the outset to start the maturation process, which can take about eight weeks. You can also obtain special bacterial cultures to add to the aquarium water. Once nitrate is dominant and both ammonia and nitrite have disappeared completely, the filter is mature. At this stage you can start to increase the stocking density of the tank. Even so, it is important to prevent the nitrate level from rising above 10 mg per liter.

▲ *A large shop display consisting of small individual units. Displays that are clean, with apparently healthy fish, are a first step to obtaining good stock.*

Choosing Healthy Stock

Although you can have fish and invertebrates shipped to you, the best option is to visit a local store and see the fish before buying them. You can then be sure that they are healthy and take them back to your aquarium with minimal delay. Importantly, in the case of species that are difficult to persuade to eat initially, such as the golden butterflyfish (*Chaetodon aureofasciatus*), you can ask to see them being fed.

Healthy fish should swim without difficulty and should be able to maintain their position in the water without bobbing around. Their fins should not have any damage, although minor injuries may heal uneventfully. It is very important that there are no abnormal swellings on the fins or the body, and you should check that both eyes appear normal and are not cloudy. The color of the fish may vary, but be wary of any that appear abnormally pale, since this is often a sign of illness, especially if the individual also has an emaciated appearance.

When it comes to purchasing invertebrates for a reef tank, it is much harder to recognize signs of ill health, although there can be some clear indicators. For example, crustaceans should not be missing claws, even though they may grow back after they molt.

Fish are normally sold in plastic bags. These are placed in paper bags to darken the environment and make the journey less stressful for them. Keep the bag in a shady place, such as the footwell, and be sure that it is adequately supported so that it cannot tip over and that nothing can fall on top of it. Once you arrive home, float the plastic bag in the aquarium with the lights off for about 15 minutes. The water in the bag will warm up gradually, reducing the stress on the fish. Then allow them to swim out into the aquarium.

INTRODUCTION AND STOCKING

The way you introduce the fish to the aquarium is important, particularly if you are planning to have a community aquarium, since some species tend to be more territorial than others. You will therefore need to place the more tolerant species in the aquarium first, allowing them to become established before putting in more assertive and potentially aggressive fish. This will reduce the likelihood of conflict and bullying. Similarly, in the case of anemones and corals, space them out well around the aquarium so there will be no risk of them stinging each other. Make sure there is enough space for them to spread without coming into contact with other sessile invertebrates.

Since most of the fish offered for sale are young, it is important to allow space for them to grow. If you stock the aquarium to maximum capacity at the outset, problems will arise, even allowing for the fact that most saltwater fish in aquarium surroundings do not grow as large as in the wild. You are soon likely to find that the filtration capacity of the tank becomes overloaded, resulting in a decline in water quality. This in turn will increase the likelihood of the fish becoming ill.

When first introduced to an aquarium, most fish will tend to seek cover and will hide away largely out of sight. They are unlikely to want to feed at this stage either, so it is pointless offering any food. Unless there are other fish already established in the aquarium, it will simply start to decompose and will have an adverse affect on the water quality. Refusing to feed initially is regular behavior and not a sign of illness. In a day or so their appetites should return to normal.

Feeding and General Care

Offering a balanced diet is critical to keeping your fish and invertebrates healthy. Happily, advances in food technology have meant that it is now possible to provide a wide range of prepared items that contain the necessary nutrients, including vitamins and minerals, to keep them in good condition. Nevertheless, you need to be aware of the individual feeding preferences of your aquarium occupants, so you can match the food to their particular needs.

Some species are much easier to cater for than others, and the shape of their face often gives a clue as to their feeding habits. Those that have broad mouths, such as lionfish, tend to be predatory by nature. Small narrow jaws, on the other hand, as displayed by many butterflyfish, suggest more elegant and specialized feeding habits.

Even with the wide range of foodstuffs now available for marine fish, some species are still challenging to wean onto artificial diets. This may also be a reflection of the way they were handled prior to reaching you. As a general guide, tank-bred specimens tend to be easier to manage because they have been accustomed to artificial diets all their life.

It will be much harder to persuade a fish to start eating again if it has evidently lost condition. Offer the widest possible range of foods available, therefore, when attempting to persuade an individual to eat. Also try to ensure a quiet environment, because some fish will not feed if surrounded by boisterous companions, which will often be keen to take the most tasty morsels in any event.

▼ *Adding food to a saltwater tank.*

Try to replicate the feeding habits of the fish in the wild. Herbivorous fish, such as tangs, tend to browse for long periods because the algae that form a large part of their diet in the wild have a low nutritional value. Tangs are much more easily established in aquarium surroundings when there is a good covering of algae on the rocks, which they can browse. Natural food of this type can also help maintain the coloration of these fish. Predatory species, in contrast, may only feed once or twice a day, while others, such as seahorses, require almost constant access to brine shrimps and similar small creatures in aquarium surroundings.

Types of Food

Fresh and frozen foods provide the most palatable feeding options, simply because they have a higher water content and a more natural texture than freeze-dried or dried foods. Frozen foods of various types are widely available and in general represent a safer feeding option than fresh food simply because they are usually gamma-irradiated to kill off potentially harmful microbes. They have the other major advantage that they can be sliced up into tiny pieces, making them suitable for small fish, whereas freeze-dried foods that have had their moisture removed no longer possess a fleshy texture.

Most frozen foods are often sold in pouches and simply need to be defrosted thoroughly before being placed in the aquarium. As with all foods, however, the amount required will depend on the size and number of fish in the aquarium as well as their individual appetites, so you may not need to use a whole pouch. Defrost just what you need because, once thawed, these foods will rapidly deteriorate.

Freeze-dried food can be used straight from the packet. It tends to float at first, which makes it suitable for fish that feed at the surface of the water. Marine flake, which comes in the form of very fine wafers, sinks slowly through the water. A heavier option, which sinks immediately and is intended for fish that live at the bottom of the aquarium, is food in the form of granules.

▲ ADDING BRINE SHRIMP CULTURES AT FEEDING TIME. THESE SHRIMPS CAN BE CULTURED IN A SEPARATE AQUARIUM.

Again, it is important not to overfeed, otherwise the uneaten granules will break down and begin to pollute the substrate.

If there are insufficient algae growing in the aquarium, herbivorous fish—rabbitfish, for example—can be offered other vegetable matter, such as organic lettuce. It can be tucked under a convenient rock to prevent the leaves from floating round the aquarium. The leaves can be blanched beforehand to improve their digestibility. A fresh, uncontaminated supply is easy to provide if you grow a lettuce plant on a windowsill and pick off the leaves as required.

Only formulated foods, such as flake, are usually supplemented with a full range of vitamins and minerals, although some prepared livefoods, such as brine shrimps, may be enhanced in this way. If you are using other types of food, you therefore need to give your fish a varied diet to ensure they do not develop deficiencies. You should also try to persuade them to eat some formulated food. Do not use these foods after the recommended use-by date on the pack, however, otherwise their vitamin and mineral content will start to decline. They should be stored in a cool dry location and kept sealed to preserve their nutritional value.

Brine shrimp cultures can be set up very easily, providing an ongoing supply of food that is suitable for certain fish and invertebrates. The brine shrimp eggs are sold in sealed airtight containers, designed to exclude moisture that would reduce their viability. It can therefore be false economy to buy a large pack.

Special hatching kits for brine shrimp eggs are available, although a small spare tank could be used for this purpose. There must be an airstone to encourage rapid water movement and to stimulate the hatching process, in addition to a heaterstat.

The eggs hatch into so-called nauplii, which can be useful for feeding invertebrates and smaller fish. You can simply sieve them out of the water. Alternatively, it may be possible to buy live brine shrimps and mysid shrimps in some outlets. Both are staple foods for seahorses. If you are hoping to breed any of your fish, then tiny planktonic creatures called rotifers are likely to be required for rearing purposes.

Although it is always interesting to watch the fish feeding, it is also important as an indicator of health—a refusal to feed can be a sign of impending illness. In some cases you may be able to tame fish sufficiently to feed from your hand, but take care not to get bitten. Choose a relatively large morsel of food for this reason. There are also special tablets of food that can be stuck onto the aquarium glass, allowing you to watch the fish eating more easily.

There are some special foods produced for certain invertebrates, but others—crustaceans and starfish, for example—will eat food provided for the fish. Bear in mind, though, that some species are nocturnal, so you need to provide food at dusk. Sessile invertebrates that are anchored on the reef need food to be delivered to their immediate vicinity, and you should therefore use a pipette if necessary.

Anemones that have zooxanthellae in their tissues will be able to obtain some nutrients from these algae. In the case of invertebrates that tend to scavenge for their food, do not simply put in extra food for the fish in the hope than any leftovers will be eaten by invertebrates. Aim to feed each group independently. It may even be possible to tame crustaceans such as crabs by developing a regular feeding routine for them.

MAINTENANCE CHECKS

You will need to continue to make regular checks on water quality, testing for nitrogenous compounds and also checking the pH. A declining pH suggests that water conditions are deteriorating. The pH should not fall below 7.8, which is still on the alkaline side of the pH scale. If it falls lower, a partial water change will be needed. You should replace about one-quarter of the total volume of the aquarium.

Using a special siphon, you can draw off water from the aquarium into a bucket without disturbing the occupants. You will have to mix up a replacement salt solution. Prepare the water the day before it is to be used. Bring it up to the correct temperature, using an aquarium heater, and ensure that it is adequately aerated. It should be possible to pour the new solution into the aquarium very carefully, so as not to upset the decor or create a strong current that will disturb the fish.

Inevitably, some of the water in the aquarium will evaporate over time. This will cause the concentration of salt to increase and will raise the SG reading. If you find the SG figure rises, then just add dechlorinated freshwater rather than a salt solution.

Any decline in water quality may be linked to a failure of the filtration system. It is important to check when you are feeding the fish that this is functioning properly. You should also glance at the thermometer to see that there is no problem with the heating. Filter systems need relatively little maintenance, although foam cartridges or filter sponges can become blocked by accumulated debris. This will not only reduce the flow rate through the unit, but it is also likely to have a detrimental effect on the population of beneficial bacteria that is established here.

Rather than replacing the medium, it is much better to service it when you carry out a partial water change. Wash the filter sponge out in some of the old tank water

MAINTENANCE CHECKLIST

DAILY

1 Check livestock—number and general health
2 Check temperature
3 Empty protein skimmer waste
4 Check all equipment is working properly
5 Remove any uneaten food after feeding
6 Enter observations into aquarium log
7 Check water flow to denitrification filter
8 Top up water in osmolator reservoir
9 Top up fresh- and saltwater reservoirs in automatic water change systems
10 Adjust ozonizer output, if required

EVERY OTHER DAY

1 Top up evaporated water
2 Remove algae from front glass
3 Clean prefilters in systemized tanks

WEEKLY

1 Clean cover glasses, if fitted
2 Remove any "salt creep"
3 Add trace elements, pH buffers, and vitamin supplements, etc., if necessary

EVERY TWO WEEKS

1 Change 15–25% of water, depending on how heavily the aquarium is stocked
2 Test for ammonia, nitrite, pH, nitrate, and specific gravity in well-established tanks (more often in new setups)
3 Replace filter floss in canister filter
4 Disconnect fluorescent tubes when cool, clean with plain water, and dry thoroughly
5 Clean probes of electronic meters and controllers with a brush; check for damage

MONTHLY

(More often if required)

1 Rake through the coral sand substrate.
2 Siphon off any detritus

BI-MONTHLY

1 Replace any airstones, including protein skimmer diffusers
2 Change carbon
3 Clean protein skimmer
4 Remove excess algae (this may need to be done more often, or less often, than bi-monthly)
5 Check all electrical connections

EVERY THREE MONTHS

1 Clean pump impellers, pipework, internal housing, and check for wear
2 Clean out all canister filter hoses with suitable hose brushes
3 Change air filter pads in air pumps and check diaphragm for wear
4 Clean the internal quartz sleeve of the ultraviolet sterilizer

EVERY SIX MONTHS+

1 Replace ultraviolet sterilizer tubes
2 Renew fluorescent tubes and other lighting if required
3 Change damaged or worn pump impeller assemblies
4 Renew nonreturn air valves

Safety: Saltwater and electricity are a lethal combination. Always fit a plug-in residual circuit breaker (RCB) to the main power supply.

▲ ▶ *REGULAR MAINTENANCE IS CRUCIAL, INCLUDING REMOVING DETRITUS FROM THE SURFACE OF GRAVEL WITH VACUUM CLEANER (ABOVE AND INSET) AND CLEANING THE GLASS SURFACE (RIGHT).*

that has been siphoned out of the aquarium and then reassemble the unit. This will preserve the bacterial population. If you rinse it under a faucet, the chlorine will destroy the bacteria and will have a potentially catastrophic effect on the functioning of the filter.

It is also important to maintain the general appearance of the aquarium. The cover glass, for example, often starts to show signs of algal growth, which will need to be wiped off at regular intervals. It is also a good idea to invest in an algal scraper with which you can clean the front of the tank. The long-handled designs are easier to use than those that rely on magnets to anchor them on each side of the glass.

Watch for blockages in the air supply from airstones, which may have to be replaced. Spray bars may need cleaning. Other equipment will need to be monitored, and the cup of the protein skimmer will have to be cleaned as necessary. Replacement of fluorescent tubes and ultraviolet lamps should be carried according to the manufacturer's instructions. Last but not least, spend time checking the fish and other tank occupants every day, to ensure they are healthy. Any deaths in the aquarium will soon result in a marked deterioration in water quality, particularly if they go unnoticed for any period of time.

ADDING MORE FISH

There is likely to be a time when you decide you want to add more fish to the aquarium. It is important to have a separate quarantine tank for this

▼ *IF WATER QUALITY DETERIORATES SO FAR THAT THE WATER MUST BE CHANGED, A SIMPLE METHOD IS ILLUSTRATED BELOW, USING AN AQUARIUM PUMP AND A LENGTH OF CANISTER FILTER HOSE.*

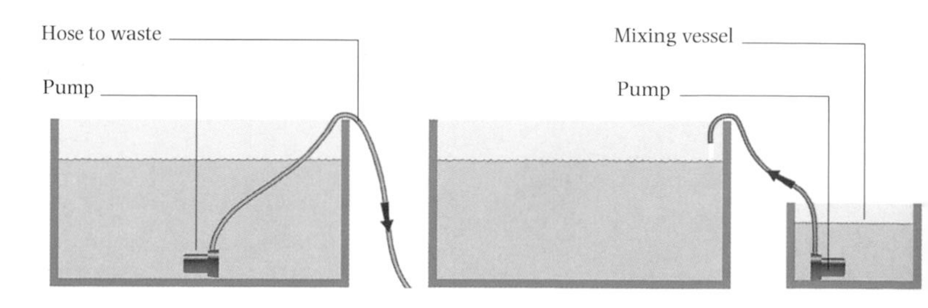

purpose initially, so you can be sure that any new fish are healthy before placing them into the aquarium. They should be kept here for at least two weeks before being transferred to the main aquarium. Otherwise, you could face major problems in dealing with an outbreak of disease, particularly in a reef tank. It is generally best to introduce more boisterous specimens at a later stage to reduce the likelihood of bullying in the aquarium.

Fish differ in their individual temperament, but the chart on the right provides a guide to species that can be housed together safely and—perhaps more importantly—indicates those that cannot be mixed safely in the same aquarium.

BREEDING

The majority of marine fish that are available originate from the wild and are caught under strict regulation in order to prevent any harm to their populations or the reef itself. Increasingly, however, a number of species are being bred in aquaria, including various clownfish, seahorses, and gobies. Much still remains to be learned about the conditions that trigger breeding. In the case of clownfish, they are hermaphrodite. Keeping two together guarantees a pair, because the dominant individual will grow into a larger female. Overall, they are probably the easiest group to breed successfully in aquarium surroundings. A pair will need an aquarium on their own, and it may take up to a year before sexual maturity is attained.

Egg laying occurs on a rock cleaned by the pair beforehand. The male fertilizes the eggs as they are laid. He then remains close by, guarding them until they hatch about a week or so later. If you cannot provide fresh rotifers, it may be possible to rear the fry on thawed rotifers, with brine shrimps and then powdered flake food being introduced as the fish grow slightly bigger. Filtration in the aquarium needs to be effective yet gentle to prevent any deterioration in water quality. Be prepared to divide the young clownfish up into separate tanks as they grow, to prevent overcrowding, which will be harmful. The young fish will start to show their characteristic coloration and patterning from the age of about three weeks.

▶ *A chart indicating which groups of fish can be kept together and which cannot. Some of the less familiar groups, such as sharks and remoras, are not covered in this book. They may, however, be available to the aquarist and are therefore included in the chart for reference.*

FISH IDEAL WITH ALL INVERTEBRATES

Anemonefish
Angelfish, Dwarf
Blennies
Cardinalfish
Damselfish
Firefish
Grammas
Gobies
Jawfish – *deep substrate required*
Mandarinfish
Pygmy Basslets
Wrasses, Dwarf

FISH GENERALLY COMPATIBLE (only with care)

Catfish – *become progressively distructive with age*
Hawkfish – *shrimps may be at risk*
Sea Basses – *not all species safe with invertebrates*
Seahorses and Pipefish – *preferably species tank*
Squirrelfish – *crustaceans at risk*
Surgeonfish – *can be difficult to treat for disease in the presence of invertebrates*

FISH NOT COMPATIBLE WITH INVERTEBRATES

Angelfish – *may be safe when very young*
Batfish
Boxfish
Butterflyfish
Filefish – *some smaller species may be suitable*
Groupers
Lionfish – *crustaceans and small fish at risk*
Moray Eels - *safe only when young*
Pine-Cone Fish
Porcupinefish
Pufferfish
Rabbitfish
Remoras
Sharks and Rays
Sweetlips – *safe when young, thereafter very destructive*
Tigerfish – *very destructive*
Trumpetfish
Wrasses – *safe when young, progressively destructive with age*

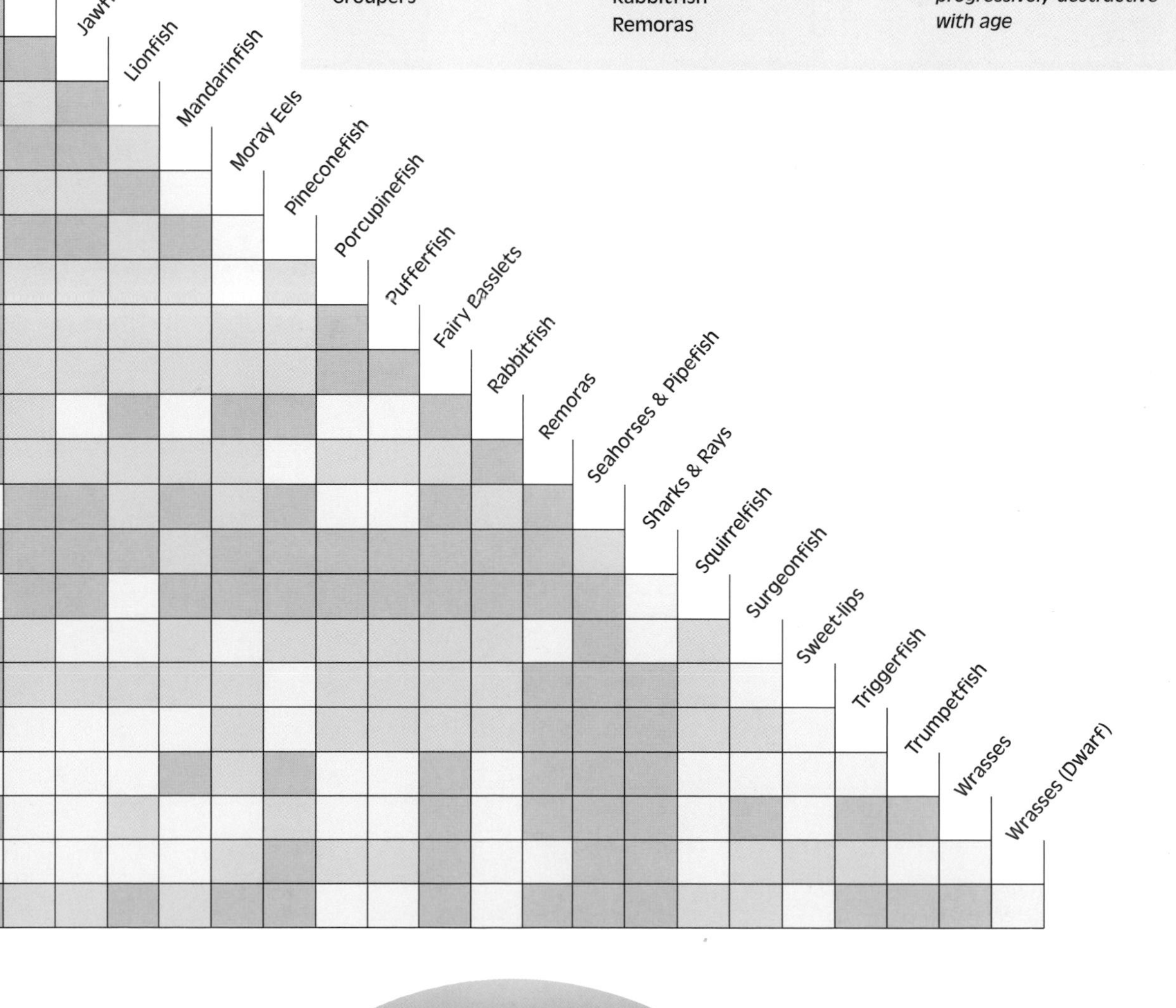

Health Matters

It is usually not too difficult to spot when a fish has a health problem. It will usually demonstrate a change in behavior, often becoming less active than usual. Its color may also alter and it may lose interest in food.

If you have a fish that you suspect is ill, you should transfer it to separate quarters as soon as possible. This should not reduce its likelihood of recovery and will protect any other fish that are sharing the aquarium. Another reason for not leaving a sick fish in the main aquarium is that some treatments will interfere with the biological activity of a filter, killing off the beneficial aerobic bacteria. On the other hand, the presence of activated carbon in the aquarium will render remedies ineffective by adsorbing them out of the water. Furthermore, it is not possible to treat fish safely in a reef tank if the remedy contains copper, because this is deadly for invertebrates.

Parasitic Ailments

Parasites are the most common causes of disease in saltwater fish, and parasitic ailments can often spread rapidly in the confines of an aquarium. **White spot**, caused by a microscopic parasite called *Cryptocaryon irritans,* is a particular problem that can be introduced to an aquarium very easily if you do not quarantine the fish for a couple of weeks before placing them in the tank. If the parasite does gain access to the main tank, all the fish are potentially at risk, although certain species such as damselfish seem more vulnerable than others to this infection.

▼ White spot, or marine itch, appears as small white spots covering the fins and body of a fish. It is caused by a parasite (*Cryptocaryon irritans*) and is highly infectious.

The signs of white spot are generally quite easy to identify. The fish's body becomes covered in a series of tiny white spots, which are cysts full of the tiny parasitic organisms. The cysts then break down, liberating huge numbers of parasites into the aquarium water and leaving tiny open wounds on the fish's body, which are then liable to become infected with bacteria or fungi. White spot can be treated successfully, but it is obviously important to remove the fish and treat them elsewhere, preferably before their spots break down.

Ultraviolet sterilization can help kill the parasites in the aquarium during the free-swimming stage in their life cycle, when they are at their most vulnerable. An ultraviolet unit of this type is particularly helpful in a reef tank, because it is possible to eliminate parasites without resorting to treatments that are likely to be harmful to invertebrates. Otherwise, if left, the white spot parasites will have no difficulty in finding further hosts, and repeated cycles of infection will prove extremely debilitating for the fish.

Cryptocaryon is one of a group of protozoan parasites, all of which are too small to be seen with the naked eye. Another relatively common parasite of this type is *Amyloodinium ocellatum*, which causes **velvet disease**. Initially this parasite attacks the gills of the fish, causing a rapid increase in its breathing rate, before giving rise to distinctive velvetlike patches over the fish's body. Velvet disease can be fatal in the early stages because of the damage it causes to the fish's gills, although—just as with white spot—the fish may later succumb to a secondary infection. Proprietary copper-based treatments are usually recommended to combat these parasites.

Head and lateral line erosion (HLLE) disease is another protozoan ailment, caused in this case by a parasite known as *Octomita necatrix*. It results in the appearance of a hole in the vicinity of the head, which is why it is sometimes also known as hole-in-the head disease. It can affect all fish, although it is most likely to be seen in tangs. There is a suggestion that they may carry the parasites in their intestinal tract. The parasites only multiply to the point where they cause disease when the fish is under stress, for example, if the nitrate level of the water is elevated or if it is suffering from a nutritional deficiency.

Offering these herbivorous fish a diet that includes blanched broccoli can be helpful both as a preventative measure and also during treatment of HLLE using a proprietary remedy. Unfortunately, even if the fish makes a successful recovery, it is likely to be left with residual scarring.

▲ *TOMINI SURGEONFISH (CTENOCHAETUS TOMINIENSIS) SHOWING HEAD AND LATERAL LINE EROSION DISEASE. IT OFTEN RESULTS IN THE DEATH OF THE FISH.*

Other, larger **external parasites** are sometimes encountered on marine fish, but the presence of flukes, for example, should usually be spotted by any dealer who quarantines his stock. That is why it is worthwhile to buy stock only from reputable sources, which should ensure that the fish are as healthy as possible. Although you may pay more in the short term, there will be long-term benefits from this approach, not least of which is the fact that the fish should be easier to establish in new surroundings.

Fish have reasonably effective immune systems, however, and their bodies are covered in a thin film of protective mucus. As a result, you should never attempt to handle a fish with dry hands, because this will tend to strip away the protective covering.

A fish's body reacts to damage or injury by producing more mucus, which is why you are often likely to see a more evident covering of mucus over the body surface when a fish has an external parasitic ailment. This is simply a normal reaction by the immune system and may help guard against any following bacterial or fungal infection. External parasitic infections often result in intense irritation, causing the fish to rub itself repeatedly against rockwork and other tank decor. Look out for such behavior, which can be an indication of this type of illness.

A few parasites, such as *Brooklynella* are very host-specific, attacking anemonefish in this case. The impact of the infection is concentrated on the gills rather than the body, and treatment can be carried out using a proprietary remedy, based perhaps on the dye malachite green. You are likely to find a range of such treatments in aquatic stores, with clear instructions for their usage provided on the accompanying packaging. You may need to immerse the fish in a concentrated solution of the medication rather than diluting it and adding to the treatment tank. This approach is likely to be more stressful to the fish in the short term, so do not exceed the recommended contact time. If the fish is showing clear signs of distress, such as very labored breathing, move it back to the treatment tank.

One advantage of this method, however, is that you can use a power filter to maintain the water quality in what then effectively becomes an isolation rather than a treatment tank, which would otherwise be much harder with medication present. The environment in which the fish is kept is often a significant factor in determining whether or not it will recover successfully.

Bacterial and Fungal Ailments

Damage to the fins or body for any reason is likely to trigger opportunistic bacterial or fungal diseases, since the fish's body defenses have been breached. The fins are especially vulnerable to bacterial infections, particularly if the water conditions are poor. Reddening of the affected area and erosion of the fins at the edges causing a ragged appearance are typical signs of **fin rot**. Improving the fish's environment will greatly enhance the likelihood of a successful recovery. Suitable treatment, which may involve the use of antibiotics, should also be provided.

Fungus is especially likely to strike a weakened fish that has not been eating properly, for example. It can gain access to any site of injury on the body, perhaps where scales have been damaged or lost. Topical treatment—applying a suitable cream to the affected area—can often be the most effective way of dealing with the problem. Treat the fish in a net, taking care to ensure that it cannot jump out. Alternatively, treatments can be added successfully to the water. Partial water changes in the main tank will help and will reduce the number of fungal spores present, lessening the likelihood of further outbreaks of disease.

Viral Disease

The most common and obvious manifestation of viral infections in fish is so-called **cauliflower disease**, caused by the *Lymphocystis* virus, which results in disfiguring cauliflower-like growths on the body. They may spread or remain localized, but they do not appear to be highly contagious or irritating to the fish itself. There is no treatment available, although in time the swelling may disappear, rather like a wart.

Life Span

Since most fish are wild-caught, it is often impossible to have any clear indication of their precise age at the time of purchase and therefore predict their life span. But they should usually live considerably longer in an aquarium, free from predators and the stress of finding food for themselves, compared with the more hostile environment in the wild.

As a general guide, larger fish such as eels have a longer life span than smaller ones. The majority will live for a maximum of between four and eight years, although some individuals may still be swimming

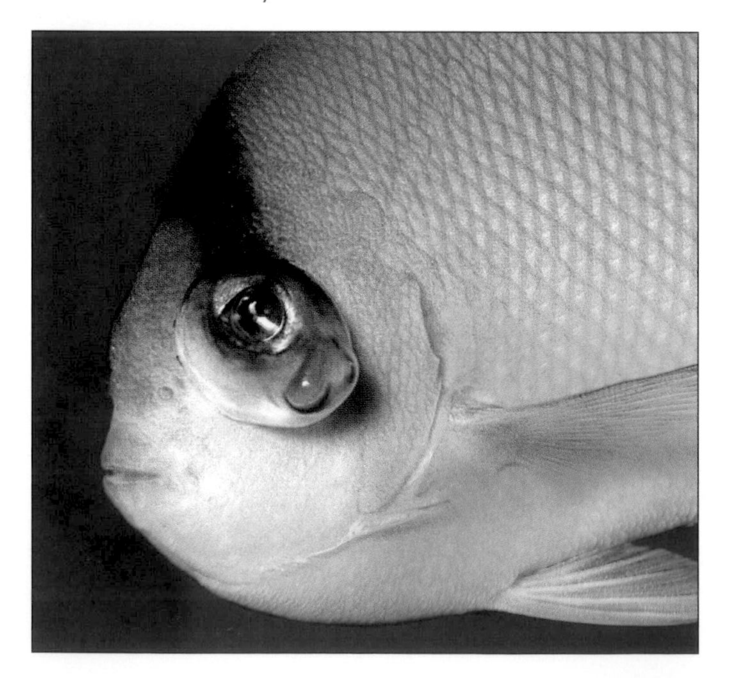

◀ *Angelfish suffering from pop-eye, which may be a sign of infection. One or both eyes can be affected.*

▶ *A sick fish often loses the ability to maintain its balance in water.*

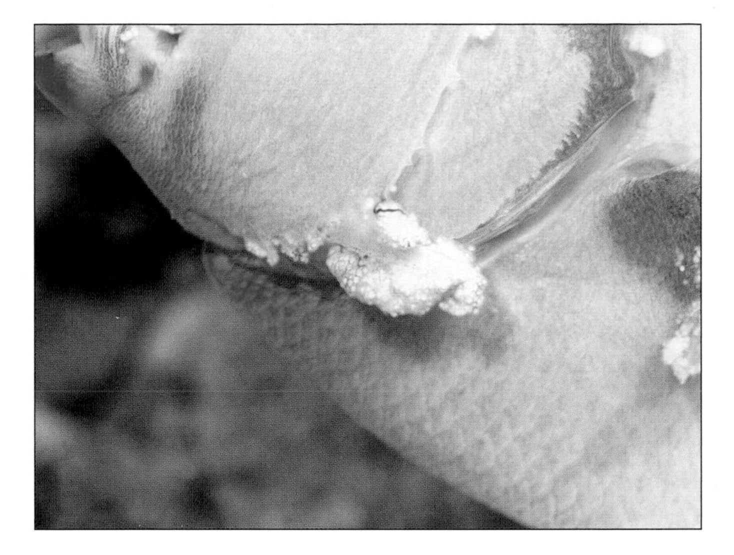

◀ *CAULIFLOWER DISEASE IS A VIRAL INFECTION THAT CAUSES CAULIFLOWER-LIKE GROWTHS TO FORM ON THE SKIN AND GILLS.*

around in the aquarium after ten years or more. Only at the very end of their lives do fish tend to show any obvious signs of age, most frequently losing the ability to swim well, and hanging in the water instead.

DEALING WITH EMERGENCIES

Emergency situations can crop up occasionally, the cause of which may be unclear at first. You may simply notice that the fish are showing serious signs of distress, gulping at the surface and swimming abnormally. Start by checking that all the aquarium equipment is functioning properly and that there are no blockages. Carry out water quality checks to see if there is a problem here. If you have just carried out a partial water change, you may have added the wrong concentration of salt, for example. There may be something in the atmosphere that has been drawn into the water and poisoned the fish. Many ordinary household sprays can be harmful to fish and should not be used, certainly not in the same room as the aquarium.

The best thing to do in this situation is to undertake a fairly large-scale water change of up to 40 percent of the tank volume without delay. Do not use water from the hot faucet to bring it up to temperature because of the risk that it may contain copper from the pipework, which would threaten the health of any invertebrates. Instead, adding water from a kettle may be safer, since time is very important if it is an environmental problem affecting the fish. Add a dechlorinator once the water in the bucket is up to the required temperature, and stir in the marine salt.

If you have a power outage, little can be done without a generator other than trying to insulate the tank as far as possible to prevent heat loss. Disconnect the equipment until the power is restored, to avoid any fire risk. You will need to monitor the water quality closely over subsequent days, because the efficiency of the biological filtration system may have been compromised. Again, more frequent partial water changes will be required until the situation has stabilized.

Tropical Saltwater Fish

This section provides practical information on the care of some of the available tropical saltwater fish. Most can be successfully accommodated within the home aquarium, providing a peaceful and amazingly colorful display. Some are ideal choices for beginners and are relatively straightforward to care for. Others are more unusual and need more specialized conditions—but the rewards for taking the extra trouble are great.

MORAY AND SNAKE EELS FAMILIES MURAENIDAE AND OPHICHTHIDAE

The serpentine shape of moray and snake eels makes them easily recognizable, and they are increasingly popular in the saltwater hobby. It is important to bear in mind, however, that some species will grow to a remarkably large size, which will make them unsuitable for home aquaria. They are also predatory by nature.

SNOWFLAKE MORAY

(STARRY MORAY; CLOUDED MORAY)

Echidna nebulosa Family Muraenidae

An attractive but highly variable patterning is a feature of this eel. Its body is broken by darker markings that form blotches and may tend to encircle it as the eel grows older. There are also yellow spots and speckling on the body, which are mostly evident in the dark areas. The jaws are usually white. The iris surrounding the pupil in the eye is orangish yellow, as are the prominent projecting nostrils located just above the upper jaw. Not only are the markings of individual snowflake morays variable, but there are also regional variations across the very wide distribution area of this species.

SYNONYMS: *E. variegata; Gymnothorax boschi.*

DISTRIBUTION: *Extends from the Red Sea and East Africa northward to southern Japan and southward as far as Lord Howe island off the east coast of Australia. Ranges across the Pacific to the Hawaiian and Society Islands, and is also represented through Micronesia. Occurs in the eastern Pacific region from the Mexican coast of Baja California southward via Costa Rica, reaching as far as Colombia in South America.*

SIZE: *39.5 in (100 cm).*

BEHAVIOR: *Occurs in relatively shallow waters in areas of the reef where there are suitable retreats. Tends to be an ambush predator, preying largely on crustaceans. As with other eels, it swims by moving its body from side to side with a rippling movement. This is made possible by a flexible vertebral column consisting of up to 123 separate bones.*

▼ *SNOWFLAKE MORAY (ECHIDNA NEBULOSA)*

▲ ZEBRA MORAY (GYMNOMURAENA ZEBRA)

DIET: *Predatory, so offer meaty foods, including those derived from crustaceans. This intelligent fish can soon be trained to take pieces of food from a feeding stick, which saves considerable disruption in the aquarium. It will come to recognize the stick and realize that it provides food. Eats very readily as a general rule, so acclimatization to artificial diets is straightforward.*

AQUARIUM: *Although secure rocky hideouts are important for this eel, it is likely to prey on the typical inhabitants encountered in a reef tank.*

COMPATIBILITY: *May agree with another eel of the same species if both are introduced to the aquarium at the same time, but they need to be of a similar size to prevent any risk of cannibalism. Has a relatively bold nature and can be housed with larger companions, which it will not attack.*

ZEBRA MORAY

Gymnomuraena zebra Family Muraenidae

The zebra moray is one of the most suitable moray eels for the saltwater aquarium. Its highly attractive patterning consists of alternating reddish brown and white bands encircling the body, with similarly colored spotted markings on the jaws. The striped body patterning is not entirely regular—as with zebras, there are variations that enable individuals to be distinguished easily.

DISTRIBUTION: *Found over a wide area from the Red Sea and the coast of East Africa across the Indian Ocean, south to Australia's Great Barrier Reef and north as far as southern Japan. Extends across the Pacific and is found around the Hawaiian and the Society Islands. Also ranges from the southern part of the Mexican coast of Baja California, south via Central America as far as Colombia. In addition, it occurs farther southwest, in the vicinity of the Galápagos Islands.*

SIZE: *59 in (150 cm).*

BEHAVIOR: *Occurs not just in rocky areas of the reef, but can also be encountered over sandy stretches after dark. However, this moray eel tends to be shy. It will retreat and hide out of sight if there is any threat of danger. Hunts crabs primarily. Also feeds on other crustaceans as well as sea urchins and mollusks, but does not prey on fish.*

DIET: *Fresh crab may be necessary to persuade a reluctant zebra eel to start feeding in aquarium surroundings. Generally, however, these eels can be weaned onto other invertebrate food such as clam and squid without too much difficulty—the use of a feeding stick will encourage the eel to strike. The stick can be of the type sold for snakes, since it needs to be soft at the tips in order not to damage the mouth of the fish when it lunges at its food.*

AQUARIUM: *Likely to prey on crustaceans and various other invertebrates in a reef tank and may also dislodge any loose rockwork. Tank decor must therefore be firmly positioned and must provide a choice of suitable retreats for the eel.*

COMPATIBILITY: *Can be housed safely with other fish, including smaller companions, without fear that it will attack any of them.*

FIMBRIATED MORAY

Gymnothorax fimbriatus Family Muraenidae

Fimbriated moray eels tend to have a very pale yellow body color, broken by a series of irregular black and brownish spots of various sizes. In juveniles the spots tend to be proportionally bigger than in adults. The spots extend into the dorsal fin that runs along much of the back, from the base of the head, although this may not be immediately obvious when the eel is resting.

Distribution: *Occurs in the Indian Ocean from the island of Madagascar off Africa's south-eastern coast northward as far as southern Japan. Also present on Australia's Great Barrier Reef off the coast of Queensland. Ranges through Micronesia in the Pacific and is found eastward to the area around the Society Islands.*

Size: *31.5 in (80 cm).*

Behavior: *As in most eels, only the head of the fimbriated moray tends to be apparent once it has chosen its lair. It will remain there and seize any prey that comes within reach rather than chasing it across the reef. It is more likely to emerge and hunt under cover of darkness, however, since it is nocturnal. It is often found close to the shore in relatively sheltered areas, and it is not uncommon to see it around harbors. Sometimes ventures into brackish waters.*

Diet: *Predatory by nature, feeding on both fish and crustaceans. Needs a similar meaty diet in an aquarium setting. Beware of this eel's sharp teeth. It is generally better to offer it food using a feeding stick to avoid the risk of being bitten—it can strike with amazing speed.*

Aquarium: *Avoid accommodating this species in a reef aquarium, because it is likely to prey on a number of the occupants.*

Compatibility: *Do not trust this moray eel with smaller companions, and bear in mind that the species can grow to a relatively large size. It may therefore be better to house it in an aquarium on its own.*

GRAY-FACE MORAY

(WHITE-EYE MORAY; YELLOW-LINED MORAY)

Gymnothorax thyrsoideius Family Muraenidae

A pale yellowish white background color with a contrasting pattern of small, sometimes dense, reddish spots is characteristic of this eel. The red mottling on the body may be more intense in some cases than others. Its eyes are also distinctively colored.

▼ Fimbriated Moray (Gymnothorax fimbriatus)

◄ BARRED-FIN MORAY (GYMNOTHORAX ZONIPECTIS)

The iris surrounding the pupil is whitish, which gives the gray-face moray an unusual appearance.

DISTRIBUTION: *Ranges from the waters around Christmas Island off the southern tip of Sumatra, northward as far as the Ryukyu Islands near Japan. Extends via the islands that form Micronesia eastward to French Polynesia and also occurs as far south as Tonga.*

SIZE: *25.5 in (65 cm).*

BEHAVIOR: *Occurs in shallow water, often in tidal pools and usually not below 10 ft (3 m) from the surface. May colonize a ship that has been wrecked on the reef, and is sometimes seen there in association with other moray eels. The gray-face moray is a relatively social species, normally encountered in pairs, but sometimes seen in larger groups.*

DIET: *Feeds on both fish and crustaceans, so requires a suitably meaty diet.*

AQUARIUM: *Will prey on various occupants in a reef tank.*

COMPATIBILITY: *If two of these eels are introduced to the aquarium at the same time, especially when young, they may agree together. Be sure to provide a choice of retreats, which can include disguised plastic tubing, hidden among firmly supported rocks. Providing a range of hiding places will minimize the risk of aggression in the confines of the aquarium.*

BARRED-FIN MORAY

(BAR-TAILED MORAY)

Gymnothorax zonipectis Family Muraenidae

There is a distinct difference in appearance between juveniles, which have a mottled brown pattern on their heads, and adults, which are recognizable by their brown heads. Although it is not immediately apparent when the eel's body is concealed, there is black barring present at the rear of the dorsal fin and on the anal fin. There is also a characteristic dark spot behind each eye, which tends to be edged above and below with white blotches. Large white spots are also evident along the lower jaws.

DISTRIBUTION: *Ranges from the coast of East Africa through the Indian Ocean, reaching as far north as the Philippine Islands and extending to Australia in the south. Occurs through the Pacific to the Society and Marquesan Islands.*

SIZE: *19.75 in (50 cm).*

BEHAVIOR: *This relatively small moray is solitary by nature. It tends to occur in deeper water than many species, lurking in caves between 65 and 130 ft (20–40 m) from the surface during the day and emerging from its hiding place at night. Its mouth contains a formidable array of sharp teeth with which the eel seizes its prey, which is thought to consist of other fish and invertebrates.*

DIET: *Meaty substitute foods are required, which must be thawed out thoroughly if frozen before being offered to the eel. Aim to provide a variety of foods.*

AQUARIUM: *Avoid housing this eel in a reef tank, partly because of its predatory nature, but also because it will not appreciate the bright illumination of a shallow-water reef aquarium.*

COMPATIBILITY: *Because of its size, this species is more easily housed in an aquarium than many of its larger relatives, but it still possesses highly developed hunting instincts. It must, therefore, never be housed with smaller fish, which it may prey on.*

RIBBON MORAY

(BLUE RIBBON EEL; BLACK RIBBON EEL)

Rhinomuraena quaesita Family Muraenidae

These particularly slender eels undergo a dramatic change in coloration as they mature. Whereas juveniles are black with a yellow dorsal fin, mature females are almost entirely yellow, apart from a blue area toward the rear of the body in some cases. Males, by contrast, are generally all-blue, with the exception of yellow markings on the head and a yellow dorsal fin. This species is highly unusual, because all the females in a population start out as males. Once they reach approximately 33 inches (85 cm), the males begin to change into females. This is possibly a result of hermaphroditism, meaning that all individuals possess both male and female sex organs.

DISTRIBUTION: *Ranges from East Africa across the Indian Ocean north as far as southern Japan. Common throughout Indonesia and extends southward in the Pacific to New Caledonia and French Polynesia, reaching as far east as the Tuamotu Archipelago.*

SIZE: *51 in (130 cm).*

BEHAVIOR: *The ribbon moray is a shy species that ambushes its prey while remaining largely buried in the substrate. It has slime glands that serve to lubricate its body and protect it from injury as it burrows into sand. Its slender shape, combined with the fanlike extensions on its nose, help disguise its presence.*

DIET: *Feeds almost entirely on small fish. Unfortunately, it can be difficult to persuade it to feed, especially in new surroundings. If it will not take fish-based meaty food, the only alternative may be to offer live mollies (such as* Poecilia velifera*), which should be able to tolerate the salty water reasonably well. However, it is important to make sure that the fish are healthy before placing them in the aquarium alongside the eel. Hopefully, it will soon be possible to persuade the eel to progress to eating inert foods.*

AQUARIUM: *May prey on small fish and crustaceans in aquarium surroundings, as well as hiding among gaps in the living rock. Be sure that airlift tubes are not left uncapped in the aquarium, because the eel might try to disappear down them.*

COMPATIBILITY: *May prey on smaller companions but can often prove quite social with its own kind, and may even share suitable retreats.*

HARLEQUIN SNAKE EEL
(BARRED SNAKE EEL)

Myrichthys colubrinus Family Ophichthidae

The characteristic bold patterning of this eel mimics that of the deadly sea snake *Laticauda colubrina*, with which it is easily confused. The harlequin snake eel is completely harmless, however. It is essentially patterned with broad bands of dark reddish brown and white, but there may also be several dark blotches apparent in the white areas of some individuals.

DISTRIBUTION: *Ranges from the Red Sea south down the eastern coast of Africa, reaching Delagoa Bay, Mozambique, and through the Indian Ocean. Extends across the Pacific as far as the Society Islands.*

SIZE: *38 in (97 cm).*

BEHAVIOR: *Occurs in shallow coastal areas of water, both where seagrass is present and over open sandy areas. It is not uncommon to find it burrowing into the sand and hiding there. It is a diurnal species, like the snake it mimics, and so will be observed hunting for prey in the daytime. Other fish, notably young tobies (*Canthigaster *species) will attempt to drive it away from an area by nipping at its tail.*

DIET: *Often proves reluctant to feed in aquarium surroundings. Its natural prey includes small fish and crustaceans. Wafting similar meaty substitutes around in the aquarium with a feeding stick and encouraging the eel to lunge at this food can trigger a regular feeding response.*

AQUARIUM: *Not to be trusted alongside small fish and crustaceans.*

COMPATIBILITY: *May be bullied by some fish. Needs a sufficient layer of sand at the base of the tank in order to be able to burrow and to feel secure. It will be more likely to feed under these circumstances. May appear to "freeze" at first when transferred to new surroundings.*

▼ *RIBBON MORAY (RHINOMURAENA QUAESITA)*

SQUIRRELFISH AND SOLDIERFISH FAMILY HOLOCENTRIDAE

Members of the family Holocentridae are typically nocturnal in their habits. This helps explain their large eyes, which enable them to see in conditions where there is little available light. They must have plenty of retreats in their aquarium and should not be kept under bright lighting. Low-intensity nighttime tubes are recommended for an aquarium that houses members of this family. Such lighting will enable the fish to be seen clearly but will not disturb them or cause them to retreat from view.

PINECONE SOLDIERFISH

Myripristis murdjan

The edging on the scales of this fish creates the impression of a pine cone. The scales are predominantly silvery pink in color, with a reddish surround. The rear edge of the operculum covering the gills is black, but there is no large spine on the gill cover itself, which is sometimes a feature of members of this family. White edging is prominent on the leading edges of most of the fins, which are suffused with red.

DISTRIBUTION: *Ranges from the Red Sea and East Africa eastward across the Indian Ocean, extending as far as the islands of Oceania in the Pacific.*

SIZE: *12 in (30 cm).*

BEHAVIOR: *By virtue of its nocturnal habits this fish tends not to be very conspicuous on the reef. It seeks out retreats such as caves on the reef, where it can remain hidden during the daytime. When it does emerge, it feeds mainly on zooplankton that is wafting in the water.*

DIET: *Can be given vitamin-enriched brine shrimps and mysid shrimps as well as other meaty foods, which should be chopped up as necessary.*

AQUARIUM: *Not to be trusted in a reef tank, since it may consume occupants, such as crustaceans, in these surroundings under cover of darkness.*

▼ *PINECONE SOLDIERFISH (MYRIPRISTIS MURDJAN)*

COMPATIBILITY: *Avoid small companions. Can usually be kept in a group with its own kind in a large aquarium, provided all the fish are introduced at the same time.*

SPOT-FIN SQUIRRELFISH
(SAMMARA SQUIRRELFISH)

Neoniphon sammara

A horizontal striped patterning is characteristic of this species. It displays narrow black lines interspersed with more widely spaced silvery pink lines. There is a network of fine black spots evident below the very large eyes, and these markings extend back along the underside of the body below the gills. A large prominent black spot is present at the front of the dorsal fin, but how far back it reaches varies from individual to individual.

DISTRIBUTION: *From the Red Sea and East Africa across the Indian Ocean to southern Japan, down to northern Australia and south to Lord Howe Island, which lies off the east coast of Australia. Extends through the Pacific to the Hawaiian, Ducie, and Marquesan Islands.*

SIZE: *12 in (30 cm).*

BEHAVIOR: *The spot-fin squirrelfish is encountered in shoals, often in areas of the reef where seagrass predominates and not uncommonly in relatively shallow water. It is also frequently seen in association with* Acropora *species coral on the reef. It is protected by a venomous spine on each side of the head and therefore needs careful handling in aquarium surroundings. In particular, when servicing the tank, avoid accidentally brushing against this sharp projection.*

DIET: *Requires a meaty diet, which should include crustaceans, cut into small pieces as necessary.*

AQUARIUM: *Not suitable for a reef tank, since it will prey on crustaceans there, particularly after dark.*

COMPATIBILITY: *Can be kept in small groups. The spot-fin squirrelfish has a surprisingly wide gape, however, and small companions are likely to be eaten.*

▲ *SPOT-FIN SQUIRRELFISH (NEONIPHON SAMMARA)*

SIZE: *6.75 in (17 cm).*

BEHAVIOR: *Like other members of the family, this species tends to become more active at night, when it emerges from its daytime hiding places on the reef. It then swims over open areas in search of invertebrate prey, which can range from small clams to crustaceans.*

DIET: *This squirrelfish requires meaty foods. It usually feeds readily in aquarium surroundings, provided ithe food is offered once it starts to become dark.*

AQUARIUM: *Cannot be housed in a reef tank, since it will feed on a variety of the invertebrates typically included in this environment.*

COMPATIBILITY: *May not agree very well with members of its own species, but can be kept successfully with other members of the family.*

CROWN SQUIRRELFISH

Sargocentron diadema

Horizontal alternating red and silver barring running down the sides of the body from behind the gills helps identify this species. Other distinguishing features include two white vertical streaks behind each eye and a diagonal white line that runs below the eyes up to the jaws.

DISTRIBUTION: *Ranges from the Red Sea and East Africa northward to the Ryukyu and Ogasawara Islands near Japan and southward to northern Australia, reaching Lord Howe Island off the east coast. Extends across the Indian Ocean via Micronesia to the Hawaiian and Pitcairn Islands in the Pacific.*

▼ *CROWN SQUIRRELFISH (SARGOCENTRON DIADEMA)*

Seahorses
Family Syngnathidae

Seahorses are among the most instantly recognizable of all marine fish and they are very appealing occupants of a marine aquarium. Unfortunately, they can prove very difficult to maintain in such surroundings, largely because of their feeding habits. Some species have undergone a marked decline in numbers, having been widely caught for use in eastern medicine. Some are also dried and sold as curios, and trade in members of this family is now being monitored to prevent overexploitation. The good news, however, is that tank breeding of seahorses is becoming more common and seems set to supply the aquarium trade with increasing numbers of these fish in the future. Pipefish have very similar requirements to seahorses but are far less commonly kept in aquaria.

▼ *Lined Seahorse (Hippocampus erectus)*

LINED SEAHORSE

Hippocampus erectus

The lined seahorse can vary widely in coloration depending on its area of origin and the background against which it finds itself. This fish can adjust its coloration to blend into its surroundings. Lined seahorses may be yellowish with orange areas on the body or very dark, almost black, with a lighter-colored belly.

Synonyms: *H. brunneus; H. fascicularis; H. hudsonius; H. kincaidi; H. marginalis; H. punctulatus; H. stylifer; H. villosus.*

Distribution: *Ranges widely through the western Atlantic from as far north as the coast of Nova Scotia in Canada, southward down the coast of North America, through the Caribbean and Gulf of Mexico to Panama and Venezuela. It may even extend down to the vicinity of Rio de Janeiro, Brazil, although the southern form may prove to be a separate species.*

Size: *6 in (15 cm).*

Behavior: *Seahorses swim vertically through the water using the dorsal fin at the rear of the body to propel themselves. The prehensile tail enables them to anchor on coral, where they can rest. Even if they are not being accommodated in a reef tank, they will need anchorage points of this type.*

Diet: *Seahorses need live brine shrimps or mysid shrimps offered for long periods regularly throughout the day. They are slow to feed and will otherwise not obtain adequate nourishment. It may be possible in time to wean them onto thawed foods of this type, but livefoods will be more palatable. Use vitamin-enriched foods as far as possible.*

Aquarium: *Ideal for a reef aquarium, but do not include anemones or corals that possess stinging tentacles alongside this seahorse.*

Compatibility: *A peaceful species and one of the most suitable for the aquarium, owing to its adaptable nature. Must be kept only with small inoffensive species, but preferably on its own or with its own kind.*

THORNY SEAHORSE

Hippocampus histrix

The coloration of this seahorse can vary dramatically among individuals. Some are virtually white in color, but they can appear yellow, red, or brown. The most obvious distinguishing features of the thorny seahorse are its relatively long snout and the prominent spiky projections running down its back.

▲ THORNY SEAHORSE (HIPPOCAMPUS HISTRIX)

DISTRIBUTION: *From the sea around Tanzania south down the east coast of Africa to South Africa. Ranges across the Indian Ocean north as far as southern Japan and south to New Calendonia. Extends as far as Hawaii and Tahiti in the Pacific, although its taxonomy is being debated, and it may prove to represent more than one species.*
SIZE: *6.75 in (17 cm).*

BEHAVIOR: *Encountered in shallow areas of the reef. Has also been seen at depths of more than 270 ft (82 m). Breeding behavior is typical of the family. After spawning, the male carries the eggs in a special brood pouch on the front of his body. They hatch about a month later as miniatures of their parents. They are ejected into the sea to fend for themselves.*

DIET: *Mysid shrimps and brine shrimps should be offered. They should be vitamin enriched if possible to ensure that the seahorse receives a balanced diet. Food needs to be provided in close proximity to this species, which explains why thorny seahorses often fare better in smaller aquaria.*

AQUARIUM: *Ideal for a reef tank in which there are no sessile invertebrates with stinging tentacles that could represent a danger to this seahorse.*

COMPATIBILITY: *Keep in a small group, since it is likely to lose out in competition for food if kept in the aquarium with other more active fish.*

SPOTTED SEAHORSE

(YELLOW SEAHORSE; BLACK SEAHORSE)

Hippocampus kuda

The network of tiny black dots over much of its body gives this species the name spotted seahorse. Its basic coloration, however, is variable. It can range from a yellowish shade to black, hence the two alternative common names. This is the species that tends to be most commonly available to marine aquarists. It will breed readily if conditions are favorable.

SYNONYMS: *H. aterrimus; H. chinensis; H. hilonis; H. horai; H. melanospilos; H. moluccensis; H. novaehebudorum; H. polytaenia; H. rhynchomacer; H. taeniops; H. taeniopterus; H. tristis.*

DISTRIBUTION: *Ranges from the central part of the Indian Ocean, from the coast of Pakistan and India, eastward via the Philippines to southern Japan and through the Pacific to the island of Hawaii and the Society Islands.*

SIZE: *12 in (30 cm).*

BEHAVIOR: *One of the larger seahorses, the spotted seahorse is sometimes encountered in brackish as well as marine habitats. It can blend in well on the reef, relying on camouflage coloration to conceal its presence.*

DIET: *Offer a typical seahorse diet based on live mysid and brine shrimps, augmented with vitamin-enriched, thawed foods of the same type. Almost constant access to food will be beneficial, because seahorses are slow to capture sufficient prey to meet their needs.*

AQUARIUM: *Avoid dangerous corals and anemones. A reef tank makes an ideal environment for this seahorse.*

COMPATIBILITY: *Will agree well together, but must not be housed with fish that will harry seahorses. Any stress may trigger the development of ick, to which seahorses can be susceptible. Check stock carefully for signs of this parasitic ailment prior to purchase.*

LIONFISH AND SCORPIONFISH FAMILY SCORPAENIDAE

In the marine aquarium hobby lionfish are probably the best-known members of this family. Lionfish and scorpionfish are spectacular, but they must have a large and relatively uncluttered aquarium that allows them to swim without difficulty, because their bodies are surprisingly bulky. In spite of their rather slow and languid swimming style, however, these fish are very effective predators, which means that any companions need to be chosen with care. It is also important to concentrate when you are attending to their needs in the aquarium, particularly any time that you have to place your hands into the water, otherwise you may end up being stung by their venomous spines. If they feel threatened, they will lunge with their spines held erect to defend themselves, so move your hand slowly in the water to avoid frightening the fish. Wear a suitable glove for extra protection. If you have the misfortune to be stung, place your hand immediately under hot running water or in a bowl of hot water. This will help neutralize the venom and make the injury less painful, but seek qualified medical advice without delay.

TWO-SPOT LIONFISH

(TWIN-SPOT TURKEYFISH; FU MANCHU LIONFISH)

Dendrochirus biocellatus

Mottled red and brown markings extend over the body of this fish, with a prominent white bar across the snout just in front of the eyes. There are conspicuous banded feelers each side of the mouth as well. Particular care needs to be taken around this fish, because it has sharp venomous dorsal spines present on the dorsal fin. The pectoral fins are enlarged into a fan shape and they display alternating light and dark banding. The most distinctive features of this lionfish, however, are the two dark eyespots (known as ocelli) that are present on the soft part of the dorsal fin. They help confuse would-be predators, which are fooled into thinking that the spots are the fish's eyes. At the same time, they

▼ *Two-spot Lionfish (Dendrochirus biocellatus)*

may serve to lure potential prey more easily within range of the fish's mouth, which is located at the opposite end of the body to the eyespots.

DISTRIBUTION: *Occurs in the southern Indian Ocean, extending from the islands of Réunion and Mauritius northeastward to the Maldives off the southwestern coast of India. It is also present around Sri Lanka. Its range extends north from there as far as southern Japan, reaching to the south as far as the Scott Reef off Australia's northwest coast and to the east as far as the Society Islands in the Pacific.*

SIZE: *5 in (13 cm).*

BEHAVIOR: *Shy by nature, this lionfish is nocturnal in its habits. It hides away in caves during the day and emerges to hunt under cover of darkness. It can use its long pelvic fins rather like legs to drag itself over the substrate, and tries to get as close as possible to its prey before launching itself with remarkable speed at its target. When two of these fish encounter each other, the dominant individual can be recognized because its eyespots turn grayish. The same change occurs in males prior to mating.*

DIET: *Meaty foods are required, although this lionfish can sometimes be reluctant to take inert food at first and is often harder to feed at this stage than other lionfish.*

AQUARIUM: *There is a risk of this fish eating both crustaceans and smaller fish in a reef aquarium.*

▲ *SHORT-FIN LIONFISH (DENDROCHIRUS BRACHYPTERUS)*

COMPATIBILITY: *Avoid boisterous companions and species such as pufferfish that have a reputation for fin-nipping. Its shy nature means that this lionfish must have adequate retreats in the aquarium, which will help it settle down. Be aware that these lionfish can become aggressive toward each other, especially as they grow larger.*

SHORT-FIN LIONFISH

(DWARF LIONFISH; SHORT-FIN TURKEYFISH)

Dendrochirus brachypterus

There is significant natural variation in the appearance of these lionfish. In individuals that are mainly brown in color, some are much redder than others. There is even a rare lime-yellow variant that is occasionally seen. Short-fin lionfish are very good at disguising themselves. Darker barred markings are evident across the large pectoral fins and, in the brown form, there is also white banding evident here. The dorsal spines vary significantly in length, with the central spines being much shorter than those occurring farther forward, nearer the head.

DISTRIBUTION: *From the Red Sea and East Africa across the Indian Ocean, ranging northward as far as southern Japan and southward as far as Lord Howe Island off the east coast of Australia. Extends through the Pacific as far as Samoa and Tonga.*

SIZE: *6.75 in (17 cm).*

BEHAVIOR: *Primarily nocturnal, this lionfish conceals itself during the day and emerges at night to seek its prey. It frequents reef flats and lagoons where there are hiding places, although it may also conceal itself among areas of seagrass. It is quite social and is sometimes seen in groups of up to ten individuals. Groups usually contain several females and smaller males, led by a dominant male. Within the confines of caves, this fish may hunt at any time, typically seizing crustaceans that come within range. In some areas it may even evolve special hunting techniques. For example, it has been seen in parts of the Red Sea seeking out sea urchins that are home to cardinalfish. The lionfish darts in and grabs the small quarry as soon as an unwary individual strays within reach.*

DIET: *Meaty foods, such as fish and crustaceans, are required.*

AQUARIUM: *This lionfish is not to be trusted in a reef tank in which there are potential prey species.*

COMPATIBILITY: *It may be possible to keep a large and a relatively small specimen together. Two individuals that are evenly matched in size are more likely to start fighting and will need to be separated before the weaker individual gets badly injured.*

ZEBRA LIONFISH
(DWARF LIONFISH; ZEBRA TURKEYFISH)

Dendrochirus zebra

Lionfish are also known as turkeyfish because the appearance of their pectoral fins are said to resemble those of a male turkey. As its name suggests, this particular species has a striped appearance with five fairly broad bands extending down its body. Edged with black, they are paler at the edges and become redder in the center. As with other lionfish, the appearance of the zebra lionfish is influenced by environmental conditions. For example, it tends to be redder in color and lighter overall during the day. The pectoral fins are webbed almost to the tips, while the dorsal spines are barred. In the mid-body region these spines are longer than the width of the body. There is also a characteristic dark spot on the operculum, covering the gills on each side of the body. It can sometimes be possible to sex these particular lionfish visually, because males generally grow larger and have more stocky heads. Females undergo a dramatic color change just before they spawn, when the front part of their body turns white.

▼ ZEBRA LIONFISH (DENDROCHIRUS ZEBRA)

DISTRIBUTION: *Ranges from the Red Sea and East Africa northward as far as southern Japan and south to Lord Howe Island off the east coast of Australia. Extends across the Pacific as far as Samoa.*

SIZE: *10 in (25 cm).*

BEHAVIOR: *The zebra lionfish is usually found in relatively shallow and sheltered stretches of water, but has been encountered down to depths of 250 ft (80 m). The males tend to be aggressive and can fight viciously, even breaking their dorsal spines and injuring their elaborate fins as a consequence. Color changes can indicate the status of individuals when they meet—the dominant lionfish becomes darker, indicating a challenge to its would-be rival. If it wants to avoid fighting, the rival will respond by turning lighter and retreating.*

DIET: *A range of meaty foods is required. It is important to provide different types to help discover which foods your fish favors. Often displays a surprisingly large appetite once settled in its quarters.*

AQUARIUM: *Cannot be trusted in a reef aquarium in which there are small crustaceans or fish.*

◀ *SPOT-FIN LIONFISH (PTEROIS ANTENNATA)*

COMPATIBILITY: *Likely to be aggressive toward its own kind and related species in the confines of an aquarium. True pairs can generally be kept together, however. Successful spawning has been achieved in aquarium surroundings, taking place just before or after dusk. Will prove shy in a brightly lit tank and should always have access to a range of retreats.*

SPOT-FIN LIONFISH
(BROAD-BARRED FIREFISH)

Pterois antennata

This species is probably the most commonly kept lionfish. The rays of its pectoral fins are largely unwebbed, although a clear spotted pattern and a dark area can be seen at the base of these fins when they are held open. The body is barred. The bars range from red to brown in color and are broken by paler intervening white-edged bands. The appearance of these lionfish can be variable, however. The venomous dorsal spines display red and white barring, and there are prominent projections over the eyes. The face is whitish apart from a dark red stripe running across the eyes, and there is some paler red spotting around the upper jaw.

DISTRIBUTION: *Extends from East Africa through the Indian Ocean, southward to the Great Barrier Reef off the southeast coast of Australia and north as far as southern Japan. Ranges through the Pacific to the Marquesan Islands and Mangareva Island.*

SIZE: *8 in (20 cm).*

BEHAVIOR: *Found in areas of the reef where there are plenty of hiding places, sometimes associating in small groups or with other lionfish, and emerging usually toward dusk to hunt over the reef. Although it is often found in shallow water, it has been reported at a depth of 165 ft (50 m).*

DIET: *Offer meaty foods, such as shrimps. Usually feeds readily but, as with other lionfish, it might be necessary to offer grass shrimps or live mollies* (Poecilia velifera) *if all else fails. It is often hard to see these fish feeding in-store prior to obtaining them because of their nocturnal feeding habits, so choose only individuals that appear plump.*

AQUARIUM: *Not to be trusted in a reef tank with crustaceans and other smaller fish.*

COMPATIBILITY: *Relatively social, but provide a choice of retreats, especially when keeping more than one of these fish in the same aquarium.*

CLEAR-FIN LIONFISH
(RADIAL FIREFISH)

Pterois radiata

With white fin spines projecting around its body, the clear-fin lionfish is a truly spectacular lionfish. It is also the only species that has two horizontal white stripes on the caudal peduncle. The membrane present at the base of the pectoral fins is not spotted. The patterning on the body itself can differ markedly among individuals. It consists of five or six alternating brownish red bands and white stripes of variable width.

▼ *CLEAR-FIN LIONFISH (PTEROIS RADIATA)*

DISTRIBUTION: *Extends from the Red Sea down the coast of Africa as far as Sodwana Bay in South Africa. Ranges across the Indian Ocean, extending northward as far as the Ryukyu Islands. Its southerly distribution is marked by New Caledonia, and it can be found through the Pacific Ocean east as far as the Society Islands.*

SIZE: *9.5 in (24 cm).*

BEHAVIOR: *Tends to favor rocky areas of the reef, frequenting areas where there are small caves and other hiding places such as rocky overhangs. Has been observed in hiding places alongside the spot-fin lionfish* (P. antennata)*. Not commonly seen and usually solitary by nature, preferring to hunt under cover of darkness. It tends to ambush its prey, which consist largely of small crabs, although sometimes shrimps may be taken as well.*

DIET: *Meaty foods, especially those of crustacean origin. May be reluctant to feed on inert foodstuffs at first, in which case you may need to offer invertebrates such as grass shrimps* (Palaemontetes *species*).

▲ *RED LIONFISH* (PTEROIS VOLITANS)

AQUARIUM: *Cannot be kept safely in a reef tank alongside crustaceans. Is also likely to eat smaller fish in these surroundings, even though they are not its natural prey.*

COMPATIBILITY: *May agree in pairs or can be housed with nonaggressive tankmates of comparable size. Most importantly, the aquarium must incorporate a suitable retreat in which the lionfish can hide at times.*

RED LIONFISH

(DEVILFISH; COMMON LIONFISH)

Pterois volitans

This lionfish is variable in appearance, with individuals from shallower water tending to be darker in coloration than those occurring elsewhere. The most colorful examples have bright red stripes offset against intervening white bands on the body. At the other extreme, the red areas on some individuals are almost black. Another characteristic is that individuals occurring in relatively clear water display less spotting on the median fins. The red lionfish is very similar to Miles's lionfish (*P. miles*), whose range lies farther west, although the latter has tubercles on its cheeks. In the red lionfish the pectoral fins are proportionately smaller in older individuals than in juveniles, and the cirri (projections) above the eyes tend to disappear with age. As with other members of the group, the red lionfish has venomous spines, which means that care needs to be taken when servicing its aquarium and when catching or moving this fish.

DISTRIBUTION: *Eastern Indian Ocean, from the Cocos (Keeling) Islands south of Sumatra south to Western Australia. Ranges as far north as southern Korea and Southern Japan, and can be found southward to Lord Howe Island, the northern coast of New Zealand, and the Austral Islands in the Pacific. Occurs eastward as far as the Marquesan and Pitcairn Islands.*

SIZE: *15 in (38 cm).*

BEHAVIOR: *This lionfish hunts crustaceans and preys more heavily than other related species on fish. It has evolved various hunting techniques for this purpose. For example it sometimes uses its long pectoral fins to drive its prey into a corner, where it can be trapped and caught more easily. It may also lurk at the water's surface, especially after dark, and catch shoaling fish that tend to occur there. During the daytime it often hides away, hanging upside down on its own or in small groups in caves, although much larger numbers have been found on wrecks. In aquarium surroundings it is quite usual for this lionfish to rest on the substrate. It tends to be more conspicuous in the aquarium than many of its relatives.*

DIET: *Meaty foods, including both crustaceans and marine fish. Always be sure to allow frozen foods to thaw out thoroughly before using.*

AQUARIUM: *Not suitable for a reef aquarium if there are smaller fish or crustaceans present, because of its predatory habits.*

COMPATIBILITY: *Cannot be kept safely with smaller companions but can agree with its own kind. Its large size, however, means that a spacious aquarium is essential.*

BEARDED SCORPIONFISH

Scorpaenopsis barbata

Scorpionfish rely on their camouflage to conceal their presence. In this species the basic body coloration is brownish, sometimes with a reddish hue. It has white and black blotches that help it blend in against the background, although the markings are highly individual. Scorpionfish can change their appearance to match their environment, which will be evident if you use a colorful substrate in the aquarium. They may even display occasional orange markings on their bodies, to represent the appearance of a sponge. When seen in profile, this species has a very prominent gap between the eyes and the snout. Its three dorsal spines increase in height progressively from the front to the back of the body, and the pectoral fins are large.

DISTRIBUTION: *Restricted to the western part of the Indian Ocean, ranging from the Red Sea south as far as the Somalian coast. Present also in the Gulf of Aden and the Arabian Gulf.*

SIZE: *8.75 in (22 cm).*

BEHAVIOR: *Sedentary in its habits, the bearded scorpionfish is a master of disguise, relying on concealment to avoid predators and obtain prey. Its ability to hide is helped by the presence of skin flaps on its head. The flaps explain why this species and its close relatives are often known collectively as tasseled scorpionfish. The bearded scorpionfish is not shy, however. It often rests in the open in exposed areas and may even perch regularly on suitable corals, from which it can seize passing invertebrates such as shrimps as well as smaller fish.*

DIET: *Requires meaty foods and can take surprisingly large pieces, thanks to its cavernous mouth. May need coaxing with grass shrimps at first.*

AQUARIUM: *An interesting occupant in a reef tank, but cannot be trusted with smaller fish or crustaceans.*

COMPATIBILITY: *Avoid keeping this fish in the company of boisterous bigger individuals or smaller companions that may fall prey to it.*

▼ BEARDED SCORPIONFISH (SCORPAENOPSIS BARBATA)

ANTHIAS AND GROUPERS FAMILY SERRANIDAE

Represented in every temperate and tropical sea throughout the world, the family Serranidae has more than 370 species. They range in size from 1 inch (2.5 cm) to 9 feet (2.7 m). Anthias (subfamily Anthiinae) are well suited to the saltwater aquarium because of their small size, their coloration—which can be very striking—and their generally inoffensive nature. The potential large size of sea basses (subfamily Serraninae) and groupers (subfamily Epinephelinae), along with their predatory nature, normally requires them to be housed alone or in small aquaria.

The ability to change sex is common in reef fish, and there instances of sex change and hermaphroditism in serranids are common. Among *Anthias* species each fish begins life as a female but has the ability to become male. This enables the species to continue even in the most difficult of circumstances. The change can be very swift, taking as little as two to three days.

THREAD-FIN ANTHIAS

Nemanthias carberryi Subfamily Anthiinae

This beautiful fish is the only member of its genus. It differs from the closely related *Pseudanthias* species by having 11 rather than 10 dorsal spines. Juveniles have 12, but they lose the smallest one at the front as they start to grow. There are two color forms, based on the predominant color on the head and underparts, which are either pink or purplish. The back and sides of the fish have a more yellowish appearance, especially toward the rear of the body. This is caused by the presence of small yellow spots on each scale. There is also a yellow stripe running from the enlarged upper jaw, extending back through the eye and curling down toward the pectoral fins. Males can be identified by the characteristic trailing reddish thread on the bottom lobe of their caudal fin

DISTRIBUTION: *The western part of the Indian Ocean from East Africa, south to South Africa, and east to the Maldive Islands off the southwestern coast of India.*

▼ *THREAD-FIN ANTHIAS (NEMANTHIAS CARBERRYI)*

▲ *Peach Anthias (Pseudanthias dispar)*

Size: *5 in (13 cm).*

Behavior: *Often seen around the slopes on the outer area of the reef, the thread-fin anthias occurs in large shoals, with males tending to associate together. They may even be encountered in the company of* Pseudanthias *species in the wild. In aquarium surroundings, however, a single male should be accommodated with a group of at least eight females to prevent bullying. Good water quality and movement—reflecting the environmental conditions on the part of the reef where this fish typically occurs—are very important in the tank.*

Diet: *Offer a diet of meaty foods, which should include vitamin-enriched brine shrimps. Livefoods can encourage this fish to eat, as can introducing it alongside other established anthias that are already feeding on a substitute diet.*

Aquarium: *Ideal occupant for a reef tank.*

Compatibility: *Should not be mixed with aggressive companions. The aquarium must include plenty of retreats to help them settle in well.*

PEACH ANTHIAS

(RED-FIN ANTHIAS)

Pseudanthias dispar Subfamily Anthiinae

The body of this anthias tends to be peach colored. The males are instantly recognizable by their red dorsal fin. Their ventral fins are also longer than those of females. In females the lower half of the body ranges from white to lavender. Both sexes have a narrow orange stripe that extends back from the snout through the eyes and curves down to the pectoral fins. The distribution of the peach anthias also influences its coloration. These fish have the ability to replicate the appearance of other related species, enabling them to join up and form large shoals. For example, peach anthias found in the waters around the Line Islands in the Pacific have yellow coloration on the top half of their body and pale purple below. In this location their coloring corresponds to that of Bartlett's anthias (*P. bartlettorum*), which is more numerous there. The only difference is that the peach anthias retains its red dorsal fin. The peach anthias may also be confused on occasion with the flame anthias (*P. ignitus*), which is found farther west in the Indian Ocean. However, it lacks the red bands along the top and bottom edges of the caudal fin.

Distribution: *Extends from Christmas Island, which lies south of Sumatra, north to the Yaeyama Islands south of Japan, and south as far as Australia's Great Barrier Reef. Ranges east across the Pacific via the Line Islands as far as Fiji and Samoa.*

Size: *3.75 in (9.5 cm).*

Behavior: *This anthias tends to occur in relatively shallow areas of the reef, down to a depth of just 60 ft (18 m). It is found in large shoals that feed on zooplankton. The males tend to congregate on the outside of the shoals where they can display more easily to females.*

Diet: *Ideally, zooplankton should be provided, especially when this fish is first introduced to an aquarium. Mysid and brine shrimps can also be offered. Because this fish is a forager in the wild, it needs to be fed several times during the day.*

Aquarium: *Ideal for a typical reef tank since, unlike species that occur in deeper water, it is not affected by the bright illumination.*

Compatibility: *This fish should be kept in large groups consisting of a single male and six or eight females. It is therefore most suitable for a single-species aquarium, although it may associate readily with other related species.*

YELLOW-BACK ANTHIAS

(YELLOW-TAIL ANTHIAS)

Pseudanthias evansi Subfamily Anthiinae

The generic name of this fish means "false anthias" and this species is often called a fairy basslet. This species has an oblique yellow area on its back, extending from the front of the dorsal fin and broadening to encompass the entire caudal fin. The remainder of the body is violet, broken by yellow dots that are especially evident on the sides of the body. A faint orange line extends from the snout through the eye to the pectoral fins on each side. It is possible to recognize males of the species because the ends of their dorsal fins are elongated, and similar extensions are evident on both the anal and pelvic fins.

DISTRIBUTION: *East Africa through the Indian Ocean, ranging into the Andaman Sea and extending as far as the Cocos (Keeling) and Christmas Islands that lie to the south of the Indonesian island of Sumatra.*

SIZE: *4.75 in (12 cm).*

BEHAVIOR: *A highly social species, encountered in areas of the reef where there is good coral growth. Hundreds of these anthias may congregate in some areas, especially where zooplankton drifts on the currents. They tend to be found over sloping areas of the reef to depths of 130 ft (40 m).*

DIET: *Feeds naturally on zooplankton. Unfortunately, it not easy to switch this species to alternative aquarium foods. Vitamin-enriched brine shrimps are a good choice, along with similar small foods.*

AQUARIUM: *Suitable for a reef tank.*

COMPATIBILITY: *Likely to settle better in a group consisting of one male and several females. Some fish, such the midas blenny* (Ecsenius midas)*, will associate with this anthias in the wild and will therefore make a natural aquarium companion.*

RED-CHEEKED ANTHIAS

(GREEN ANTHIAS; THREAD-FIN ANTHIAS)

Pseudanthias huchtii Subfamily Anthiinae

This species is sometimes described as the thread-fin anthias—the same name as *Nemanthias carberryi*. However, the two species are easy to tell apart. The red-cheeked anthias is much duller in terms of its overall coloration. The males display a greenish body coloration, and the underparts in particular have an overlying purplish hue. A very distinctive threadlike projection is present at the front of the dorsal fin, which is yellowish green at its base. A thin blue border extends along the top and bottom of the caudal fin. The most characteristic feature is the relatively broad red stripe present on each side of the face, extending from the eye to the pectoral fins. Females are an olive-green shade overall, becoming more silvery on the underparts. Like the males, females possess stripes, but they are not especially conspicuous and are orange in color rather than red.

▲ *RED-CHEEKED ANTHIAS (PSEUDANTHIAS HUCHTII)*

DISTRIBUTION: *Occurs in western and central parts of the Pacific, from around the Indonesian Moluccas north to the Philippines and south to Australia's Great Barrier Reef. Often encountered around New Guinea and also present in the vicinity of Vanuatu and Palau, which lie to the southeast of the Philippine island of Mindanao.*

SIZE: *4.75 in (12 cm).*

BEHAVIOR: *Males are territorial and aggressive and they maintain harems of females. When purchasing this species, try to ensure that none of the apparent females are beginning to develop into males. (If no male is present the dominant female will develop into a biological and phenotypic male. Hormonal changes in the female's body trigger a change in appearance and the associated development of the dormant male gonads.)*

DIET: *Usually takes readily to substitute diets such as vitamin-enriched brine shrimps. Naturally feeds on zooplankton, so offer a range of similar meaty types of food, including mysid shrimps. Food should be offered in small amounts several times during the day.*

AQUARIUM: *Suitable for a reef aquarium, although males must be not be housed together.*

COMPATIBILITY: *Likely to bully related species and others that compete for its food. It is a relatively easy member of the group to keep in aquarium surroundings, usually settling in without any significant problems.*

YELLOW-STRIPE ANTHIAS

(PURPLE ANTHIAS)

Pseudanthias tuka Subfamily Anthiinae

The yellow-stripe anthias is a particularly beautiful member of the group, but unfortunately it is usually very difficult to keep in aquarium surroundings. Males have an overall purple coloration, with a darker spot close to the base at the rear of the dorsal fin. There is a relatively inconspicuous area of yellow that extends from the throat along the underside of the body. Females display more evident yellow markings, with a stripe of this color running along the base of the dorsal fin and extending along the top lobe of the caudal fin. The lower edge of the caudal fin is also yellow. A narrow yellow stripe runs from the snout through the eyes to the base of the pectoral fins.

Distribution: *Extends from Mauritius in the Indian Ocean south to Rowley Shoals off Australia's northwestern coast and around the Great Barrier Reef on the east coast of Australia. Ranges north to the Philippines from Indonesia and via Palau east as far as the Solomon Islands in the Pacific.*

Size: *4.75 in (12 cm).*

Behavior: *Likely to be encountered in groups in the company of species such as the similarly colored purple queen anthias* (P. pascalus). *Males of the latter species lack the yellow area seen on the underside of the yellow-stripe anthias and have a red rather than a purple spot on the dorsal fin. Females do not display the yellow stripes on the sides of the body, making them easier to distinguish. Young blue-headed tilefish* (Hoplolatilus starcki) *may also join up with groups of yellow-stripe anthias in some areas, developing yellow stripes that help them blend in alongside the anthias.*

Diet: *The diet of this anthias in the wild consists of zooplankton and fish eggs. It is now possible to culture zooplankton in special chambers, known as refugiums, alongside the main aquarium. They offer the best possibility of being able to maintain the yellow-stripe anthias, but mysid shrimps will also be acceptable. Food of this type contains the natural coloring agents that help preserve the vivid appearance associated with this and many other species of anthias.*

Aquarium: *This species is suitable for a reef tank. It will form a close bond with cleaner shrimps* (Lysmata species.) *in this environment and will encourage their attentions.*

Compatibility: *Ideal companions for this species include the blue-headed tilefish as well as the purple queen anthias and other inoffensive species that feed on zooplankton.*

SLENDER GROUPER

(WHITE-LINED ROCK COD)

Anyperodon leucogrammicus Subfamily Epinephelinae

Groupers, also known as rock cod, are significantly larger in size than anthias and other members of the family. Their large size, and the fact that they are predatory in their feeding habits, tends to preclude them from being popular occupants of the saltwater aquarium. In spite of their size, however, many species are very effective at camouflaging their appearance on the reef. In the slender grouper, which is one of the more colorful species, there is a marked difference in coloration between juveniles and adults. The young fish display a series of horizontal yellow and turquoise-blue stripes running down the length of the body. The stripes on the head and the back are narrower than those lower down on the sides of the body. In contrast, adults display a spotted patterning, although they usually retain the outline of the white stripe under each eye, and the stripe extends back over the gills. The background body color of the adults tends to be whitish or pale grayish, and the spots are orange.

Distribution: *From the Red Sea south as far as the coast of Mozambique and through the Indian Ocean northward as far as Japan and southward to Australia. Extends across the Pacific as far as the Phoenix Islands that lie to the north of Samoa.*

▼ *Yellow-stripe Anthias (Pseudanthius tuka)*

▼ *Slender Grouper (Anyperodon leucogrammicus)*

SIZE: *25.75 in (65 cm).*

BEHAVIOR: *Tends to be encountered in areas of clear water in which coral predominates. The unusually colorful appearance of the young fish resembles that of the wrasse* Halichoeres purpurescens. *The wrasse is not seen as a threat by other small fish on the reef, so by disguising itself as a wrasse, the young grouper can approach its prey more easily. It is possible, however, to confuse the two in the aquarium trade at this stage, and you may find that inoffensive tank companions begin to disappear in the company of one of these supposed "wrasse."*

DIET: *This species feeds on fish and crustaceans in the wild, so offer meaty foods in aquarium surroundings.*

AQUARIUM: *Not suitable for a reef aquarium because it will prey on a number of the occupants typically kept in these surroundings.*

COMPATIBILITY: *The predatory nature and likely adult size of the slender grouper means that accommodating it with other fish in aquarium surroundings can be difficult.*

ARGUS GROUPER
(PEACOCK HIND)

Cephalopholis argus Subfamily Epinephelinae

The patterning of this grouper is very distinctive, the main features being the five or six pale vertical bars that extend down the sides at the rear of its body. There is also a pale area at the base of the pectoral fins, the tips of which are often brownish. The spines at the front of the dorsal fin are orange-gold in color, although they are evident only when this area of the fin is raised. A series of relatively small blue spots with black edging extends over the entire body, including the lighter areas, and over the fins.

DISTRIBUTION: *From the Red Sea down to the coast off Durban, South Africa, across the Indian Ocean to the Ryukyu and Ogasawara Islands near Japan. Its southerly distribution extends around the Australian coast to Lord Howe Island in the southeast. Ranges through the Pacific as far as the islands of French Polynesia and Pitcairn Island to the south.*

SIZE: *23.75 in (60 cm).*

BEHAVIOR: *Most likely to be seen in relatively shallow water over the reef, but adults can range down to depths of 130 ft (40 m). Juveniles prefer to frequent areas of coral growth that provide them with places to hide and hunting opportunities. Fish is the predominant item in their diet, although they will also prey on crustaceans. They tend to become more active at dusk and may hunt at night in some parts of their range.*

DIET: *Requires meaty foods, replicating its natural diet.*

AQUARIUM: *Not suitable for a reef aquarium alongside other fish or crustaceans.*

COMPATIBILITY: *Its large size means that it requires suitably spacious accommodation and can be housed safely only with fish of similar size. Most easily maintained on its own.*

▼ *ARGUS GROUPER (CEPHALOPHOLIS ARGUS)*

▲ Blue-spotted Rock Cod (Cephalopholis cyanostigma)

BLUE-SPOTTED ROCK COD

(BLUE-SPOTTED HIND)

Cephalopholis cyanostigma Subfamily Epinephelinae

Young examples of this particular rock cod undergo a significant change in color as they mature. At first they appear dark brown with contrasting bright yellow fins. Then, once they have grown to about 3.5 inches (9 cm), the juveniles begin to lose their yellow appearance, which is replaced by a blackish brown coloration. They also start to develop a spotted blue pattern on their bodies. Their background body coloration lightens and becomes pale reddish, but there are still brown areas evident, particularly toward the rear of the body.

Distribution: *Ranges from Dampier Land off the north coast of Western Australia, around the coast as far as the Capricorn Islands, which lie at the southern end of the Great Barrier Reef. Extends north to the Philippines and through the western Pacific via Palau and New Britain as far east as the Solomon Islands.*

Size: *15.75 in (40 cm).*

Behavior: *Can be encountered in relatively shallow coastal areas, particularly when young, frequenting seagrass beds. Also often seen among coral. Older individuals especially may wander more widely into deeper areas of water, down to depths of 165 ft (50 m) or so.*

Diet: *Hunts both fish and crustaceans. Needs a correspondingly meaty diet in aquarium surroundings. Usually has a healthy appetite, like its relatives. An effective filtration system is therefore needed to prevent the build-up of nitrogenous waste.*

Aquarium: *Do not mix with fish or crustaceans in a reef tank.*

Compatibility: *Although it may be possible to house young of this species together without any problems, they will need to be separated as they grow older. Avoid choosing smaller companions, since they are likely to end up being eaten by the rock cod.*

LEOPARD ROCK COD

(LEOPARD HIND)

Cephalopholis leopardus Subfamily Epinephelinae

This species displays a highly individual blotched patterning on its body, with darker, more brownish red areas extending over its upperparts. The background color tends to be a dirty white shade, with the reddish markings broken into more of a spotted pattern on the underside of the body. The eyes are relatively large and prominent, as in the case of related species, which suggests that these fish rely largely on vision to detect potential prey.

Distribution: *Occurs in the Indian Ocean around the coast of East Africa and nearby islands, and ranges northward to the Ryukyu Islands near Japan. Extends southward as far as northern Australia and through the Pacific as far east as the Society Islands.*

Size: *9.5 in (24 cm).*

Behavior: *Ranges over areas of the reef in which coral is present and where there are plenty of suitable refuges, such as caves. It may be a crepuscular fish, emerging from its hiding place as darkness falls. Its crepuscular habits may also explain its relatively large eyes. It ranks as one of the smaller of the rock cod, however, which makes it more vulnerable to predators. It is known to feed on crustaceans and also probably preys on small fish. It is solitary by nature.*

Diet: *Requires a meaty diet in aquarium surroundings, including shrimps. Be sure to defrost frozen foods thoroughly before offering them to the fish because, like other groupers, individual leopard rock cod can feed on large chunks of food. A partially frozen item of food could lead to a digestive upset.*

Aquarium: *Will prey on the typical occupants of a reef aquarium.*

Compatibility: *Its size means that this species is better suited to the home aquarium than many of its relatives, provided it is kept on its own in a tank that offers suitable retreats.*

▼ Leopard Rock Cod (Cephalopholis leopardus)

▲ Coral Rock Cod (Cephalopholis miniata)

CORAL ROCK COD
(CORAL HIND)

Cephalopholis miniata Subfamily Epinephelinae

The coral rock cod is one of the most colorful and attractive members of the family. Its body is a bright shade of orange-red, overlaid with a series of regular bright blue spots across the entire body and the fins. The spots have narrow black borders. This fish is conspicuous when swimming over the reef, although its coloration may vary in intensity, appearing more orange than red in some conditions.

Distribution: *Extends from the Red Sea down to the vicinity of Durban off the coast of South Africa. Ranges across the Indian Ocean and is encountered around the islands throughout this region as well as north to southern Japan and south to Australia. Occurs east through the Pacific as far as the Line Islands.*

Size: *17.75 in (45 cm).*

Behavior: *A prominent member of the reef fauna where it is present, tending to prefer clear areas of water in more exposed areas of the reef. It has a well-defined social structure, with a single male patrolling a large area of up to 570 square yards (475 sq m). This area is occupied by up to 12 females, each of which maintains its own smaller territory within this area. They hunt individually rather than as a shoal, preying mainly on other reef fish, particularly jewel lyre-tailed anthias* (Pseudanthias squamipinnis) *as well as crustaceans.*

Diet: *Offer meaty foods derived from both fish and crustaceans.*

Aquarium: *Will prove to be predatory in a reef aquarium in which there are other fish and crustaceans.*

Compatibility: *Not suitable for keeping alongside smaller companions. Visual sexing is impossible, and mature males are likely to fight. It is better to house them on their own rather than attempting to keep them as a group, although this may be possible at first with small specimens.*

TOMATO ROCK COD
(TOMATO HIND)

Cephalopholis sonnerati Subfamily Epinephelinae

The coloration of the tomato rock cod varies significantly throughout its range, with the most colorful individuals originating from the Indian Ocean. Its color ranges from orange-red to reddish brown, often with small white areas on the body. The fins are darker than the body and may display blackish brown markings toward the tips. The head may be darker too, having purplish red coloration broken by an orange spotted patterning. Tomato rock cod occurring in the Pacific are far less brightly colored, and have a much more brownish appearance overall with brownish red spots over the entire body, including the fins.

DISTRIBUTION: *Ranges from East Africa southward to the vicinity of Durban off the coast of South Africa. Found through the Indian Ocean north to southern Japan and south as far as Australia's Great Barrier Reef off the coast of Queensland. Extends through the Pacific as far as the Line Islands.*

SIZE: *22 in (57 cm).*

BEHAVIOR: *This relatively deepwater species of rock cod is occasionally found down to depths of 490 ft (150 m). The young tend to occupy shallower water, however, often occurring in the vicinity of coral and sponges. It is quite a secretive fish by nature—the adults prefer rocky areas in which there are suitable retreats for them and where they can hunt for crustaceans such as cleaner shrimps. The tomato rock cod appears to be solitary by nature.*

DIET: *Preys on fish and a variety of crustaceans, so it requires similar meaty foods in aquarium surroundings.*

AQUARIUM: *Not to be trusted in a reef tank—even when young—because of its predatory nature.*

COMPATIBILITY: *Avoid mixing with smaller companions that could become its prey. Its large size demands a suitably spacious aquarium.*

DARK-FIN ROCK COD

(BANDED-TAILED HIND)

Cephalopholis urodeta Subfamily Epinephelinae

Vivid red coloration is often a feature of members of this genus. As one of the smaller species, the dark-fin rock cod is easier to house in an aquarium than its larger relatives. It tends to be a fiery orange-red color, often with a blackish suffusion on the head, creating a spotted appearance. This darker area extends along the base of the dorsal fin to the caudal fin, which tends to be blacker than the body, giving rise to the fish's common name.

SYNONYM: *Epinephelus urodeta.*

DISTRIBUTION: *East Africa, ranging from the coast of Kenya southward as far as South Africa. Extends across the Indian Ocean (but not as far north as India), and through the Pacific as far as French Polynesia and Pitcairn Island.*

SIZE: *11 in (28 cm).*

BEHAVIOR: *Favors relatively shallow areas of water with good visibility, which helps it spot and catch potential prey in the open. The bulk of its diet is made up of small fish, but it will also feed on crustaceans.*

DIET: *Meaty foods of various kinds, replicating the types of food it prefers to eat in the wild.*

▼ DARK-FIN ROCK COD (CEPHALOPHOLIS URODETA)

AQUARIUM: *Not to be trusted with many of the occupants that are likely to be included in a reef aquarium.*

COMPATIBILITY: *Solitary by nature. Any companions must be large and sufficiently robust not to be harassed by or fall prey to this species.*

ROCK HIND

Epinephelus adscensionis Subfamily Epinephelinae

The coloration of this sea bass can vary dramatically among individuals. The young generally have a brown background coloration, with pale yellow markings evident along the dorsal fin and the caudal fin. There are black spots of variable size over the body, and whitish blotches of varying size also apparent. As the fish grows older, the black spots tend to be replaced by a rusty-red spotted patterning with a lighter brown background, although between one and four black blotches are still likely to be evident along the base of the dorsal fin. The reddish spots are generally larger in the ventral region than elsewhere on the body. In some cases the background patterning will appear much lighter, almost white.

DISTRIBUTION: *Found on the North American coast from Massachusetts, extending south via Bermuda to the Caribbean and the Gulf of Mexico. Continues around the coast of South America south as far as Brazil. This fish can also be found in the eastern Atlantic and is present around St. Helena and Ascension Island. Its range may extend east as far as parts of western Africa.*

SIZE: *24 in (61 cm)*

▼ *ROCK HIND (EPINEPHELUS ADSCENSIONIS)*

▲ *BLUE-AND-YELLOW GROUPER (EPINEPHELUS FLAVOCAERULEUS)*

BEHAVIOR: *Tends to be found in rocky, shallow areas where it can hide away, being shy by nature. Often drifts along close to the bottom with its tail pointing downward. The rock hind is solitary and predatory by nature. Crabs are the major items in its diet, although it also hunts fish. The Ascension Island population preys on hatchling sea turtles as well.*

DIET: *Meaty foods of suitable size will be taken readily.*

AQUARIUM: *Not suitable for a reef aquarium alongside crustaceans.*

COMPATIBILITY: *Cannot be mixed safely with smaller companions in view of its predatory nature.*

BLUE-AND-YELLOW GROUPER

Epinephelus flavocaeruleus Subfamily Epinephelinae

Unfortunately, this attractively colored species grows to a very large size, making it unsuitable for the home aquarium unless you have room for a correspondingly spacious tank. Young blue- and-yellow groupers look more attractive than adults, since they display more bright yellow coloration. In the case of adults, the body tends to be a pale shade of blue, with yellow coloration covering much of the area of the fins, including the pectorals. There may be some black markings also evident on the tips of the fins.

DISTRIBUTION: *Occurs widely through the Indian Ocean, southward as far as the area of Port Alfred, South Africa, and around islands such as Rodriguez. Extends eastward as far as northwestern Sumatra.*

SIZE: *35.5 in (90 cm).*

Behavior: *The young of this species of grouper are found in relatively shallow waters, while adults occur down to depths of 490 ft (150 m). This fish is highly predatory by nature and will seize a variety of invertebrate and vertebrate prey in its large, powerful jaws.*

Diet: *Meaty foods of various types. Larger individuals display healthy appetites and can be fed big chunks of food.*

Aquarium: *Not suitable for a reef tank, since it preys on a variety of the creatures normally occurring there, including cephalopods such as octopus.*

Compatibility: *Should be kept alone in view of its predatory behavior. Young individuals soon become quite tame.*

BROWN MARBLED GROUPER

(BLOTCHY GROUPER)

Epinephelus fuscoguttatus Subfamily Epinephelinae

Spotted patterning is a characteristic of this species when young. The spots themselves are larger on the fins and the rear of the body. Blotched areas start to develop as the fish matures, but this patterning appears to be highly individual. It tends to be brown but may be more brownish black. The so-called disruptive camouflage helps obscure the shape of this potentially very large fish, particularly when it is lurking among rocks. The appearance of the first part of the dorsal fin is very evidently serrated.

Distribution: *From the Red Sea and East Africa, ranging northward to southern Japan and south to the Australian coast. Extends out through the Pacific to Samoa and the Phoenix Islands. However, it is absent from some areas in this region, including French Polynesia.*

Size: *47.25 in (120 cm).*

Behavior: *While young brown marbled groupers prefer to hunt among beds of seagrass on the reef, adults wander more widely—often into deeper water down to about 200 ft (60 m). This species is crepuscular by nature. It tends to become active at dusk, when it will begin hunting from the shadows.*

Diet: *Feeds naturally on fish and various invertebrates, notably crustaceans and cephalopods. Meaty foods should therefore be offered.*

Aquarium: *Will prey on various inhabitants likely to be found in a typical reef tank. Its large size means that housing this fish will be difficult.*

Compatibility: *Cannot be housed with other smaller fish that may become its prey.*

▼ *Brown Marbled Grouper (Epinephelus fuscoguttatus)*

▲ *MALABAR GROUPER (EPINEPHELUS MALABARICUS)*

STAR-SPOTTED GROUPER

(WHITE-SPECKLED ROCK COD)

Epinephelus hexagonatus Subfamily Epinephelinae

As with related species, the mottled patterning of this grouper helps break up the outline of its body, enabling it to merge into the background and avoid being spotted by its prey. The markings of this fish are highly individual, and consist of darker brownish black areas on a lighter background. There are typically between four and six such areas extending along the base of the dorsal fin. White spotted patterning occurs over the entire body, except for the underparts, and extends into the dorsal and caudal fins. The white markings are more definite on the lower sides of the body, creating a hexagonal pattern. There may also be occasional red markings close to the head.

DISTRIBUTION: *This grouper is found in the vicinity of most tropical islands in the Indian Ocean and the western Pacific .*

SIZE: *11 in (28 cm).*

BEHAVIOR: *Tends to be encountered predominantly in shallow waters down to depths of 100 ft (30 m) and is often seen on the more exposed areas of the reef, where its disruptive camouflage pattern serves it well. Although it is a good swimmer, it will ambush its prey whenever possible.*

DIET: *Hunts crustaceans, such as crabs, and fish. Will readily eat a range of similar meaty foods in the aquarium.*

AQUARIUM: *Not suitable for a reef tank where there are crustaceans or small fish present.*

COMPATIBILITY: *Tends to be solitary by nature, so best kept alone. As one of the small groupers, this species is more easy to accommodate in aquarium surroundings than its larger relatives.*

MALABAR GROUPER

(MALABAR ROCK COD)

Epinephelus malabaricus Subfamily Epinephelinae

Young of this species display a spotted appearance. They have a combination of brownish black and white markings evident over a background body color of pale brown. Like other groupers, they have large mouths with angled jaws that are directed upward. The front part of the dorsal fin is also heavily serrated. As it grows older, the appearance of the fish changes and develops a more definite pattern of dark blotches.

DISTRIBUTION: *In spite of its name, this species has a wide distribution and is not confined to the vicinity of the Malabar Islands that lie off the southwest coast of India. Its range extends from the Red Sea and East Africa in the west, through the Indian Ocean north to southern Japan and southward as far as Australia. It also extends across the Pacific as far as Tonga.*

SIZE: *92 in (234 cm).*

BEHAVIOR: *A highly adaptable species, this grouper can be found in shallow waters close to the shore but has also been encountered down to depths of 490 ft (150 m) in the ocean. Juveniles in particular often wander into mangrove areas and even into the brackish water of estuaries. It is very important to bear in mind the size this fish can ultimately attain once it matures. It is really a species suitable only for a public aquarium, unless it can be housed in a saltwater pond.*

DIET: *Predatory, so must be offered a meaty diet of fish and crustaceans.*

AQUARIUM: *Will prey on various occupants likely to be encountered in a reef aquarium, even when young.*

COMPATIBILITY: *The huge potential size and predatory nature of this fish means that finding suitable companions for it is likely to be very difficult.*

HONEYCOMB GROUPER

Epinephelus merra Subfamily Epinephelinae

Young individuals have a light gray body color with dense but variable spotted patterning over the body and the fins. The spots on the face are smaller in size. Each fish has a highly individual patterning, even when young. Subsequently, as they grow older their patterning alters and takes on more of a hexagonal design—the brownish markings often merge together to create blocks that appear piled on top of each other. The remaining spots then form a honeycomb pattern and are especially evident at the rear of the dorsal fin and on the caudal fins.

DISTRIBUTION: *Ranges through the southern part of the Indo-Pacific from South Africa to French Polynesia, but is absent close to mainland Asia.*

SIZE: *17.75 in (45 cm).*

BEHAVIOR: *Young are often encountered in close proximity to* Acropora *species coral. Adults, on the other hand, wander farther in shallow lagoons and over other areas of the reef. This may reflect differences in their feeding habits. Whereas the young hunt invertebrates rather than other fish, adult honeycomb groupers are more predatory in their feeding habits.*

DIET: *Meaty items should be offered. Adults can swallow relatively large pieces of food.*

AQUARIUM: *Not suitable for a reef tank, because this fish will prey on many of the other inhabitants.*

COMPATIBILITY: *Its predatory habits mean that this fish must not be kept with smaller companions. Because of its size, it requires a large aquarium and is best housed individually.*

CAMOUFLAGE GROUPER

Epinephelus polyphekadion Subfamily Epinephelinae

The dull appearance of the camouflage grouper enables it to merge well into the background of the reef. Its basic coloration is brownish. It sometimes has a reddish hue, especially around the jaws and on the underparts, where a darker brown spotted patterning is evident. A series of grayish blotches can be seen down the sides of the body. Variable white spots and speckling are especially evident on the caudal and dorsal fins.

SYNONYM: *E. microdon.*

DISTRIBUTION: *Ranges from the Red Sea down the eastern coast of Africa and across the Indian Ocean. Extends north to southern Japan and*

▼ HONEYCOMB GROUPER (EPINEPHELUS MERRA)

◀ *FOUR-SADDLE GROUPER (EPINEPHALUS SPILOTOCEPS)*

southward to Queensland and Lord Howe Island off the east coast of Australia. Can be found through the Pacific as far east as French Polynesia.

SIZE: *35.5 in (90 cm).*

BEHAVIOR: *This grouper is found in the vicinity of coral atolls throughout its range, especially in areas where there are caves and other hiding places. Unlike many members of the family, it occurs in small groups rather than alone. It feeds predominantly on fish and crabs but also eats other crustaceans as well as small octopuses and gastropods.*

DIET: *Requires a meaty diet and will benefit from different types of food.*

AQUARIUM: *Not suitable for a reef aquarium, since many of the occupants are likely to be its natural prey.*

COMPATIBILITY: *Can be kept with its own kind in a suitably spacious aquarium, which must include sufficient retreats.*

FOUR-SADDLE GROUPER

Epinephalus spilotoceps — Subfamily Epinephelinae

Like a number of related species, this grouper has a mottled appearance. Its markings vary widely among individuals. It has a dense patterning of brown or reddish brown spots as well as olive ones, which can create the impression of solid coloration. There are intervening white areas on the body, and smaller dark spots may be present on the snout, extending up between the eyes. Its most distinctive feature, however, and one that sets it apart from its relatives, is the presence of three blackish areas at the base of the dorsal fin. These blotches—together with a fourth blackish area farther back on the body, above the caudal peduncle—give the fish its common name.

DISTRIBUTION: *East Africa, down the coast to Mozambique and across the Indian Ocean into the Pacific, reaching as far as the Line Islands. Not found around Sri Lanka, the Philippines, or farther north. Its southerly range extends to Rowley Shoals off Australia's northwest coast.*

SIZE: *14 in (35 cm).*

BEHAVIOR: *Largely unstudied, but certainly shows an affiliation with islands throughout its range. Not generally found close to major landmasses.*

DIET: *Offer meaty types of food of appropriate size.*

AQUARIUM: *Not to be trusted in a reef tank alongside invertebrates and small fish, which are its natural prey.*

COMPATIBILITY: *Although one of the smaller members of the group, it is safer to house this fish on its own, particularly once it starts to mature.*

GREASY GROUPER
(GREASY ROCK COD)

Epinephelus tauvina — Subfamily Epinephelinae

The young of this species are predominantly brown in color. They display some prominent white areas on their body, usually extending to the fins, and some darker, blackish spots as well. Adult fish tend to appear much whiter, with the entire body displaying more evenly spaced spotted markings that can vary in color from orange-red through to brown. The underlying coloration of the rest of the body can vary from greenish gray through to brown, with a prominent black area usually evident at the base of the rear spines on the dorsal fin. There may also be dark barring on the body. The unusual name of this fish comes from its scale pattern, which changes from being ctenoid (meaning that the rear edge of the scale is serrated) when young to cycloid (in which the scales have a wavy back edge) as adults. The only vestige of ctenoid scales in adults is in the vicinity of the pectoral fins on each side of the body.

◀ *GREASY GROUPER (EPINEPHELUS TAUVINA)*

▲ BUTTER HAMLET (HYPOPLECTRUS UNICOLOR)

SYNONYMS: *E. chewa; E. elongatus.*

DISTRIBUTION: *Ranges from the Red Sea down the coast of Africa as far as South Africa. Extends northward to southern Japan and can be found as far south as the coast of New South Wales, Australia, and around Lord Howe Island to the east of Australia. Extends eastward through the Pacific to Ducie Island, which lies to the west of Easter Island.*

SIZE: *29.5 in (75 cm).*

BEHAVIOR: *Young individuals frequent relatively shallow areas of the reef, even venturing into tide pools. Adults, on the other hand, prefer deeper water, and are found at depths of 985 ft (300 m). These groupers are solitary by nature and predatory in their feeding habits, preying largely on other fish. In the wild their diet often seems to include angelfish.*

DIET: *Requires meaty foods of an appropriate size.*

AQUARIUM: *Unsuitable for a reef aquarium, because smaller fish and invertebrates will be at risk of being eaten by this grouper.*

COMPATIBILITY: *Its large size and predatory nature means that this species should be housed individually.*

BUTTER HAMLET

Hypoplectrus unicolor Subfamily Serraninae

An attractive pale silvery yellow coloration predominates in this fish, with bluish silver coloration becoming apparent along the upper part of the body and the dorsal fin, extending to the caudal fin. There is a very distinctive black blotch present on the upper part of the caudal peduncle, while the pelvic fins are yellow and the pectoral fins are clear. A narrow blue line encircles each eye and extends down across the cheeks. There may be two black spots evident on the snout, and also blue spotted patterning extending up between the eyes.

DISTRIBUTION: *Ranges from the North American coast of Florida south to the Bahamas. Also occurs throughout the Caribbean, although this species is not present in the Gulf of Mexico.*

SIZE: *5 in (13 cm).*

BEHAVIOR: *The butter hamlet often displays variable coloration and can modify its body coloring to appear darker or lighter depending on its surroundings and mood. Yellow coloration is often most evident along the underside of the body. It is a common species in the waters around Florida, and it can be approached relatively closely as long as it is not frightened by sudden movements.*

DIET: *Predatory by nature, this fish requires a diet of meaty food, including crustaceans and fish.*

AQUARIUM: *Not to be trusted in a reef tank, because it is likely to prey on crustaceans and other invertebrates and possibly small fish too.*

COMPATIBILITY: *Should be housed on its own. Can be nervous at first, so ensure that there are adequate retreats in the aquarium. May instinctively spend more time close to the substrate than many fish. Younger individuals tend to adapt more easily to these surroundings. Can be housed as part of a group with other fish of similar size that will not be aggressive.*

BLACK-SADDLED CORAL GROUPER
(BLACK-SADDLED LEOPARD GROUPER)

Plectropomus laevis Subfamily Epinephelinae

This species has very striking patterning, although its markings vary significantly from individual to individual. A black band extends over the head down to the eyes, and a further series of (typically) four distinct black bands, which can be quite broad, extends over the top of the body. Similar markings may also be present on the underparts and there may be black spots on the sides of the body as well, creating a sharp contrast with the white areas over the rest of the body. The caudal peduncle is often bright yellow. This color extends into the caudal fin, and there may be a yellow mark on the head and elsewhere on the body and fins. The iris surrounding the pupil in the eye is orange and is especially evident in younger individuals.

DISTRIBUTION: *East Africa, extending southward from Kenya down to Delagoa Bay, Mozambique. Ranges through the Indian Ocean, extending north as far as the Ryukyu Islands and south to Australia's Great Barrier Reef. Also present off the coast of Queensland and ranges out across the Pacific as far as the Tuamotu Archipelago.*

SIZE: *49 in (125 cm).*

BEHAVIOR: *The young of this species can resemble the black-saddled toby* (Canthigaster valentini). *This gives them protection, since the toby is recognized by others on the reef as being toxic if eaten, and the young coral groupers are therefore essentially ignored by predators. They grow into aggressive predators themselves, however. They have a vicious array of teeth and can hunt down large prey, including other groupers. When breeding, black-saddled coral groupers may migrate over short distances to communal spawning grounds. At this time they may form shoals, although they are otherwise solitary by nature. The young frequent shallow waters, lurking around coral rubble to seize prey such as crustaceans and small fish.*

DIET: *Requires a meaty diet based on fish and crustaceans.*

AQUARIUM: *Cannot be included in a reef tank.*

COMPATIBILITY: *Its aggressive and essentially solitary nature means that this fish should be housed on its own in a suitably spacious aquarium.*

PAINTED COMBER

Serranus scriba Subfamily Serraninae

This member of the family displays highly variable patterning. It has a series of reddish brown vertical bars running down

▼ BLACK-SADDLED CORAL GROUPER (PLECTROPOMUS LAEVIS)

each side of the body, with intervening whitish areas. There is usually a vermiculated (wavy-line) patterning evident on the sides of the face below the eyes, and this can also extend over the lips. The variations in appearance among painted combers from different parts of the species' range may reflect local conditions, and it is possible that their markings help the fish conceal their presence in different habitats.

DISTRIBUTION: *Found in the eastern Atlantic, extending south from the Bay of Biscay off the coast of France to the coast of Mauritania in West Africa. Also found around islands in this region, including the Canaries, Madeira, and the Azores, which mark the western point of its range. Extends eastward through parts of the Mediterranean to the Black Sea.*

SIZE: *14 in (36 cm).*

BEHAVIOR: *A fairly sedentary species that lives close to to the seabed. It conceals its presence there, lurking under rocks and among seaweed, looking for opportunities to feed. Both fish and crustaceans feature in its diet. It has a surprisingly wide gape.*

DIET: *Offer meaty foods that correspond to those it eats in the wild.*

AQUARIUM: *Not to be trusted in the company of smaller fish or crustaceans.*

COMPATIBILITY: *Its size and feeding habits dictate that this fish should be housed on its own.*

▲ *PAINTED COMBER (SERRANUS SCRIBA)*

HARLEQUIN BASS

Serranus tigrinus Subfamily Serraninae

The distinctive markings of the harlequin bass are less apparent in juveniles than in adults. In addition, the young fish tend to display horizontal rather than vertical stripes on their bodies. These are still evident in the adult fish, but less prominent. In adults the broader vertical stripes are the most obvious feature, even though they are not of a consistent width and do not therefore create an even pattern. Black spots are more evident in the clear fins, and although white is the predominant background color, there may be yellow areas, especially running horizontally down the midline of the body. Yellow may also be evident in the serrated part of the dorsal fin and at the tips of the caudal fin.

DISTRIBUTION: *Occurs in the western Atlantic, extending from the coast of Florida eastward to the Bahamas and Bermuda. Also ranges southward through the Caribbean down to the coast of northern South America.*

SIZE: *4 in (10 cm).*

BEHAVIOR: *This species is most commonly seen in shallow waters, often where there is coral rubble present, but also in areas where seagrass predominates. It has developed an unusual and seemingly effortless method of hunting—it frequently drifts over the substrate, watching keenly for small fish or crustaceans in the vicinity. When it spots a suitable victim, it darts down and seizes it.*

DIET: *Offer meaty items of an appropriate size.*

AQUARIUM: *Cannot be trusted in a reef tank that contains crustaceans and small fish, because these animals are its natural prey.*

COMPATIBILITY: *If two individuals are introduced to the aquarium at the same time, there is a strong possibility that they will get on together, particularly if they have plenty of space and there are sufficient retreats. It also helps if they have been kept together beforehand. Two males, however, are likely to fight ferociously and will need to be separated. This species is hermaphrodite, which means that if two females are placed in the aquarium, one will develop into a male fish. Harlequin bass are often encountered in pairs in the wild.*

WHITE-EDGED LYRETAIL

(CRESCENT-TAILED GROUPER)

Variola albimarginata Subfamily Epinephelinae

This grouper gets its name from the narrow white band that is evident along the rear edge of the caudal fin. The young are particularly brightly colored, with orange-red coloration over

▼ *HARLEQUIN BASS (SERRANUS TIGRINUS)*

◀ WHITE-EDGED LYRETAIL (VARIOLA ALBIMARGINATA)

most of their bodies—apart from the throat and the area around the jaws, which is paler in color. A more pronounced spotted patterning develops with maturity, as do several pale blotches along the base of the dorsal fin. There is also a series of smaller pale purplish or white spots extending over the body.

DISTRIBUTION: *Ranges from East Africa through the Indian Ocean. Extends northward as far as the Ryukyu Islands and south to the northern part of Australia's Great Barrier Reef off the coast of Queensland. Found through the Pacific as far east as Samoa.*

SIZE: *25.75 in (65 cm).*

BEHAVIOR: *Young fish favor shallow stretches of water, while adults occupy deeper water where there are caves that can serve as retreats. This species is not very common anywhere in its range, but it can occasionally be encountered in groups and is not entirely solitary in its habits. It is quite bold by nature.*

DIET: *Preys on both fish and crustaceans, so requires a meaty diet incorporating items of this type.*

AQUARIUM: *Cannot be kept in a reef tank because it is likely to prey on a number of the other occupants typically included in these surroundings.*

COMPATIBILITY: *Has a surprisingly capacious mouth and will readily attempt to swallow smaller companions. Grows fast, so young fish will quickly need much larger accommodation. Its large size tends to preclude keeping this species in groups.*

LYRETAIL GROUPER

(YELLOW-EDGED LYRETAIL; CORONATION GROUPER)

Variola louti Subfamily Epinephelinae

One of the most evident features of juveniles of this species is the presence of a white stripe that extends from the snout back between the eyes to the base of the dorsal fin. The body color itself is effectively divided in the midline, with a reddish area above and and a pale whitish yellow area beneath. Pale dark-edged bluish spots extend over the entire body and into the caudal fin as well. Some individuals also display some blackish areas on the sides of the body. The coloration of adults is very different, however. They have a variable reddish gray background color, and red spots and stripes over the body. Their caudal fin displays the typical lyre-shaped appearance, which is less conspicuous in juveniles.

SYNONYMS: *V. longipinna; V. melanotaenia; V. punctulatus.*

DISTRIBUTION: *Extends from the Red Sea down to South Africa and east across the Indian Ocean, north to southern Japan, and south as far as the southern part of Australia's Great Barrier Reef off the coast of New South Wales. Its easterly range in the Pacific reaches to the waters around Pitcairn Island.*

SIZE: *32.75 in (83 cm).*

BEHAVIOR: *Most commonly encountered around islands rather than around continental landmasses, this grouper is more conspicuous on the reef and in the aquarium than many of its relatives. It is a highly effective predator, and can seize a variety of other fish and crustaceans. Crabs often feature in the diets of larger individuals.*

DIET: *Feed meaty items of the appropriate size.*

AQUARIUM: *Not to be trusted in a reef aquarium. Will soon become too large for the average reef setup.*

COMPATIBILITY: *This is not a species that should be acquired without considerable thought as to its long-term care. While juveniles are attractive, they usually grow quickly, and tankmates easily become their prey. Unlike some of its relatives, this species also needs plenty of swimming space in the aquarium as well as some retreats.*

◀ LYRETAIL GROUPER (VARIOLA LOUTI)

CARDINALFISH FAMILY APOGONIDAE

The Apogonidae is a fairly large family of more than 200 species. Although cardinalfish are relatively easy to maintain in a saltwater aquarium, only a few species are well known in the hobby. This may be linked to the fact that they tend to be nocturnal by nature and are therefore difficult to observe.

Their reproductive habits are especially interesting. Cardinalfish are mouthbrooders. The adult fish collect the eggs after spawning and keep them in the relative safety of their mouths until they hatch. This task is usually the responsibility of the male. Such parental care means that, compared with many other groups, cardinalfish can potentially be bred more easily in the home saltwater aquarium. Their relatively small size also makes them an ideal choice for such an environment.

SHORT-TOOTH CARDINALFISH

(GOLD-BELLY CARDINALFISH)

Apogon apogonoides

The front part of the body of the short-tooth cardinalfish is golden, while the area behind has a reddish hue. Two vibrant blue horizontal stripes run from the upper jaw to behind the eyes, separated by a black bar in front of the eyes. In common with other related species, this cardinalfish has a relatively slimline body shape. The upper part of the body is darker but still retains a golden hue. Young cardinalfish are paler than adults, and the blue area is restricted to the area in front of the eyes.

DISTRIBUTION: *East Africa through the Indo-Pacific. Found in the waters around Indonesia and the Philippines, extending northward via Taiwan to Japan. Ranges southward to Australia's Great Barrier Reef.*

▼ *SHORT-TOOTH CARDINALFISH (APOGON APOGONOIDES)*

SIZE: *4 in (10 cm).*

BEHAVIOR: *Ranges from coastal areas to open areas of the reef and to depths of 200 ft (60 m). It is not uncommon to see it around coral, hiding among the branches or seeking out other refuges during the daytime. The social structure displayed by this species can vary markedly. These cardinalfish may be observed on their own, in pairs, or even as part of large shoals, sometimes associating with other related species as well.*

DIET: *Crustaceans of various types and small fish form the basis of its diet, so offer similar meaty foods in aquarium surroundings.*

AQUARIUM: *Not entirely trustworthy in a reef tank, particularly where crustaceans are present.*

COMPATIBILITY: *Can be kept in groups, which look attractive, but cannot be housed safely with smaller companions. If kept in groups, introduce all the fish to the aquarium at the same time to prevent bullying.*

RING-TAILED CARDINALFISH

Apogon aureus

As its name suggests, the most distinctive feature of this species is the prominent black band present on the caudal peduncle at the base of the tail. The body color overall is golden yellow. There is also a blue stripe running from the upper jaw underneath the eye, a black bar present between the eye and snout, and a blue line above. What sets this species apart is the blue stripe that extends down each side of the face from the lower jaw. Juveniles can be identified easily, since they have a spot on each side of the caudal peduncle rather than an encircling band, according to some reports. However, this may be a regional variation.

DISTRIBUTION: *From the Red Sea and East Africa through the Indo-Pacific, north as far as Miyakejima Island in southern Japan. Ranges south to the coast of southeast Australia and New Caledonia and east to Tonga.*

▼ RING-TAILED CARDINALFISH (APOGON AUREUS)

SIZE: *6 in (15 cm).*

BEHAVIOR: *This is a relatively shy species that often hides under ledges on the reef or even in crevices during hours of daylight. Where they form shoals, these cardinalfish tend to venture farther into the open, presumably because this provides a greater opportunity to spot any threat and take avoiding action. At certain times of the year only, they may form mixed shoals with the short-tooth cardinalfish* (A. apogonoides).

DIET: *Feed meaty foods, including items such as chopped shrimps, since this cardinalfish is predatory by nature.*

AQUARIUM: *Not suitable for a reef tank where there are crustaceans, but will not harm sessile invertebrates.*

COMPATIBILITY: *Can be kept in aquarium surroundings in small groups if they are introduced together, so as to prevent bullying. Avoid mixing with smaller companions, which are likely to be eaten.*

YELLOW-STRIPED CARDINALFISH

Apogon cyanosoma

The taxonomy surrounding this species is controversial. It has been suggested that the fish previously classified under this scientific name may actually represent from four to eight separate species. These divisions tend to be based essentially on differences in their markings, such as the background color and the width and number of stripes. In terms of its basic coloration, this cardinalfish has a silvery body, intersected by a variable series of horizontal orangish yellow stripes, with a small narrow white stripe above the eye. The orangish yellow stripes do not extend into the caudal fin, but they create pinkish spots in this region that appear redder under conditions of darkness.

DISTRIBUTION: *From the Red Sea down the eastern coast of Africa as far as Mozambique. Ranges northward up to the Ryukyu and Bonin Islands, and*

▼ YELLOW-STRIPED CARDINALFISH (APOGON CYANOSOMA)

▲ *CARDINALFISH (APOGON IMBERBIS)*

southward around the eastern coast of Australia to the southern part of the Great Barrier Reef and New Caledonia. Also ranges across the Pacific Ocean to Fiji and Tonga. Current taxonomic proposals suggest that the true yellow-striped cardinalfish's range may be restricted to the western area of the Pacific.

SIZE: *3 in (8 cm).*

BEHAVIOR: *Remains close to cover during the day, sometimes seeking out the protection provided by the stinging tentacles of anemones or the sharp spines of sea urchins. May be seen on its own or in larger aggregations. Males collect and brood the eggs in their mouths once spawning has occurred. This paternal care is appreciated by* Thalassoma *species wrasses on the reef, which then try to harass individual males in a bid to persuade them to spit out their eggs. If successful, the wrasses proceed to consume the cardinalfish eggs.*

DIET: *Will need a range of meaty foods as a substitute for the crustaceans that form the bulk of its diet in the wild.*

AQUARIUM: *Not to be trusted in an aquarium alongside crustaceans.*

COMPATIBILITY: *Can be kept on its own as part of a community aquarium with other, nonaggressive species. If you hope to spawn these cardinalfish successfully in the aquarium, they will need to be housed in a larger group of about five individuals. Obtain all the fish at the same time and place them in the aquarium simultaneously in order to prevent any territorial disagreements or bullying.*

CARDINALFISH

Apogon imberbis

The red coloration of this fish explains the common name that has been given to the members of this group—it resembles the color of a cardinal's robe. The distribution of this fish in regions of the Old World meant that it was one of the first species to be classified (in 1758). The coloration of individuals varies, with some being a brighter shade of red than others. The area from the head along the upper part of the back appears dusky compared with the lower part of the body. A particular feature of the cardinalfish is that it has relatively large eyes, a characteristic found in fish that become more active at dusk—the enlarged eyes enhance their ability to see under these conditions.

DISTRIBUTION: *Occurs in the eastern part of the Atlantic and the western Mediterranean, ranging from the coast of Portugal via Morocco and the Azores down as far as the Gulf of Guinea off the west coast of Africa.*

SIZE: *6 in (15 cm).*

BEHAVIOR: *Usually occurs in areas where visibility is poor, lurking close to caves into which it can retreat in the event of danger. It has been found down to depths of 650 ft (200 m), but it often occurs in much shallower water. May be encountered individually, although it is sometimes observed in large shoals.*

DIET: *Feeds naturally on both crustaceans and small fish, so requires meaty foods in the aquarium.*

▲ LARGE-TOOTHED CARDINALFISH (CHEILODIPTERUS MACRODON)

AQUARIUM: *Not safe to accommodate in the reef aquarium with potential prey species.*

COMPATIBILITY: *May be kept in a small group together, but certainly cannot be trusted with smaller companions.*

LARGE-TOOTHED CARDINALFISH
(EIGHT-LINED CARDINALFISH)

Cheilodipterus macrodon

These fish are so called because of the relatively large sharp pointed teeth present in their jaws. The species' coloration alters with age, although the narrow patterning of horizontal lines is consistent. Juveniles have a characteristic black spot at the base of the caudal fin, which fades in intensity with age. They also tend to be paler in color overall, with a yellowish hue on the head, and blue coloring evident on the body. In contrast, adults tend to be of a more consistent black shade, but there are some regional variations in terms of both color and size, especially among individuals found in the Indian and Pacific Oceans. Those that occur in the Pacific attain a larger size.

DISTRIBUTION: *Ranges from the Red Sea and East Africa eastward and reaches the Ryukyu Islands to the north and Lord Howe Island off the east coast of Australia in the south. Extends across the Pacific Ocean as far as Pitcairn Island.*

SIZE: *10 in (25 cm).*

BEHAVIOR: *This is a relatively solitary species. Adults are usually encountered in pairs or small groups rather than large shoals. They frequent caves and ledges, which provide them with cover. These vantage points may also make it easier for the fish to seize prey, which consist largely of smaller fish.*

DIET: *Provide meaty substitutes such as lancefish, allowing frozen foods to defrost thoroughly before offering them to the aquarium occupants.*

AQUARIUM: *May prey on crustaceans in a reef tank and will certainly take small fish, for example, blennies, that may also be present in these surroundings.*

COMPATIBILITY: *Companions need to be chosen carefully, in view of the predatory habits of this species. Best kept singly rather than in a group, unless the aquarium is suitably large and has plenty of retreats. The fish must all be introduced to the tank at the same time.*

BANGGAI CARDINALFISH

Pterapogon kauderni

In a very short space of time the Banggai cardinalfish has become one of the most popular of all saltwater fish. Originally discovered in 1933, its exact locality remained a mystery until it was found again and introduced to the hobby in 1994. It is not just the fish's stunning appearance that has led to its popularity but also its fascinating breeding habits, which have enabled it to be bred repeatedly in aquarium surroundings.

The greatest difficulty in achieving success can be obtaining a pair in the first place, because there is no way of distinguishing visually between the sexes. It may be possible just before spawning occurs, however, because the male's jaw darkens at this time and the female's body swells with eggs. These cardinalfish have a black stripe running through the eyes and two more vertical black stripes running across their bodies from their split dorsal fins down through the fins on the ventral part of the body. The intervening areas can vary from yellowish to cream, depending on the individual and its surroundings, with these areas broken by white dots. The caudal fin is narrow and elongated. It is black with white spots along its edges, and the lower fins are also

▼ BANGGAI CARDINALFISH (PTERAPOGON KAUDERNI)

spotted. Juveniles lack any spotting, however, allowing them to be distinguished without difficulty.

DISTRIBUTION: *Restricted as far as is currently known to 16 islands in the Banggai Archipelago and in the vicinity of Luwuk harbor, central Sulawesi. There is also another population apparently thriving in Lembeh Strait farther north on the Sulawesi coast. The ancestors of this fish were introduced there when a number were inadvertently released in the area. They are now quite commonly reported by divers.*

SIZE: *3 in (8 cm).*

BEHAVIOR: *Tends to prefer relatively shallow water near the coast, often occurring around jetties as well as in mangroves, and favoring areas where seagrass is thriving. Has formed a close relationship with the sea urchin* Diadema setosum*—young fish in particular dart in readily among its spines if threatened. These cardinalfish may also be found in association with certain sea anemones that give them protection against predators. The best way of acquiring a pair is to keep a number of these cardinalfish together and allow them to choose their own mates. The pair should then be transferred to an aquarium on their own. Keep them well fed to encourage spawning and build up the male's bodily condition, since he will fast for a month during the period in which the eggs incubate and the fry continue to develop in his mouth. Plenty of cover in the tank is essential, otherwise the male will eat the young fry once they leave his mouth. A gentle foam filter is also needed, because a power filter would suck up the young fish. Around 40 young are likely to result from a single spawning. They can be reared on brine shrimp nauplii with the addition of unsaturated fatty acids and vitamins. Young are mature by the time they measure just over 3 in (7.5 cm) in total length. In-breeding must be discouraged subsequently, since it can lead to fin and mouth abnormalities.*

DIET: *Provide meaty foods chopped into small pieces as well as vitamin-enriched brine shrimps.*

AQUARIUM: *Ideal for a reef tank, since it lives in harmony with both invertebrates and other peaceful fish such as firefish.*

COMPATIBILITY: *Avoid housing this cardinalfish with companions such as tobies (which will nip its fins) or aggressive species such as small hawkfish.*

PAJAMA CARDINALFISH

Sphaeramia nematoptera

The pajama cardinalfish can be kept easily and tends to breed readily. Unfortunately, however, it is not so easy to rear the fry. Juveniles are more colorful than adults, although they all display similar patterning. The front half of the body is mainly yellow, with most of the pupil in the eyes being bright red. A broad black vertical band encircles the body and extends to the fins as well. The rear part of the body is relatively transparent in young fish and has reddish brown spots. In adults, this area turns milky, and the spotted patterning becomes correspondingly paler. Their black banding is also less pronounced. Sexing may be possible, since males reputedly have a longer filament at the rear of the dorsal fin compared with that of the females.

▲ *PAJAMA CARDINALFISH (SPHAERAMIA NEMATOPTERA)*

DISTRIBUTION: *Occurs in the western Pacific from the vicinity of Java, ranging to Papua New Guinea and eastward to Tonga and Fiji. Extends as far north as the Ryukyu Islands and south to Australia's Great Barrier Reef.*

SIZE: *3 in (8 cm).*

BEHAVIOR: *Encountered in groups among stony coral, which provides protection during the day. Will venture out to feed under cover of darkness. Thrives in small groups in which a definite pecking order is established. Violent conflicts are therefore avoided, provided the fish are all introduced to the aquarium at the same time.*

DIET: *Chopped meaty foods such as shrimps should be given, along with vitamin-enriched brine shrimps. Naturally hunts small invertebrates close to the substrate.*

AQUARIUM: *Safe for a reef aquarium, making an attractive addition, although it may possibly feed on polychaete and other worms in these surroundings.*

COMPATIBILITY: *Avoid mixing with aggressive fish.*

BUTTERFLYFISH FAMILY CHAETODONTIDAE

The elegant butterflyfish are very attractive, but unfortunately they are not always easy to maintain in the saltwater aquarium because of their feeding habits. There are an estimated 114 species, whose distribution is centered on the Indian and Pacific Oceans, although the family is also represented in the Atlantic Ocean.

It is very important to examine the feeding habits of the individual species, because those that are termed obligate corallivores (meaning that they are essentially dependent on coral to form the basis of their diet) are extremely difficult to wean onto alternative food sources, and should be avoided. On the other hand, in the case of so-called facultative corallivores, it is possible to persuade them to take other foods. Once fully established in an aquarium, butterflyfish can enjoy a long life. Some have been known to live for up to 25 years.

Young butterflyfish can differ in appearance from adults. They often display on their dorsal fin a protective spot, called an ocellus, which looks like an eye. Its purpose it to confuse predators, and it is particularly effective if the real eye is concealed with a dark band running through it.

Butterflyfish in the wild tend to be active during the day. At night they retreat to a protected area of the reef, hopefully avoiding predators. This type of behavior can be seen in aquarium fish, too. They will seek out a safe refuge and return there to spend the night.

The family Chaetodontidae is not a group of marine fish that has been regularly bred in the aquarium, but pairs of male and female butterflyfish usually develop a strong bond. There are no differences in markings between the sexes to aid identification, although in some cases males may attain a larger size than females. The eggs laid by the females float and are carried by the current after spawning.

▼ *PHILIPPINE BUTTERFLYFISH (CHAETODON ADIERGASTOS)*

▲ *ASIAN BUTTERFLYFISH (CHAETODON ARGENTATUS)*

PHILIPPINE BUTTERFLYFISH
(PANDA BUTTERFLYFISH)

Chaetodon adiergastos

The body coloration of the Philippine butterflyfish is whitish with a prominent black band encircling the eyes (hence the alternative common name of panda butterflyfish). There is also a black spot present on the fish's forehead. Diagonal brown stripes run across the sides of the body, and there is yellow coloration on both the snout and the fins.

DISTRIBUTION: *Extends from northwest Australia northward via Indonesia and the Philippines to Taiwan and the Ryukyu Islands near Japan.*

SIZE: *8 in (20 cm).*

BEHAVIOR: *Young of this species tend to be solitary and may venture into estuaries, but adults range over reefs and are seen either in pairs or in small groups.*

DIET: *Provide a varied diet, based on finely chopped meaty foods.*

AQUARIUM: *Cannot be housed safely in a reef aquarium.*

COMPATIBILITY: *Quite tolerant of its own kind, especially if all the individuals are introduced to the aquarium at the same time.*

ASIAN BUTTERFLYFISH
(BLACK PEARL-SCALED BUTTERFLYFISH)

Chaetodon argentatus

One of the less colorful butterflyfish, it has a series of bold black bands extending a variable distance down the sides of the body. On the sides its body scales are highlighted with black edges.

DISTRIBUTION: *Restricted to an area from the Philippines northward to the Chinese coast and Taiwan. Reaches as far north as southern Japan. Also present around the Ryukyu and Izu Islands.*

SIZE: *8 in (20 cm).*

BEHAVIOR: *Occurs on rocky reefs, sometimes in the company of other butterflyfish. Hybridizes in the wild in Japanese waters with the pearl-scale butterflyfish (*C. xanthurus*).*

DIET: *Introducing this species to an aquarium where filamentous algae are growing well will help it become established.*

AQUARIUM: *Not to be trusted in a reef tank.*

COMPATIBILITY: *Reasonably tolerant of other butterflyfish.*

GOLDEN BUTTERFLYFISH

(GOLDEN-STRIPED BUTTERFLYFISH)

Chaetodon aureofasciatus

As its name suggests, the coloration of the golden butterflyfish tends to be yellowish. There is a broad dark grayish area on the upper part of the body. One orangish yellow horizontal stripe runs along the same line as the pectoral fin, and another passes through the eyes.

DISTRIBUTION: *Extends from western Australia north to New Guinea. Also encountered around Australia's Great Barrier Reef.*

SIZE: *5.5 in (14 cm).*

BEHAVIOR: *Typically found in areas of the reef where there is staghorn coral (*Acropora *species). Will retreat to the relative safety of branching corals at night, but may be found occasionally in brackish waters.*

DIET: *Mainly consists of stony coral, but will eat algae. Hard to cater for.*

AQUARIUM: *Not suitable for a reef aquarium.*

COMPATIBILITY: *A difficult species to maintain. May be persuaded to sample foods other than stony corals if kept with more adaptable butterflyfish or in small groups of its own kind.*

THREAD-FIN BUTTERFLYFISH

(AURIGA BUTTERFLYFISH)

Chaetodon auriga

A black band encircles the eyes in this species. The area behind the head and the underparts has a white background, overlaid with dark bands that intersect each other at 90 degrees. The remainder of the body is yellowish, and there is a prominent black spot on the dorsal fin. A threadlike filament extends from this area to beyond the caudal fin.

DISTRIBUTION: *Ranges widely from the Red Sea and East Africa across the Pacific to the Hawaiian and Marquesan Islands. Reaches as far east as the Ducie Islands. Northward, it extends to southern Japan; southward to Lord Howe Island off the east coast of Australia and Rapa Island in the South Pacific.*

SIZE: *9 in (23 cm).*

BEHAVIOR: *Very adaptable, particularly in terms of its diet. Feeds on a wide range of invertebrate prey and algae, often ripping off parts of sessile species such as corals and anemones, and often hunts worms.*

DIET: *Will eat prepared foods of all types and small livefoods.*

▼ *THREAD-FIN BUTTERFLYFISH (CHAETODON AURIGA)*

AQUARIUM: *Definitely not a species suitable for a reef aquarium.*

COMPATIBILITY: *More aggressive than some butterflyfish, even to other fish of similar coloring. Can be kept with others of the same species, however, in a large aquarium. One of the easiest species of butterflyfish, it has been recorded as living up to 12 years in aquarium surroundings.*

BLACK-TAIL BUTTERFLYFISH
(EXQUISITE BUTTERFLYFISH)

Chaetodon austriacus

A particularly beautiful species, the black-tail butterflyfish has black coloration evident on the fins at the rear of the body and a broad white band over the dorsal area. The body color is vibrant yellow and has a series of bluish horizontal lines. The snout is dark, and there is a vertical dark bar running through the head, with another one behind.

DISTRIBUTION: *Restricted to the Red Sea and the Gulf of Aden.*

SIZE: *5 in (13 cm).*

BEHAVIOR: *Invariably seen in pairs when adult, the black-tail butterflyfish is highly territorial, guarding the hard corals that form the bulk of its diet.*

DIET: *Without access to hard coral this particular species is impossible to maintain in aquarium surroundings.*

AQUARIUM: *Will destroy hard coral, being an obligate corallivore.*

COMPATIBILITY: *Generally intolerant by nature.*

▲ *BLACK-TAIL BUTTERFLYFISH (CHAETODON AUSTRIACUS)*

▼ *EASTERN TRIANGULAR BUTTERFLYFISH (CHAETODON BARONESSA)*

EASTERN TRIANGULAR BUTTERFLYFISH

Chaetodon baronessa

Dark orange-black bands alternating with white (or sometimes yellow) stripes over the head give way to alternating blue and pale yellowish white bands that meet like the apex of a triangle over the body of this butterflyfish. It has, unfortunately, proved to be another exceedingly difficult species to maintain in aquarium surroundings, because of its feeding habits.

DISTRIBUTION: *Extends eastward from the Cocos (Keeling) Islands south of Sumatra and around Australia to northern New South Wales and northward to southern Japan. Its easterly range reaches to Fiji and Tonga.*

SIZE: *6 in (15 cm).*

BEHAVIOR: *Occurs in association with* Acropora *species corals, using its beak to bite off the polyps. Usually encountered in pairs.*

DIET: *An obligate corallivore, which means that it is very difficult to feed without constant access to suitable live coral.*

▲ BENNETT'S BUTTERFLYFISH (CHAETODON BENNETTI)

AQUARIUM: *Needs to be kept in an aquarium with coral.*

COMPATIBILITY: *Generally aggressive and intolerant, although a strong bond exists between true pairs.*

BENNETT'S BUTTERFLYFISH
(BLUE-LASHED BUTTERFLYFISH)

Chaetodon bennetti

Rich yellow coloration predominates in this species, and there is a prominent black spot edged with blue beneath the dorsal fin. Two pale blue stripes pass either side of the pectoral fins, and a dark blue-edged stripe runs across each eye.

DISTRIBUTION: *From East Africa across the Pacific Ocean as far east as Pitcairn Island, as far south as Lord Howe Island off the east coast of Australia, and as far north as the waters of southern Japan.*

SIZE: *8 in (20 cm).*

BEHAVIOR: *Bennett's butterflyfish is closely allied with colonies of* Acropora *species coral, which provide retreats and food for this species. While young, these fish are sometimes seen in groups, but adults invariably occur in pairs.*

DIET: *Feeds mainly on hard coral but will eat some algae. Very difficult to wean onto other foods.*

AQUARIUM: *Needs to be housed alongside suitable coral.*

COMPATIBILITY: *Becomes increasingly territorial as it matures.*

BLACKBURN'S BUTTERFLYFISH
(BROWNBURNIE)

Chaetodon blackburnii

The majority of the body is dark, and a blackish stripe runs through the eyes. The intervening areas are yellowish white, with a white area also present at the base of the caudal fin. The blue bands that run into the dark areas are less apparent in young fish than in more mature individuals.

DISTRIBUTION: *Restricted to the western part of the Indian Ocean, extending from Kenya down to Madagascar and eastward to Mauritius.*

SIZE: *5 in (13 cm).*

BEHAVIOR: *Becomes more common in southern areas during the summer, suggesting that some population movements occur.*

DIET: *Thought to eat soft coral as well as worms and small crustaceans.*

AQUARIUM: *Something of a rarity, so its habits are not well understood.*

COMPATIBILITY: *Tends to be solitary if unpaired. It is important to choose companions very carefully.*

BURGESS'S BUTTERFLYFISH

Chaetodon burgessi

There are two well-documented color variants in this species. In both types there is a broad black stripe through the eye and another extending from the back down to the pectoral fin. There is also a broad diagonal black patch across the rear of the fish's body. The rest of the body is usually either yellow or white, but individuals displaying variable patches of both colors are occasionally documented. The yellow color variant found in waters around the Philippines is commonly available.

DISTRIBUTION: *Ranges from northeastern Borneo north to the Philippines and south to Flores and Pohnpei, Indonesia, and east to Tonga.*

SIZE: *5.5 in (14 cm).*

BEHAVIOR: *Often occurs in relatively deep water, below 130 ft (40 m); usually found in pairs.*

DIET: *Quite adaptable, taking a variety of foods in the aquarium, including flake. Believed to feed on worms, small crustaceans, and coral in the wild.*

AQUARIUM: *Can be housed with certain invertebrates, but this butterflyfish may prey on them.*

▼ BURGESS'S BUTTERFLYFISH (CHAETODON BURGESSI)

▲ FOUR-EYE BUTTERFLYFISH (CHAETODON CAPISTRATUS)

COMPATIBILITY: *Relatively tolerant by nature toward its own kind and not usually aggressive toward other fish.*

FOUR-EYE BUTTERFLYFISH

Chaetodon capistratus

A bold black spot with a white border located just in front of the caudal peduncle on each side of the body gives this butterflyfish its common name. There is also a dark brown stripe running through the eyes and a radiating pattern of dark lines across the body. The underlying body color is yellowish white. Young individuals have two spots on each side as opposed to one.

DISTRIBUTION: *An Atlantic species, sometimes wandering as far north as Massachusetts, but generally ranging from the Bahamas and Bermuda through the Gulf of Mexico and the Caribbean to northern South America.*

SIZE: *6 in (15 cm).*

BEHAVIOR: *Known to form both true and same-sex pairings, with both fish collaborating to defend a feeding area and drive away other fish.*

DIET: *Stony coral polyps are a major food item, as well as gorgonian polyps in some areas. Worms and crustaceans are sometimes eaten as well.*

AQUARIUM: *Can be difficult to persuade this fish to eat at first. Food needs to be chopped into tiny pieces, reflecting the size of its mouth, and offered frequently throughout the day in small quantities.*

COMPATIBILITY: *Should be kept only with quiet companions. This butterflyfish is likely to be territorial.*

CITRUS BUTTERFLYFISH
(SPECKLED BUTTERFLYFISH)

Chaetodon citrinellus

This is another butterflyfish that displays very variable coloration, with either a white or a yellow base color. A dark stripe runs through the eyes, and rows of distinct spots, which may be blue or orange or sometimes a combination of both, extend over the sides of the body.

DISTRIBUTION: *From East Africa across the Pacific to the Hawaiian, Marquesan, and Tuamotu Islands. Its northerly distribution reaches to southern Japan, while to the south it can be found around Lord Howe Island off Australia's east coast.*

SIZE: *5 in (13 cm).*

BEHAVIOR: *Tends to be found in relatively shallow stretches of water on the reef, seeking food over sandy areas. Usually observed in pairs.*

DIET: *One of the easier butterflyfish to feed. Takes a varied diet, but is most easily established if there is a good supply of filamentous algae growing in the aquarium.*

AQUARIUM: *Not really suited to a reef aquarium, since it will attack coral polyps.*

COMPATIBILITY: *Keep singly or in pairs. The citrus butterflyfish is one of the more aggressive members of the family.*

RED-TAIL BUTTERFLYFISH
(PAKISTAN BUTTERFLYFISH)

Chaetodon collare

Alternate blackish and white banding is present on the head. The body has a speckled appearance, with the background being whitish or ocher yellow. The base of the caudal fin is reddish in adults. Young individuals display more reddish brown color overall and have a blackish spot at the rear of the dorsal fin.

DISTRIBUTION: *Extends from the southeast region of the Arabian Peninsula through the Persian Gulf and the coast of Pakistan down to the Maldives in the Indian Ocean and eastward to Bali, Indonesia.*

SIZE: *7 in (18 cm).*

BEHAVIOR: *Although often seen in pairs, this species can also occur in large aggregations of up to 20 individuals. The black bands on the face may turn gray when confronting a rival.*

▼ *RED-TAIL BUTTERFLYFISH (CHAETODON COLLARE)*

DIET: *Feeds largely on coral polyps, but will eat filamentous algae. Young fish can be weaned most easily onto substitute diets.*

AQUARIUM: *Unsuitable for a reef tank alongside live coral. A quiet environment increases the likelihood that the fish will feed well.*

COMPATIBILITY: *Not especially aggressive, but introduce a pair to the aquarium at the same time. Can sometimes take a dislike to other butterflyfish.*

INDIAN VAGABOND BUTTERFLYFISH

Chaetodon decussatus

Pale diagonal striping against a white background, with a prominent black area at the back of the body, helps identify this species. The caudal fin is yellow with a black band close to the tip.

DISTRIBUTION: *Ranges from the Maldives off India's southwest coast to Sri Lanka, north to the Andamans, and east to Indonesia.*

SIZE: *8 in (20 cm).*

BEHAVIOR: *Occurs not just on coral reefs but over a variety of other habitats, both rocky and sandy. Most commonly seen in pairs.*

DIET: *A facultative corallivore, eating coral polyps in preference to other foods but, in fact, very adaptable in terms of its feeding habits.*

AQUARIUM: *Cannot be kept safely with corals in a reef aquarium.*

▲ *SADDLED BUTTERFLYFISH (CHAETODON EPHIPPIUM)*

COMPATIBILITY: *Small juveniles are more quarrelsome than adults. Can be kept in pairs in a large aquarium, and will settle quite easily.*

SADDLED BUTTERFLYFISH

(SADDLE-BACK BUTTERFLYFISH)

Chaetodon ephippium

The body color is predominantly yellowish gray. A broad white band over the upper back separates the background color from a large black area. Undulating blue bands are present on the lower part of the body, with yellow coloration around the mouth area. A trailing filament extends back from the dorsal fin. Young individuals are mainly white and they retain the black mark seen in adults. They also have yellowish fins.

DISTRIBUTION: *From Sri Lanka and the Cocos (Keeling) Islands across the Pacific to the Hawaiian, Marquesan, and Tuamotu Islands. Ranges north to southern Japan and south as far as Australia's southeastern coast.*

SIZE: *12 in (30 cm).*

BEHAVIOR: *Ranges widely over the reef, but particularly common in clear water where there is abundant growth of hard coral. The young tend to be solitary, but adult pairs form a strong bond.*

DIET: *Omnivorous. Eats coral polyps, filamentous algae, invertebrates, and even fish eggs.*

◀ *SADDLE-BACK BUTTERFLYFISH (CHAETODON FALCULA)*

AQUARIUM: *Not suitable for a reef tank. Juveniles will often start eating more readily than adults, particularly if there is filamentous algae present.*

COMPATIBILITY: *Can be kept in a mixed community of fish, but it is best to avoid boisterous companions.*

SADDLE-BACK BUTTERFLYFISH
(BLACK-WEDGED BUTTERFLYFISH)

Chaetodon falcula

A yellowish saddlelike area across the top of the back, bordered by two large black blotches, is characteristic of this butterflyfish. Dark stripes run vertically up the white sides of the body to join with these colored areas. There is a black stripe extending through the eyes and another at the base of the caudal fin.

DISTRIBUTION: *Ranges across the Indian Ocean from East Africa to Indonesia, extending as far north as India and south to Mauritius.*

SIZE: *8 in (20 cm).*

BEHAVIOR: *The long jaws of this butterflyfish help it seize the tentacles of featherduster worms and enable it to grab small crustaceans.*

DIET: *Acclimatize this fish by offering it brine shrimps and finely chopped seafood such as squid. Reluctant individuals may need to be tempted with an anemone.*

AQUARIUM: *Not safe for inclusion in a reef setup, since it feeds on both corals and anemones. Provide other hiding places in the aquarium.*

COMPATIBILITY: *Relatively tolerant of its own kind.*

RED SEA RACCOON BUTTERFLYFISH
(DIAGONAL BUTTERFLYFISH)

Chaetodon fasciatus

Yellow is the predominant color of this butterflyfish, and it has a prominent white area in the vicinity of the forehead. There is a dark area above, at the base of the dorsal fin, and relatively thick stripes curling up across the body.

DISTRIBUTION: *Restricted to the Red Sea and the Gulf of Aden in the western Indian Ocean.*

SIZE: *8.75 in (22 cm).*

BEHAVIOR: *A common, relatively large, and highly territorial species, usually encountered in individual pairs that may drive away all other butterflyfish.*

DIET: *Feeds largely on coral polyps but will also eat other invertebrates and some filamentous algae. The latter is useful for settling an individual into an aquarium environment.*

AQUARIUM: *Should not be included in a reef aquarium.*

COMPATIBILITY: *Highly aggressive, so keep singly unless you acquire a bonded pair. Likely to bully other butterflyfish.*

BLACK BUTTERFLYFISH
(DUSKY BUTTERFLYFISH)

Chaetodon flavirostris

Adults are a rich dusky blue in color with yellow edging around the rear of the body and on the snout. There is a distinctive black bump on the forehead. Young black butterflyfish are paler and

▼ *RED SEA RACCOON BUTTERFLYFISH (CHAETODON FASCIATUS)*

▲ *BLACK BUTTERFLYFISH (CHAETODON FLAVIROSTRIS)*

have a white band extending down behind the eyes and an orange stripe on a whitish background at the rear of the body.

DISTRIBUTION: *Occurs in the southern Pacific down Australia's east coast, reaching as far south as Lord Howe Island. Also ranges eastward via Fiji and Samoa to Pitcairn Island.*

SIZE: *8 in (20 cm).*

BEHAVIOR: *Adults roam widely over lagoon areas of the reef, whereas juveniles seek more protection and even venture into estuaries.*

DIET: *Coral features prominently in the diet of this butterflyfish, but it may also eat zooplankton and nibble at algae.*

AQUARIUM: *Although shy, this butterflyfish cannot be trusted in a reef aquarium.*

COMPATIBILITY: *Keep this butterflyfish separate and be sure to provide plenty of retreats. It can be very difficult to acclimatize this species to aquarium surroundings.*

SPOTTED BUTTERFLYFISH
(PEPPERED BUTTERFLYFISH)

Chaetodon guttatissimus

Dusky brownish spots of variable size on the sides of the body, resembling specks of pepper, are a feature of this butterflyfish.

The background body color is yellowish, and the fins often have a bluish tinge. An unusual bright orange area is present above the caudal peduncle, and there is a dark stripe with narrow edging running through the eyes.

DISTRIBUTION: *Occurs in the Red Sea to East Africa, across the Indian Ocean to Christmas Island, ranging north to western Thailand and east as far as Bali.*

SIZE: *4.75 in (12 cm).*

BEHAVIOR: *Often seen in pairs and sometimes in small groups, typically in areas of the reef where there is hard coral, which forms a significant part of its diet.*

DIET: *Although it may be persuaded to take filamentous algae, this butterflyfish has a reputation for being difficult to wean onto alternative diets.*

AQUARIUM: *Cannot be included safely in a reef aquarium.*

COMPATIBILITY: *The young of this species often settle better in aquarium surroundings than adults and feed more readily. Allowing a pair to grow up together in the aquarium also lessens the risk of aggression.*

BLACK-LIP BUTTERFLYFISH

(BROWN BUTTERFLYFISH; SUNBURST BUTTERFLYFISH)

Chaetodon kleinii

Among the characteristic features of this species are black markings on the lips. Behind the black bar that passes through the eye is a white band, followed by a variable brownish black area and another white band. The rear part of the body is yellowish.

SYNONYMS: *C. melammystax; C. melastomus; C. virescens.*

DISTRIBUTION: *From the Red Sea and East Africa across the Indian and Pacific Oceans to the Galápagos Islands off the coast of Ecuador. Also occurs in Hawaiian waters.*

SIZE: *5.5 in (14 cm).*

BEHAVIOR: *A deepwater species, usually found from 33 ft (10 m) down to 400 ft (122 m) on the reef, but it sometimes feeds on zooplankton near the surface. Shoals may raid damselfish nesting grounds to eat their eggs.*

DIET: *One of the easiest species to feed in aquarium surroundings. Takes a wide variety of foods, especially once established.*

AQUARIUM: *Will prey on coral, so not really trustworthy in a reef tank.*

COMPATIBILITY: *Can be housed with more assertive fish, but introduce it to the aquarium first and provide access to a choice of hiding places.*

▼ BLACK-LIP BUTTERFLYFISH (CHAETODON KLEINII)

HOODED BUTTERFLYFISH

(ORANGE-FACED BUTTERFLYFISH)

Chaetodon larvatus

This unmistakable species can be identified by its orange face with a white area behind. It has a bluish gray body traversed by yellowish stripes that meet in the midline. The upper part of the body to the rear and the caudal fin are both black, and the whole body is edged with a sky-blue color.

DISTRIBUTION: *Restricted to the Red Sea and the Gulf of Aden.*

SIZE: *5 in (12 cm).*

BEHAVIOR: *Lives in close association with* Acropora *coral, usually in pairs.*

DIET: *Feeds almost exclusively on coral polyps, making it virtually impossible to keep without access to a regular supply.*

AQUARIUM: *Cannot be included in a reef tank, since it will destroy coral.*

COMPATIBILITY: *Pairs will agree together, but otherwise territorial.*

RACCOON BUTTERFLYFISH

Chaetodon lunula

The incomplete broad black mask across the face and the white area behind give the impression of a raccoon, which explains this fish's common name. There is another yellow-edged broad

▲ RACCOON BUTTERFLYFISH (CHAETODON LUNULA)

black band running up to the dorsal fin, and a black spot with a similar border on the caudal peduncle. The upper body is dark, and the lower part of the body is yellow and striped.

DISTRIBUTION: *From East Africa north to southern Japan, south to Lord Howe Island off Australia's east coast, and east to the Hawaiian and Marquesan Islands.*

SIZE: *8 in (20 cm).*

BEHAVIOR: *Unlike most butterflyfish, this is a nocturnal species, which is indicated by its relatively large eyes. It feeds at night and hides during the day. The young often venture very close to the shoreline.*

DIET: *Will eat a wide range of foods and may even feed during the day in the aquarium.*

AQUARIUM: *Do not keep with invertebrates other than large crustaceans.*

COMPATIBILITY: *Not aggressive with other fish, and easily maintained.*

BLACK-BACK BUTTERFLYFISH

Chaetodon melannotus

Yellow edging around the body extending to the fins and a darker area on the upper part of the body help identify this species. Dark streaks run obliquely from the flanks, and there are black stripes running through the eyes as well as at the base of the caudal fin.

SYNONYM: *C. abhortani.*

DISTRIBUTION: *East Africa and the Red Sea across the Pacific to Samoa. Extends north to southern Japan and south to Lord Howe Island off Australia's east coast.*

SIZE: *6 in (15 cm).*

BEHAVIOR: *As in a number of other reef fish, it changes color significantly at night, at which time black coloration extends over much of the upper part of the body, with the exception of two white patches.*

DIET: *Tends to feed on soft coral polyps, but usually adapts very easily and can be persuaded to take a range of foods, occasionally even flake.*

AQUARIUM: *Not suited to a reef aquarium.*

COMPATIBILITY: *Tends not to agree with others of its own species or with other butterflyfish that have similar markings.*

MERTEN'S BUTTERFLYFISH
(SEYCHELLES BUTTERFLYFISH)

Chaetodon mertensii

A white body with V-shaped black lines that meet in the midline are features of this butterflyfish, as well as a broad orange band on the rear of the body and a similar band across the caudal fin. An incomplete eyebar enables the Indian Ocean form to be differentiated from its Pacific counterpart.

SYNONYM: *C. madagaskariensis (used for the Indian Ocean form).*

DISTRIBUTION: *Down the east coast of Africa across the Indian and Pacific Oceans to the Tuamotu Archipelago. North to southern Japan and south to Lord Howe Island off the east coast of Australia.*

SIZE: *5 in (13 cm).*

BEHAVIOR: *Adults live in pairs, usually on the outer fringe of the reef.*

DIET: *Not especially difficult to wean onto a diet of prepared foods and marine algae.*

AQUARIUM: *Will damage hard corals and will also eat invertebrates in a reef tank.*

▼ BLACK-BACK BUTTERFLYFISH (CHAETODON MELANNOTUS)

▲ *MILLET-SEED BUTTERFLYFISH (CHAETODON MILIARIS)*

COMPATIBILITY: *May be aggressive toward its own kind and also toward other members of the family.*

SCRAWLED BUTTERFLYFISH

(MEYER'S BUTTERFLYFISH)

Chaetodon meyeri

The strikingly marked scrawled butterflyfish has a series of relatively broad black bands on its predominantly white body. Black markings also swirl across the dorsal, caudal, and anal fins. There is striped patterning on the face, too, and all these areas are yellowish. Even juveniles are boldly marked, although their stripes are straighter than those of the adults.

DISTRIBUTION: *East Africa across the Pacific to the Galápagos Islands, north to the Ryukyu Islands near Japan and south to Australia's Great Barrier Reef.*

SIZE: *8 in (20 cm).*

BEHAVIOR: *Occurs in areas of the reef where there is abundant growing coral. The young hide below branches of coral.*

DIET: *An obligate corallivore that eats only coral polyps.*

AQUARIUM: *Can be housed only with its natural food source and can be kept only in areas of the world where a supply of suitable stony coral can be adequately maintained.*

COMPATIBILITY: *Tends to be territorial.*

MILLET-SEED BUTTERFLYFISH

(LEMON BUTTERFLYFISH)

Chaetodon miliaris

While the predominantly underlying color of this butterflyfish is yellowish, individuals are variably spotted with markings that resemble grains of millet. There is a black stripe running through the eye, and the caudal peduncle is also black.

DISTRIBUTION: *Restricted to the vicinity of Johnston Atoll and the Hawaiian Islands in the eastern Pacific.*

SIZE: *5 in (13 cm).*

BEHAVIOR: *May extend as far as 820 ft (250 m) down into the depths on the reef. Often seen in shoals much closer to the surface during the day, feeding on zooplankton.*

DIET: *Feeds naturally on copepods and other items ranging from mysid shrimps to fish eggs. Easy to adapt to aquarium foods such as finely chopped bivalve meat and brine shrimps.*

AQUARIUM: *Reasonably safe to include in a reef setup, although it may occasionally eat coral, so keep it well fed with other foods.*

COMPATIBILITY: *Can be kept in groups if all the fish are introduced to the aquarium at the same time.*

INDIAN BUTTERFLYFISH
(HEAD-BAND BUTTERFLYFISH)

Chaetodon mitratus

A broad black band through the eye and two more sloping back across and down the body typify this butterflyfish. Its underlying color varies among individuals, from white to shades of yellow. Young have an additional, much smaller black spot on the upper part of the dorsal fin.

DISTRIBUTION: *Various Indian Ocean islands, known to include Madagascar, Mauritius, and Réunion, extending eastward via the Maldives to Cocos (Keeling) and Christmas Islands.*

SIZE: *5.5 in (14 cm).*

BEHAVIOR: *A deepwater species, often seen hiding under ledges or in caves by divers. Associates in small groups.*

DIET: *An adaptable species. Will take chopped meaty foods of various types.*

AQUARIUM: *Will also feed on invertebrates, so not suitable for a reef tank.*

COMPATIBILITY: *An aggressive nature means that it is best housed on its own rather than in the company of other butterflyfish, unless the aquarium is very large.*

▼ INDIAN BUTTERFLYFISH (CHAETODON MITRATUS)

SPOT-FIN BUTTERFLYFISH

Chaetodon ocellatus

The body of the spot-fin butterflyfish is unmarked. It is whitish with some yellow parts, especially in the area around the caudal peduncle. As its name suggests, there is a characteristic black spot at the rear of the dorsal fin. There is also a black stripe running through the eye. Young individuals display a second spot on the anal fin, and a dark line runs between the two spots.

SYNONYM: *C. bimaculatus.*

DISTRIBUTION: *Occurs in the western Atlantic, ranging from Florida through the Gulf of Mexico and the Caribbean to the coast of Brazil.*

SIZE: *8 in (20 cm).*

BEHAVIOR: *Swims over sandy areas in search of food. Larvae sometimes swept north, which explains the fact that juveniles have been recorded as far north as Nova Scotia.*

DIET: *Tends to feed on coral but can be switched to meaty seafood diets, provided the food is chopped into small pieces.*

AQUARIUM: *Will destroy both hard and soft corals in a reef tank.*

COMPATIBILITY: *It is possible to keep two spot-fins together if they are placed in a spacious aquarium at the same time.*

ORNATE BUTTERFLYFISH

Chaetodon ornatissimus

Yellow and pale gray oblique banding extends across the body of this fish. It has black stripes on the face and tail, and a narrower black band extends along the dorsal fin. Juveniles are even more striking, displaying a vivid yellow and white body.

DISTRIBUTION: *Sri Lanka north to southern Japan, south to Lord Howe Island off Australia's east coast, and eastward to the Marquesan, Ducie, and Hawaiian Islands in the Pacific.*

SIZE: *8 in (20 cm).*

BEHAVIOR: *Found in reef waters where coral growth is abundant. Young fish hide among branching corals.*

DIET: *Feeds exclusively on hard corals, including those in the genera* Porites *and* Pocillopora.

AQUARIUM: *Can be kept only in areas in which its natural food—hard coral—is readily obtainable.*

COMPATIBILITY: *Not aggressive toward other species, but territorial with its own kind.*

SPOT-NAPE BUTTERFLYFISH

Chaetodon oxycephalus

As its name suggests, this species has a prominent black spot lying above and behind the eyes. Narrow vertical lines extend up the sides of the body, and there is a solid black blotch toward the rear, beneath the dorsal fin. The dark blotch is surrounded by an area of yellow, and yellow coloring also extends down to the ventral surface of the body.

▼ *ORNATE BUTTERFLYFISH (CHAETODON ORNATISSIMUS)*

▲ *SPOT-NAPE BUTTERFLYFISH (CHAETODON OXYCEPHALUS)*

DISTRIBUTION: *Ranges from the Maldives and Sri Lanka east to Papua New Guinea, north to the Philippines, and south to the Great Barrier Reef off Australia's eastern coast.*

SIZE: *10 in (25 cm).*

BEHAVIOR: *Often encountered in pairs, inhabiting areas where coral is abundant.*

DIET: *Eats coral polyps as well as anemones, but younger individuals especially can usually be weaned relatively easily onto other foods.*

AQUARIUM: *Not safe to include in a reef aquarium.*

COMPATIBILTY: *The large size and dominant nature of this fish means that companions need to be chosen carefully. Avoid mixing with other butterflyfish, especially those with similar markings, such as* C. lineolatus.

RED-BACK BUTTERFLYFISH

(ERITREAN BUTTERFLYFISH; CROWN BUTTERFLYFISH)

Chaetodon paucifasciatus

The red-back butterflyfish is one of the most easily identifiable butterflyfish. It has a stunning rich red area extending from the rear of the dorsal fin and tapering down to just below the caudal peduncle. The caudal fin also has a matching red band near its tip, while the rest of the fin is yellow. Brownish and yellow angular stripes extend backward across much of the body, and a white, finely spotted area divides them from the red coloring.

DISTRIBUTION: *The Red Sea and the Gulf of Aden.*

SIZE: *5.5 in (14 cm).*

▲ RED-BACK BUTTERFLYFISH (CHAETODON PAUCIFASCIATUS)

BEHAVIOR: *Pairs display a strong bond. Spawning takes place between August and October, occurring soon after dusk falls. It is the female that initiates courtship.*

DIET: *Relatively adaptable, eating coral polyps, polychaete worms, and filamentous algae among other foods in the wild, as well as small crustaceans. Will usually take chopped meaty foods readily, but can be less inclined to sample flake.*

AQUARIUM: *Not suitable for a reef tank.*

COMPATIBILTY: *Reasonably tolerant of the company of its own kind if the individuals are introduced to the aquarium at the same time.*

BLUE-BLOTCH BUTTERFLYFISH

Chaetodon plebius

This is another beautiful and popular butterflyfish. It is characterized by a bluish stripe running down the body just above the midline, with some other indistinct narrow stripes running above and through this area. The rest of the body is yellow, apart from a black spot with pale blue edging on the caudal peduncle, and a similar dark eye stripe.

DISTRIBUTION: *From the Andaman Sea off India's east coast north to southern Japan and south to Lord Howe Island off Australia's east coast. Extends across the Pacific to New Caledonia, Fiji, and Tonga.*

SIZE: *6 in (15 cm).*

BEHAVIOR: *Free-ranging in small numbers during the day. At night much larger aggregations form in* Acropora *coral, and disputes often break out.*

▶ BLUE-BLOTCH BUTTERFLYFISH (CHAETODON PLEBIUS)

DIET: *Polyps of hard coral feature prominently in its diet in the wild, but it also acts as a cleaner fish, removing and eating external parasites from the bodies of other fish. Not one of the easiest species to adapt to artificial diets but, once established, individuals may live for more than a decade.*

AQUARIUM: *Not suitable for a reef tank.*

COMPATIBILTY: *Relatively tolerant of the company of its own kind. Housing it with other butterflyfish may encourage feeding, although the fish should all be placed in the aquarium at the same time.*

SPOT-BANDED BUTTERFLYFISH

Chaetodon punctatofasciatus

Individuals are variable in appearance, as with many butterflyfish. Some display more spotting than banding on the sides of their bodies. The area at the base of the caudal fin is orange, with an adjacent whitish band on the tail in the case of adults. In juveniles the band is yellowish green.

DISTRIBUTION: *Extends from Christmas Island in the Indian Ocean, south to Rowley Shoals off Australia's northwestern coast, east to the northern part of the Great Barrier Reef, and out to the Line Islands. Its northernmost range is the Ryukyu Islands near Japan.*

SIZE: *4.75 in (12 cm).*

BEHAVIOR: *Occurs in relatively shallow water. Known to hybridize naturally on occasion with closely related species such as the spotted butterflyfish* (C. guttatissimus). *This behavior can sometimes lead to unidentifiable individuals cropping up in the trade.*

DIET: *Takes a variety of food. Krill and brine shrimps can be provided, along with algae growing in the tank or a* Spirulina *species alga substitute. Its small mouth means that it may not be able to ingest large brine shrimp, so chop up any meaty food very finely.*

▲ LATTICED BUTTERFLYFISH (CHAETODON RAFFLESII)

AQUARIUM: *Not recommended for a reef aquarium.*

COMPATIBILTY: *Rather shy, so choose companions carefully. Can be housed in pairs in a large aquarium.*

LATTICED BUTTERFLYFISH

Chaetodon rafflesii

The latticed butterflyfish is predominantly yellowish with faint bluish latticelike markings over its body. The patterning is most evident in adults. There is a black band running through the eyes and another on the caudal fin.

SYNONYMS: *C. dahli; C. sebae.*

DISTRIBUTION: *Ranges fro Sri Lanka off India's southeast coast northward to southern Japan and southward to Australia's Great Barrier Reef. Extends across the Pacific to the Tuamotu Archipelago.*

SIZE: *7 in (18 cm).*

BEHAVIOR: *Its color may darken temporarily after being moved to a new aquarium. This is normal. A similar change is often observed at night, too.*

DIET: *Provide brine shrimps and finely chopped meaty foods, along with plenty of algae.*

AQUARIUM: *Not to be trusted in a reef aquarium.*

COMPATIBILTY: *Tolerant and somewhat shy by nature, so ensure that the tank incorporates hiding places in which the fish can shelter.*

RAINFORD'S BUTTERFLYFISH

Chaetodon rainfordi

A banded butterflyfish, Rainford's has two predominantly yellow vertical bands on its body. The first one runs through the eyes, and the second lies just behind. These yellow bands are followed by three broader blackish bands with yellow edging. The last of the black bands extends around the white-bordered dark spot on the caudal peduncle. The underlying body color varies from pale yellowish to white.

DISTRIBUTION: *Ranges from Papua New Guinea down the Great Barrier Reef and east to Lord Howe Island off the east coast of Australia.*

SIZE: *6 in (15 cm).*

BEHAVIOR: *Studies suggest that some groups may be made up of females only, possibly depending on the habitat. The young may join with other butterflyfish when feeding, possibly seeking safety in numbers.*

DIET: *This butterflyfish survives by eating corals in the wild and finds it difficult to adjust to a new diet. Only keep in aquaria with well-established colonies of filamentous algae, a vital food source that will provide the opportunity of introducing other items to the diet.*

AQUARIUM: *Not suitable for a typical reef tank.*

COMPATIBILTY: *Often does better in small groups, particularly if you can obtain juvenile specimens at the outset.*

MAILED BUTTERFLYISH

Chaetodon reticulatus

The unusual name of this butterflyfish stems from the appearance of its body markings, which resemble the armor known as chain mail. The dorsal fin is a pale bluish gray, and there is a similarly colored area behind the black eye band, although the grayish areas are whiter in juveniles.

DISTRIBUTION: *Extends from the Philippines north to the Ryukyu Islands near Japan. Ranges east across the Pacific to the Hawaiian, Marquesan, and Ducie Islands, and south to the Great Barrier Reef off the eastern coast of Australia.*

SIZE: *7 in (18 cm).*

BEHAVIOR: *Often seen in pairs, but may form larger aggregations in some areas.*

DIET: *Feeds almost entirely on hard coral, but may also eat a little filamentous algae.*

AQUARIUM: *This butterflyfish will destroy hard coral in a reef tank, but it cannot be weaned onto other foods and should therefore be kept only where a guaranteed supply is available.*

COMPATIBILTY: *Can prove to be territorial and reaches a relatively large size when mature.*

▼ *MAILED BUTTERFLYFISH (CHAETODON RETICULATUS)*

▲ *BLUE-CHEEK BUTTERFLYFISH (CHAETODON SEMILARVATUS)*

BLUE-CHEEK BUTTERFLYFISH

(GOLDEN BUTTERFLYFISH)

Chaetodon semilarvatus

Bright yellow is the dominant color, with a bluish patch on the cheeks that partially encircles the eyes. There are a number of wavy vertical orange stripes extending down the side of the body.

DISTRIBUTION: *Restricted to the Red Sea and the Gulf of Aden.*

SIZE: *9 in (23 cm).*

BEHAVIOR: *Usually encountered in areas where coral predominates. Sometimes retreats under* Acropora *species corals, apparently to rest.*

DIET: *May need livefoods such as brine shrimps at first, but can be introduced to inanimate foods, which must be chopped into small pieces.*

AQUARIUM: *Needs a large 100-gallon (380-l) tank, because of the likely adult size of this butterflyfish. Do not include as part of a reef aquarium.*

COMPATIBILTY: *Can be housed in small groups, provided the aquarium is sufficiently spacious.*

MIRROR BUTTERFLYFISH

(OVAL-SPOT BUTTERFLYFISH)

Chaetodon speculum

A dark, prominent, rather oval-shaped spot is apparent on each side of the body, between the dorsal and caudal fins. There is also a dark stripe running through each eye, and the remainder of the body is yellow.

DISTRIBUTION: *Ranges from Christmas Island north to Japan, east to Tonga, and south to Lord Howe Island off Australia's east coast. Also recorded from some southern Indian Ocean islands, notably Madagascar, Réunion, and Mauritius.*

◀ BANDED BUTTERFLYFISH (CHAETODON STRIATUS)

SIZE: *7 in (18 cm).*

BEHAVIOR: *Tends to be solitary by nature. The young hide among the coral branches.*

DIET: *Based largely on hard coral polyps, anemones, and filamentous algae. It is hard to wean this butterflyfish without using livefoods. Older juveniles may be acclimatized with the least difficulty, but it is very tricky without natural foods being provided.*

AQUARIUM: *Not to be trusted in a reef aquarium.*

COMPATIBILTY: *Accommodating this fish with another docile species of butterflyfish may aid the acclimatization process.*

BANDED BUTTERFLYFISH

Chaetodon striatus

Although it is not especially colorful and is not necessarily the most interesting fish to look at, the banded butterflyfish will adapt well to aquarium life. It has four black vertical bands running down the side of its body. The first of these passes through the eyes, and the rear one extends down through the caudal peduncle. In between the two bands is a large area of silvery white coloration.

DISTRIBUTION: *An Atlantic species ranging from Bermuda and the northern part of the Gulf of Mexico through the Caribbean down to the vicinity of Rio de Janeiro, Brazil.*

SIZE: *6.25 in (16 cm).*

BEHAVIOR: *Hunts polychaetes (e.g., fan worms) and may form groups of as many as 20 fish, all feeding on plankton. May also remove external parasites from other fish, such as parrotfish, that are attracted to feed alongside it.*

DIET: *Offer a variety of foodstuffs to maintain its appetite, although persuading this fish to take unfamiliar foodstuffs is often easier than with certain other species.*

AQUARIUM: *Cannot be included safely in a reef aquarium.*

COMPATIBILTY: *May be kept in pairs or possibly larger groups in a sufficiently spacious aquarium.*

HAWAIIAN BUTTERFLYFISH
(TINKER'S BUTTERFLYFISH)

Chaetodon tinkeri

A yellow beak with a purple band behind and yellow surrounding the eyes are all features of this species, which is greatly sought after in the aquarium trade. The sides of its white body are decorated with spots and there is contrasting black diagonal marking extending across the rear of the body. Some regional variations of this coloration have been documented.

DISTRIBUTION: *Originally thought to be restricted to the Hawaiian Islands, but now known also from Johnston Atoll and the Marshall and Cook Islands.*

SIZE: *7 in (18 cm).*

BEHAVIOR: *Primarily a deepwater butterflyfish, which explains why it is difficult to obtain and expensive. It is bold by nature, but it must have hiding places in the aquarium.*

DIET: *Feeds naturally on plankton and worms. Tends to adapt readily, often taking substitute foods within a day of being moved to a new aquarium.*

AQUARIUM: *Not suitable for a reef setup, partly because its natural deepwater habitat is not as brightly lit. Subdued lighting is recommended.*

COMPATIBILTY: *If two individuals are introduced to the aquarium at the same time they are likely to agree reasonably well.*

▼ HAWAIIAN BUTTERFLYFISH (CHAETODON TINKERI)

▲ TRIANGULAR BUTTERFLYFISH (CHAETODON TRIANGULUM)

TRIANGULAR BUTTERFLYFISH

Chaetodon triangulum

A series of alternate dark and light chevron-shaped stripes help identify this species. A broad dark band runs through the eye, and there is a broad black triangle on the caudal fin.

DISTRIBUTION: *Ranges from Madagascar in the Indian Ocean northward to the Andaman Sea and westward as far as Java.*

SIZE: *6 in (15 cm).*

BEHAVIOR: *The triangular butterflyfish is usually seen in the vicinity of* Acropora *species corals, which provide food as well as shelter, in particular for juveniles of the species.*

DIET: *Its dependence on coral means that this species rarely takes other foods and should be kept only in areas where its natural food can be provided easily.*

AQUARIUM: *Destroys coral, so it is not recommended for reef tanks.*

COMPATIBILITY: *Aggressive toward its own kind and also toward other members of the family.*

MELON BUTTERFLYFISH
(INDIAN OCEAN RED-FIN BUTTERFLYFISH)

Chaetodon trifasciatus

Blue stripes run horizontally down the body, which is yellowish on the lower half, especially near the head, and a lighter shade of blue above. Unusually, this species can be sexed relatively easily, since its first anal spine is red in males but pink in females.

SYNONYMS: *C. bellus; C. layardi; C. leachii; C.ovalis; C. striangulus.*

DISTRIBUTION: *East Africa across the Indian Ocean to Bali. A similar species—the Pacific red-fin butterflyfish (C. lunulatus)—occurs in the Pacific Ocean.*

SIZE: *6 in (15 cm).*

BEHAVIOR: *Seen in pairs swimming over the reef, usually in the vicinity of* Pocillopora *and* Acropora *species corals.*

DIET: *A specialist corallivore; impossible to maintain without access to a plentiful supply of this type of food.*

AQUARIUM: *Destroys hard coral.*

COMPATIBILITY: *Pairs are aggressive and territorial and will drive away any other butterflyfish they encounter.*

PACIFIC DOUBLE-SADDLE BUTTERFLYFISH

Chaetodon ulietensis

Two prominent black saddlelike patches on the sides of the body, separated by a white area crossed by narrow dark lines, help identify this butterflyfish. The rear of the body is yellow with just a small black area located at the base of the caudal fin.

SYNONYM: *C. aurora.*

▼ PACIFIC DOUBLE-SADDLE BUTTERFLYFISH (CHAETODON ULIETENSIS)

DISTRIBUTION: *From the Cocos (Keeling) Islands south of Sumatra across the Pacific to the Tuamotu Archipelago. To the north it ranges via the Philippines up to southern Japan and to the south it extends as far as Lord Howe Island off the east coast of Australia.*

SIZE: *6 in (15 cm).*

BEHAVIOR: *Juveniles frequent estuaries and harbors, whereas adults seek out areas of the reef where there is growing coral. As a result, juveniles tend to settle most easily in aquarium surroundings.*

DIET: *Eats a very wide range of foods in the wild and adapts well to aquarium substitutes. However, may prove shy about feeding at first.*

AQUARIUM: *Will prey on many of the inhabitants of a reef aquarium.*

COMPATIBILITY: *Can be kept in small groups in suitably spacious surroundings or alongside other more tolerant butterflyfish.*

TEARDROP BUTTERFLYFISH

Chaetodon unimaculatus

A prominent tear-shaped black spot lies just behind the midpoint along the top part of the body, below the dorsal fin. Another characteristic of this fish is a bold black stripe passing through the eye. A narrower black stripe is present at the rear end of the dorsal fin, extending down beyond the caudal peduncle. The body is yellow and often silvery on the underside.

DISTRIBUTION: *From East Africa across the Pacific Ocean to the Hawaiian, Marquesan, and Ducie Islands. To the north as far as southern Japan and to the south as far as Lord Howe Island and Rapa Island in the South Pacific.*

SIZE: *8 in (20 cm).*

BEHAVIOR: *Juveniles are often encountered around the Hawaiian coast in the summer. Unlike some butterflyfish, they do not differ markedly in appearance from adults at this age.*

DIET: *Varies through their range, but invariably involves coral. Also eats filamentous algae, which should be present in the aquarium. Generally takes a range of other foods quite readily.*

AQUARIUM: *Will destroy almost anything in a reef tank, sometimes even seeking to prey on larger crustaceans.*

COMPATIBILITY: *Relatively social with its own kind and other butterflyfish, but requires a spacious aquarium.*

VAGABOND BUTTERFLYFISH

Chaetodon vagabundus

Dark stripes running from the back of the head up to the dorsal fin meet others extending up from the lower part of the body. There is a broad black band at the rear of the body and another one passing through the eyes. The overall body coloration is primarily silvery.

SYNONYM: *C. mesogallicus.*

DISTRIBUTION: *East Africa northward up to southern Japan and south as far as Lord Howe Island off the east coast of Australia and the Austral Islands. Its easterly range reaches the Line and Tuamotu Islands.*

▼ VAGABOND BUTTERFLYFISH (CHAETODON VAGABUNDUS)

▲ YELLOW-HEAD BUTTERFLYFISH (CHAETODON XANTHOCEPHALUS)

SIZE: *9 in (23 cm).*

BEHAVIOR: *Often found close to the shore, even venturing into brackish water occasionally. Pairs tend to return to the same retreats every night. Their large size means they need a spacious aquarium, partly because they are active fish by nature.*

DIET: *Eats a wide variety of invertebrates as well as filamentous algae and therefore usually adjusts easily to aquarium life, especially if this type of plant matter is available.*

AQUARIUM: *Not suitable for a reef tank.*

COMPATIBILITY: *Agrees well in pairs. Relatively tolerant of the company of other butterflyfish.*

YELLOW-HEAD BUTTERFLYFISH

Chaetodon xanthocephalus

This beautiful species is not encountered as regularly as some other butterflyfish, but its appearance is unmistakable. As its name suggests, its head is primarily yellow, but the yellow areas are broken by a small area of blue in front of the eyes. There is also yellow evident in front of the pectoral fins. The body is a whitish color with irregular pinkish mauve stripes running across it. The dorsal fin is ocher, while the caudal fin is dark with bright edging. Juvenile yellow-head butterflyfish are predominantly white with a black eye bar and a spot on the caudal peduncle, but these markings fade with age.

SYNONYMS: *C. nigripinnatus; C. nigripinnis.*

DISTRIBUTION: *Ranges down much of the East African coast and across to the Maldives and Sri Lanka in the Indian Ocean, reaching as far as Mauritius in a southeasterly direction.*

SIZE: *8 in (20 cm).*

BEHAVIOR: *Often lives on its own, but can occasionally be seen on the reef in small groups. Can be very territorial.*

DIET: *The exact diet of this butterflyfish in the wild is unclear, but it is known to include coral polyps. However, it will adapt to a diet based on finely chopped meaty foods.*

AQUARIUM: *Not safe to include in a reef aquarium.*

COMPATIBILITY: *Best kept singly. Unlikely to agree with other butterflyfish or its own kind. Will benefit from having a retreat in the aquarium, which will help it settle.*

PEARL-SCALE BUTTERFLYFISH

(YELLOW-TAILED BUTTERFLYFISH)

Chaetodon xanthurus

In this species an orange area extends from above to below the caudal peduncle, and there is another orange area on the caudal fin. The background body color is blue below, turning browner above, and overlaid with a meshlike patterning. A blue-bordered dark stripe runs through the eye and a similarly colored ocellus (eyespot) is located higher up in the center of the head.

DISTRIBUTION: *Ranges from the vicinity of Indonesia north via the Philippines to the Ryukyu Islands.*

SIZE: *5.5 in (14 cm).*

BEHAVIOR: *Tends to be observed at depths below 49 ft (15 m). Seems to prefer relatively subdued aquarium lighting and will feed more readily under such conditions.*

DIET: *The exact feeding habits of this butterflyfish are unclear, but it certainly eats small invertebrates and algae.*

▲ *PEARL-SCALE BUTTERFLYFISH* (CHAETODON XANTHURUS)

AQUARIUM: *Not a good choice for a reef tank.*

COMPATIBILITY: *Will not thrive in the company of more dominant butterflyfish, but it can be housed with its own kind and with gentler species. Place the fish together at the same time to minimize the risk of any bullying.*

MARGINED CORALFISH

(MARGINED BUTTERFLYFISH)

Chelmon marginalis

Members of the genus *Chelmon* have much longer jaws than the chaetodont butterflyfish (members of the genus *Chaetodon*). The basic coloration of the margined coralfish is silvery. Two black-edged yellowish bands, which may have a more orange hue in some cases, pass through and behind the eye. In juveniles the rear part of the body has a less distinct yellowish bar, and there is also a dark spot on the dorsal fin.

DISTRIBUTION: *Ranges along the northern coast of Australia up to Papua New Guinea and along the Great Barrier Reef.*

SIZE: *7 in (18 cm).*

BEHAVIOR: *Uses its jaws to pick at polychaete worms and to grab small crustaceans. Usually seen in pairs, sometimes in larger feeding aggregations.*

DIET: *Will take finely chopped meaty foods, and it is relatively easy to convert this fish to an inanimate diet.*

AQUARIUM: *Not suitable for an invertebrate setup because of its feeding habits.*

COMPATIBILITY: *Keep individually unless you can be sure that you have a true pair. Bullying can be a trigger for* Lymphocystis *infection, to which these fish are susceptible.*

BLACK-FIN CORALFISH

Chelmon muelleri

Ocher-brown banding set against a silvery body color is a feature of this species. As they mature, these coralfish develop a swelling on the forehead. They use it as a weapon to settle territorial disputes, forcing an opponent to give ground as they batter their heads together in combat.

DISTRIBUTION: *Ranges from the northwest coast of Australia eastward to Queensland.*

SIZE: *8 in (20 cm).*

BEHAVIOR: *Often encountered in estuaries and on coastal reefs, where the water may be murky owing to the presence of mud. Juveniles are shy and tend to hide away. Adults are most commonly seen in pairs.*

DIET: *Can be persuaded to eat finely chopped meaty foods without great difficulty. Normally feeds on small invertebrates.*

AQUARIUM: *Likely to cause some damage in a reef tank.*

COMPATIBILITY: *Keep singly unless you are certain that you have a compatible pair.*

▼ BLACK-FIN CORALFISH (CHELMON MUELLERI)

▲ COPPER-BAND BUTTERFLYFISH (CHELMON ROSTRATUS)

COPPER-BAND BUTTERFLYFISH

Chelmon rostratus

This is a particularly beautiful species, with four black-edged orange stripes running vertically down the body. There is a prominent blue-bordered black spot high up near the rear of the body and a black band at the base of the caudal fin.

DISTRIBUTION: *Andaman Sea off India's east coast to the seas around Papua New Guinea. Reaches to the Ryukyu Islands close to Japan in the north and extends southward along the northwestern coast of Australia to the Great Barrier Reef.*

SIZE: *8 in (20 cm).*

BEHAVIOR: *Encountered relatively close to the shore and ventures into estuaries as well. Territorial disputes are often settled by head butting.*

DIET: *Thought to feed on tubeworms and small crustaceans. Can sometimes be difficult to wean onto aquarium foods, but livefoods such as brine shrimps can help. Juveniles measuring about 3 in (7.5 in) generally seem to be most adaptable.*

AQUARIUM: *May not inflict too much damage if kept in a reef aquarium. By and large it ignores corals but will prey on polychaete worms.*

COMPATIBILITY: *Territorial, so do not attempt to keep more than one individual together unless they are a compatible pair.*

HIGH-FIN CORALFISH

Coradion altivelis

The genus *Coradion* is a small genus, consisting of just four species. The high-fin coralfish is easily recognized by the elongated shape of its dorsal fin. There is a brownish black streak running through the eyes, and two others edged with yellow running down the body. These streaks merge together on the ventral surface. A broader brownish streak is evident at the rear of the body, and there is a prominent black mark at the base of the caudal fin. The young can be differentiated by means of an ocellus (eyespot) located toward the rear of the dorsal fin. It is represented by a dark area with a lighter border.

DISTRIBUTION: *Ranges from the Andaman Sea via Sumatra and neighboring islands to Papua New Guinea. Extends northward as far as southern Japan and southward around the northern Australian coast down to the Great Barrier Reef off Australia's east coast.*

SIZE: *7 in (18 cm).*

BEHAVIOR: *Often ranges quite close to the shore, sometimes in fairly murky waters. The young may retreat into the relative safety of sponges for protection.*

DIET: *Can usually be weaned onto substitute diets quite easily.*

AQUARIUM: *Could be housed in a reef aquarium, but be prepared for some damage to sponges and corals in particular.*

COMPATIBILITY: *A pair can usually be kept without serious disputes in the company of other members of the family.*

GOLDEN-GIRDLED CORALFISH

(ORANGE-BANDED CORALFISH)

Coradion chrysozonus

This species displays the typical markings of the genus, consisting of four vertical bands that are orange-brown in color, in this case set on a white background. It has a distinctive black spot at the back of the dorsal fin. There is also a black band at the base of the caudal fin, which is otherwise clear.

DISTRIBUTION: *Extends from Thailand and Malaysia to parts of Indonesia and the Philippines northward to the Ryukyu and Bonin Islands. Reaches Tonga in the east and is found around the northern and eastern coasts of Australia.*

SIZE: *6 in (15 cm).*

BEHAVIOR: *Often found in areas where sponges rather than corals predominate, reflecting its natural feeding preferences.*

DIET: *Also eats small crustaceans. Can be persuaded to sample aquarium foods quite easily, although brine shrimps can help in the case of a reluctant individual.*

▼ GOLDEN-GIRDLED CORALFISH (CORADION CHRYSOZONUS)

AQUARIUM: *Less destructive than some butterflyfish in a reef aquarium, but you should not mix this species with sponges, which will be attacked.*

COMPATIBILITY: *Reasonably tolerant both of its own species and of other members of the family.*

YELLOW LONG-NOSE BUTTERFLYFISH

Forcipiger flavissimus

The upper area of the head, including the eyes, is dark—the black triangle serving to disrupt the contours of the head and confuse predators. The lower part of the head is white. The body is a stunning shade of yellow and it has blue edging running down the back. There is a small black spot just below the caudal peduncle, while the caudal fin is clear.

DISTRIBUTION: *From the Red Sea and East Africa across the Pacific as far as southern Baja California, Mexico. Also around the Hawaiian and Revilla Gigedo Islands off the coast of Mexico and down to the Galápagos and Easter Islands in the eastern Pacific. Its northern range extends to Japan, with its southern limit being Lord Howe Island off Australia's east coast.*

SIZE: *8.5 in (22 cm).*

BEHAVIOR: *It uses its long and powerful jaws to rip into invertebrates, but will also pick up food items such as fish eggs without difficulty.*

▲ YELLOW LONG-NOSE BUTTERFLYFISH (FORCIPIGER FLAVISSIMUS)

DIET: *Highly adaptable, usually settling to feed very readily and quickly in aquarium surroundings.*

AQUARIUM: *Not recommended as a resident of a reef aquarium.*

COMPATIBILITY: *Keep individuals separate, especially in view of their strong dorsal spines, because conflicts will break out. Do not mix with aggressive fish.*

BIG LONG-NOSE BUTTERFLYFISH

Forcipiger longirostris

This species has a very similar appearance to its relative, the yellow long-nose butterflyfish (*F. flavissimus*), but it can be distinguished by the bluish gray hue to the white area under the mouth. It also has significantly longer jaws and a much smaller mouth. A few individuals, however, display brown rather than yellow coloration. They tend to crop up in certain localities, but their coloration is not apparently fixed, because they sometimes change into the much more widely seen yellow variant.

DISTRIBUTION: *East Africa extending across the Pacific to the Hawaiian, Marquesan, and Pitcairn Islands. Its northerly range extends to Bonin Island, while in the south it is found around New Caledonia and the Austral Islands.*

SIZE: *8.5 cm (22 cm).*

BEHAVIOR: *It roams the reef, often singly or in pairs. It uses its long jaws to probe around coral for tiny crustaceans, including minute hermit crabs that may be sheltering there.*

DIET: *Its tiny mouth means that it cannot feed on large pieces of food. Items such as mysid and brine shrimps can be offered, but any meaty foods must be cut into very small pieces.*

AQUARIUM: *Ideal for a reef aquarium, since it does not harm coral. If kept with living rock, it should be able to forage naturally for some food there.*

COMPATIBILITY: *It is important to keep individuals of this species separate, but they can be mixed with inoffensive companions.*

PYRAMID BUTTERFLYFISH

Hemitaurichthys polylepis

The color pattern of this fish is striking. On the body a pale whitish area stands out from its yellow surroundings, while the head is blackish. At night the white area turns black, leaving just a small white spot evident on the side of the body and giving the fish better camouflage in the dark.

DISTRIBUTION: *Extends from Christmas Island in the Indian Ocean via Indonesia to the Hawaiian, Line, and Pitcairn Islands. Ranges north to southern Japan and south to Rowley Shoals off Australia's northwest coast.*

SIZE: *7 in (18 cm).*

BEHAVIOR: *Associates in large groups in areas of the reef where there is a strong current, seeking plankton. As darkness falls, individuals then look for crevices in which to hide overnight.*

DIET: *Will take a wide range of substitute foods, including brine shrimps, mysid shrimps, and meaty items, which should be chopped up very finely.*

▼ *PYRAMID BUTTERFLYFISH (HEMITAURICHTHYS POLYLEPIS)*

▲ *BROWN-AND-WHITE BUTTERFLYFISH (HEMITAURICHTHYS ZOSTER)*

AQUARIUM: *Generally can be trusted in a reef aquarium, although some individuals may be tempted by soft coral.*

COMPATIBILITY: *Adjusts well to aquarium life and can be kept in groups if space permits.*

BROWN-AND-WHITE BUTTERFLYFISH

(BLACK PYRAMID BUTTERFLYFISH)

Hemitaurichthys zoster

This is a relatively dull butterflyfish. A white band encircles its body, which is brownish black in color. Both juveniles and adults are similarly marked. The caudal fin is white, with some yellow highlights evident in the central area of the dorsal fin.

DISTRIBUTION: *Ranges from East Africa to the Andaman Sea and India. Its southern distribution extends to Mauritius.*

SIZE: *7 in (18 cm).*

BEHAVIOR: *Lives in large shoals that can number more than 200 individuals. Often seen in association with other plankton eaters.*

DIET: *Can be persuaded to sample a range of items ranging from algae to finely chopped meaty foods and livefoods such as brine shrimps. May even be persuaded to take flake food in due course.*

AQUARIUM: *Relatively safe to include in a reef tank.*

COMPATIBILITY: *Can be housed singly or in large numbers, depending on the available space. An easily managed species.*

LONG-FIN BANNERFISH
(PENNANT CORALFISH)

Heniochus acuminatus

This majestic fish is one of the easiest members of the entire family to maintain in aquarium surroundings. A dark area runs down the center of the face, and two black bands encircle the body. These features, together with yellow coloration at the rear of the dorsal and caudal fins, help identify the species. Its most striking feature, however, is the long white pennant that trails upward way beyond the end of its body.

DISTRIBUTION: *East Africa eastward to the Society Islands, ranging north into the Arabian Sea and up to southern Japan. Extends as far south as Lord Howe Island off the east coast of Australia.*

SIZE: *10 in (25 cm).*

BEHAVIOR: *Very active by nature. Juveniles tend to act as cleaner fish, picking parasites off other fish. Beware of individuals whose black markings appear lighter than normal, however, since this can be an early indicator that they are suffering from a parasitic infection themselves.*

DIET: *Will take virtually all types of marine aquarium food and adapts readily to prepared diets.*

AQUARIUM: *Not ideal for a reef tank, since it is likely to prey on coral and other occupants.*

COMPATIBILITY: *These bannerfish will look magnificent in a shoal in a large aquarium. Members of the shoal should be placed in the aquarium at the same time. They will then form a pecking order. Should a dispute arise, the order of dominance is reinforced as necessary by head banging, with the victorious fish driving its weaker opponent back.*

▼ LONG-FIN BANNERFISH (HENIOCHUS ACUMINATUS)

▲ RED SEA BANNERFISH (HENIOCHUS INTERMEDIUS)

RED SEA BANNERFISH

Heniochus intermedius

This fish has a fairly chunky profile and a significantly shorter banner than the long-fin bannerfish (*H. acuminatus*). In the Red Sea bannerfish it barely extends back to the caudal fin. A brownish black band curls up through the eyes, and another one angles forward from the base of the caudal fin up toward the dorsal fin.

DISTRIBUTION: *Restricted to the Red Sea and the Gulf of Aden.*

SIZE: *7 in (18 cm).*

BEHAVIOR: *Juveniles are far more social than adults and may sometimes associate with the schooling bannerfish* (H. diphreutes) *that occurs in the same area. Adults tend to become more solitary after they have paired up.*

DIET: *Adapts well to a wide range of aquarium foods.*

AQUARIUM: *Not a good choice for a reef tank, since it is likely to prey on stony coral polyps and polychaete worm tentacles.*

COMPATIBILITY: *Can be housed in pairs and even with other bannerfish provided the aquarium is large enough.*

PHANTOM BANNERFISH
(INDIAN BANNERFISH)

Heniochus pleurotaenia

The diffuse coloration of this bannerfish provides excellent camouflage, breaking up its outline most effectively. There are black areas on the face, sides of the body, and at the rear, with a white stripe extending back from the dorsal fin down to the caudal fin and another on the head. The rest of the body is yellowish brown, with an orange hump on the forehead. Only juveniles have a slight white banner, which disappears as they mature.

DISTRIBUTION: *Ranges eastward through the Indian Ocean from the Maldives and Sri Lanka, extending northward to the Andaman Sea and eastward as far as Java.*

SIZE: *6.5 cm (17 cm).*

BEHAVIOR: *This bannerfish regularly associates in relatively large groups, even when adult. It often occurs on areas of the reef where there is a noticeable current.*

DIET: *May need livefoods at first to help it acclimatize to aquarium life, because it tends not to feed as readily as related species. Adult brine shrimps are recommended for this purpose.*

AQUARIUM: *Will feed on sessile invertebrates, so it is not a good choice for a reef tank.*

COMPATIBILITY: *This fish is social by nature, but introduce the group at the same time to avoid the risk of subsequent conflicts.*

SINGULAR BANNERFISH

Heniochus singularius

There are black bands across the jaws of this fish, extending through the eyes and passing through the pectoral fins, as well as at the rear of the body. In the central body area there are black spots in the middle of each individual white scale. A white pennant extends back from the fourth dorsal fin spine, and the rest of the dorsal fin and the caudal fin are both yellow.

DISTRIBUTION: *Extends from the Maldive and Andaman Islands eastward through the Pacific to Samoa, ranging as far north as southern Japan, southward to Rowley Shoals off Australia's northwest coast, and down to New Caledonia.*

SIZE: *12 in (30 cm).*

BEHAVIOR: *This is a rarely encountered species. Adults are most likely to be observed singly on areas of the reef where there is good coral growth, while juveniles usually hide away.*

DIET: *Stony coral polyps feature prominently in its natural diet, and this species may not be as easy to wean across to standard aquarium fare as other* Heniochus *species.*

AQUARIUM: *Not suitable for a reef tank.*

▼ *PHANTOM BANNERFISH (HENIOCHUS PLEUROTAENIA)*

◀ HORNED BANNERFISH (HENIOCHUS VARIUS)

SIX-SPINE BUTTERFLYFISH

Parachaetodon ocellatus

Vertical yellowish stripes with relatively wide brown edges extend down the sides of these fish. The body itself is whitish. There is some yellow at the top of the dorsal fin as well as a dark spot in this area, blending in with one of the bands.

DISTRIBUTION: *Extends from the coast of India and Sri Lanka northward to the Ryukyu Islands and Bonin close to Japan and eastward as far as Fiji. Also ranges southward down the eastern coast of Australia and is found on the Great Barrier Reef.*

SIZE: *7 in (18 cm).*

BEHAVIOR: *Tends to be encountered over flat sandy areas of reef, sometimes forming quite large shoals. Juveniles often prefer areas with seagrass* (Posidonia *species*).

DIET: *Eats algae, small crustaceans, and other invertebrates. Wean at first onto artificial diets, using livefoods such as brine shrimps and black worms. Gradually add prepared foods, which need to be finely chopped beforehand. May even eat flake in due course.*

AQUARIUM: *Can create problems for some invertebrates in a reef aquarium, although coral should be unharmed.*

COMPATIBILITY: *These fish may do better if kept in pairs or in larger numbers, provided they are introduced to the tank at the same time and they have sufficient space.*

▼ SIX-SPINE BUTTERFLYFISH (PARACHAETODON OCELLATUS)

COMPATIBILITY: *Usually proves more territorial toward its own kind and other bannerfish, so mixing can be problematic. In addition, being the largest member of the genus, it requires spacious accommodation.*

HORNED BANNERFISH
(HUMP-HEAD BANNERFISH)

Heniochus varius

Once adult, these fish have distinctive curved horns above each eye. They also develop a pronounced lump on the forehead, having shed the elongated dorsal filament. The body is blackish, especially on the head and underparts, with a white stripe around the neck and another running from the base of the caudal fin up to the spiny front section of the dorsal fin.

DISTRIBUTION: *Extends from Indonesia north to southern Japan. In the south it reaches as far as Rowley Shoals and New Caledonia. Its easterly distribution reaches to the Society Islands.*

SIZE: *8 in (20 cm).*

BEHAVIOR: *Adults use the hump on their foreheads to battle with rivals. Naturally shy, they are not conspicuous on the reef, darting back to hide among sponges and corals. Must have suitable hiding places in the aquarium.*

DIET: *Can be very nervous at first, so a calm, tranquil aquarium is important to encourage this bannerfish to start eating after a move. Livefood can also be helpful at this stage.*

AQUARIUM: *Eats sponges and stony coral polyps in particular, as well as other invertebrates, so not suitable for a reef aquarium.*

COMPATIBILITY: *Somewhat territorial, so house individually.*

ANGELFISH
FAMILY POMACANTHIDAE

The angelfish form a large group that is of considerable significance in the saltwater hobby. There are nine separate genera of angelfish, and currently about 75 recognized species. They have a worldwide distribution, occurring in the Indian, Pacific, and Atlantic Oceans. Many are popular as food fish in various parts of their range, but as far as the saltwater aquarium is concerned, it is better to start off with older juveniles, because they usually prove easier than adults to adapt to this environment. The coloration of older juvenile angelfish is often very different to that of mature specimens, however, and accurate identification can be made more difficult by the fact that these fish sometimes hybridize in parts of their range.

Being relatively large in size, angelfish need an extremely efficient filtration system that will maintain water quality in the aquarium. Unfortunately, they are generally aggressive, even while young, which means that they often need to be housed individually. In any event, the potential adult size of many species prohibits keeping them together in larger numbers in the typical home aquarium.

It is sometimes possible to distinguish between the sexes by differences in their appearance. All juveniles are female, and some change into males. Feeding is reasonably straightforward, but angelfish must be given regular supplies of greenstuff, preferably in the form of fresh marine algae growing in the aquarium, although this can be supplemented with other similar foods, such as spinach. If not, they can suffer from a deficiency of vitamin A and ultimately go blind.

The natural intelligence of angelfish means that they soon come to recognize their owner and await mealtimes eagerly. Records show that they can live for more than 20 years in the relative safety of the saltwater aquarium. However, it is worthwhile bearing in mind that some species are harder to acclimatize than others, and they can all be at risk from a range of parasitic illnesses when first acquired.

▼ *Three-spot Angelfish (Apolemichthys trimaculatus)*

THREE-SPOT ANGELFISH

(FLAG-FIN ANGELFISH)

Apolemichthys trimaculatus

Both juveniles and adults are a rich shade of yellow, but their spotted appearance alters with age. Only juveniles have a black spot on each side of the body, just above the base of the caudal fin. The undulating body markings are more evident in juveniles than in adults, and at this early stage in life they lack the prominent blue lips, too. There can be regional variations in coloring.

Distribution: *East Africa eastward, reaching to southern Japan and southward to New Caledonia. Its easterly range extends to Samoa.*

Size: *10 in (25 cm).*

Behavior: *Often favoring areas of the reef that fall away sharply, this species is frequently observed in small groups. Tunicates and sponges form the bulk of the diet of adult fish, whereas juveniles feed on algae.*

Diet: *The feeding habits of juveniles mean that they can be acclimatized more easily than adults to aquarium surroundings, provided there is good established growth of algae here. In addition, chopped meaty foods and* Spirulina *species of algae should be provided.*

▲ Gold-spotted Angelfish (*Apolemichthys xanthopunctatus*)

Aquarium: *Juveniles may not cause disruption in a reef aquarium until they grow older.*

Compatibility: *Relatively peaceful, although disagreements with their own kind are not unknown. Should be kept only with placid tankmates.*

GOLD-SPOTTED ANGELFISH

Apolemichthys xanthopunctatus

Golden spotting on the flanks is a feature of this angelfish. This extends widely over the greenish brown background color of its body. The dorsal, caudal, and anal fins are black with a fine blue edge. There is a black bar on the forehead, and the lips are purplish blue. Juveniles are very different in appearance, being yellow with a black stripe extending down to the eyes. They also have an eyespot, or ocellus, at the rear of the dorsal fin.

Distribution: *Restricted to various islands in the central Pacific, from the Caroline Islands eastward via Gilbert, Phoenix, and the Line Islands south to the Cook Islands.*

Size: *10 in (25 cm).*

BEHAVIOR: *Adults are relatively conspicuous over shallower areas of the reef, whereas juveniles live at significantly greater depths, often below 100 ft (30 m).*

DIET: *In this species young adults can often be transferred more easily to artificial diets than juveniles can, since in the wild young adults tend to feed on both corals and sponges. Finely chopped meaty foods serve as a good substitute, not forgetting algae, too.*

AQUARIUM: *Adults will cause damage in a reef aquarium.*

COMPATIBILITY: *Likely not to agree with its own kind and can bully other species, especially ones that are introduced to the aquarium later.*

CHERUBFISH

Centropyge argi

As a group the pygmy, or dwarf, angelfish that make up the genus *Centropyge* vary significantly in coloration. The cherubfish is deep blue in color, with a variable orange-yellow area on the sides of the face and the throat. Some individuals are much bluer overall than others. While the pectoral fins are yellowish, the other fins match the body color but have paler edging.

DISTRIBUTION: *Restricted to the western Atlantic, extending from Bermuda, Florida, and the Bahamas southward across the Gulf of Mexico and the Caribbean Sea to the coast of French Guiana.*

SIZE: *3 in (8 cm).*

BEHAVIOR: *Often found on rubble-strewn areas of the reef, living in groups of one larger male, several females, and juveniles. Males are territorial.*

DIET: *Filamentous green algae are important components of its diet, and chopped meaty foods are also required in the aquarium.*

▼ *CHERUBFISH (CENTROPYGE ARGI)*

▲ *BICOLOR ANGELFISH (CENTROPYGE BICOLOR)*

AQUARIUM: *Cannot be trusted in a reef tank, since it sometimes preys on various occupants, including corals, clams, and anemones.*

COMPATIBILITY: *Can be kept in pairs, but males will fight savagely. It is better to buy two specimens that are already living together rather than try to introduce two strangers, because the larger males can only be identified in terms of size difference within an existing group.*

BICOLOR ANGELFISH
(ORIOLE ANGELFISH)

Centropyge bicolor

The typical coloration of this angelfish is a combination of yellow on the front half of the body and rich deep blue behind. There is a blue band over the head, extending between the eyes, and the caudal fin is yellow. A striking blue and white color form, in which the yellow areas are replaced by white, has been recorded from the Great Barrier Reef and around Fiji. These fish may develop regular coloration in due course. Juveniles have darker barring on the blue areas, which disappears with age.

DISTRIBUTION: *East Africa, extending northward as far as southern Japan, reaching New Caledonia in the south. Ranges across the Pacific as far as the Samoan and Phoenix Islands.*

SIZE: *6 in (15 cm).*

BEHAVIOR: *The larger male has a harem of females within its territory, which often extends over the rubble area of the reef. The fish use retreats here if they feel threatened.*

DIET: *Must transfer to an aquarium where there is well-established filamentous algal growth. Offer chopped meaty foods. Feed twice a day but more frequently at first, when the fish is eating well.*

AQUARIUM: *Will prey on a variety of invertebrates found in a reef tank.*

COMPATIBILITY: *Try to obtain pairs. Can prove to be a bully, since it is larger than many related species.*

TWO-SPINED ANGELFISH

(CORAL BEAUTY; DUSKY ANGELFISH)

Centropyge bispinosa

This angelfish occurs in a remarkable range of color variations. Its typical coloration is orangish with some darker markings on the sides of the body, and the fins around the edge of the body are a rich blue color. However, some individuals are predominantly blue, while others may be mainly orange, yellow, or even white. It is thought that their coloration may not be a geographical feature. It may be an indication of the depth at which the fish are living, since they become paler at greater depths.

DISTRIBUTION: *Ranges from East Africa across the Pacific to the Tuamotu Archipelago. Extends northward to the Izu Islands near Japan and south as far as Lord Howe Island off the east coast of Australia.*

SIZE: *4 in (10 cm).*

BEHAVIOR: *Lives in groups consisting of a male fish and several females. Rather shy by nature, often inhabiting areas of the reef where protection is provided by coral. The two-spined angelfish spawns much closer to the substrate than many angelfish.*

DIET: *Algae feature prominently in its diet in the wild, so introduce this fish to an aquarium where growth of filamentous algae is well established. Prepared* Spirulina *species of algae can be offered alongside other foods.*

AQUARIUM: *Causes less damage in a reef tank than most angelfish, particularly if kept well fed. It is not entirely trustworthy in these surroundings, however, since it feeds on coral polyps.*

COMPATIBILITY: *Disputes are most likely to break out in relatively small aquaria. Try to pick pairs, choosing the largest and smallest individuals on offer, assuming they appear healthy.*

BLUE MAURITIUS ANGELFISH

(DEBELIUS'S ANGELFISH)

Centropyge debelius

The front lower area of the body of the blue Mauritius angelfish is yellowish with variable purple spotting. By contrast, the rear part is a consistent shade of purple—apart from the caudal fin, which is yellow.

DISTRIBUTION: *Restricted to the vicinity of islands in the western part of the Indian Ocean, occurring around Seychelles, Réunion, and the Aldabras, as well as Mauritius.*

▼ *TWO-SPINED ANGELFISH (CENTROPYGE BISPINOSA)*

SIZE: *3.5 in (9 cm).*

BEHAVIOR: *Found in deepwater areas where the reef falls away sharply, down to depths of 295 ft (90 m). Appears solitary by nature and shy, often hiding away in suitable crevices.*

DIET: *Eats algae, which should be established in its tank to assist the acclimatization process. Will also take prepared meaty foods.*

AQUARIUM: *May damage tridacnid clams and hard coral in a reef tank.*

COMPATIBILITY: *Probably safest to keep this rarely seen angelfish apart from its own kind.*

RUSTY ANGELFISH

Centropyge ferrugata

In spite of its name, the coloration of the rusty angelfish can prove to be variable, ranging from a brownish shade on the upper parts to a brighter shade of orange on the lower surface of the body, although in some individuals this area may even appear a creamy color. There is also a variable vertical pattern of dark spots and bars running down the sides of the body.

DISTRIBUTION: *Ranges from the southern Philippines northward via southwest Taiwan to the Ryukyu Islands and Tanabe Bay in southern Japan, and eastward to the Bonin Islands.*

SIZE: *4 in (10 cm).*

BEHAVIOR: *Found in areas of the reef where algae growth is widespread, often quite close to the surface. A male has a large territory, embracing that of several females. As in most related species, they tend to spawn around sunset.*

DIET: *Eats algae and most other aquarium foods.*

AQUARIUM: *Relatively trustworthy in a reef aquarium, especially if well fed.*

COMPATIBILITY: *Not very aggressive; pairs have even spawned successfully in aquarium surroundings.*

ORANGE ANGELFISH

(FISHER'S ANGELFISH)

Centropyge fisheri

This angelfish is yellowish orange overall, but with a dark patch on each side of the body just above the base of the pectoral fin. There is also some darker speckling evident on the dorsal and anal fin, while the latter is clearly edged with blue, as is the pelvic fin. The caudal fin is yellow, becoming transparent toward its tip.

DISTRIBUTION: *The Hawaiian Islands and Johnston Atoll southwest of Honolulu. Some people consider this species to be synonymous with the*

▼ *RUSTY ANGELFISH (CENTROPYGE FERRUGATA)*

◀ LEMONPEEL ANGELFISH (CENTROPYGE FLAVISSIMUS)

much more widely distributed whitetail angelfish (C. flavicauda).

SIZE: *2.75 in (7 cm).*

BEHAVIOR: *Tends to occur in areas of rubble on the seaward side of the reef at depths below 33 ft (10 m) down to at least 280 ft (85 m). This angelfish regularly hides away in crevices.*

DIET: *Feeds on algae and will also take meaty foods.*

AQUARIUM: *Not especially destructive, but may attack clams as well as corals in a reef aquarium.*

COMPATIBILITY: *Usually suitable for a community aquarium. Tends not to bully other occupants, but is likely to disagree with others of its own kind.*

LEMONPEEL ANGELFISH

Centropyge flavissimus

There is some variation in the appearance of these angelfish, depending on both their age and their origins. Young fish have a black eyespot edged with blue on each side of the body, and a more prominent area of blue encircling each eye. As in adults, their basic body coloration is lemon yellow, with blue stripes around the gill area and similar edging on the fins around the rear of the body.

DISTRIBUTION: *Occurs around Cocos (Keeling) and Christmas Islands in the Indian Ocean northward to the Ryukyu Islands and across the Pacific to the Line, Marquesan, and Ducie Islands. Occasionally recorded around Easter Island, but not found in Hawaiian waters.*

SIZE: *5.5 in (14 cm).*

BEHAVIOR: *This secretive species tends to occur in areas where coral is abundant, around small oceanic islands rather than bigger landmasses. Females undergo a dramatic color change prior to spawning, turning almost completely white.*

DIET: *Marine algae and meaty foods such as mysid shrimps are recommended as well as a good supply of live rock. Brine shrimps can induce a reluctant individual to feed.*

AQUARIUM: *Likely to damage both coral and the mantles of clams in a reef tank.*

COMPATIBILITY: *Usually proves somewhat aggressive toward other members of its genus.*

FLAME ANGEL

Centropyge loricula

The flame angel is one of the most striking of all the pygmy angelfish, and its fiery orange coloration certainly justifies its name. There is a dark spot of variable size behind the gills, and there are typically a variable number of dark stripes on the sides of the body (which are absent in individuals originating from the Marquesan Islands). At the rear of the dorsal and anal fins there is a series of alternating purple and black bars. Males from the same area of origin have more striping than females, allowing them to be distinguished by sight.

DISTRIBUTION: *Ranges from Belau in the western Pacific southward to the Great Barrier Reef and the Pitcairn Islands and extends eastward across the Pacific to the Hawaiian, Marquesan, and Ducie Islands.*

SIZE: *6 in (15 cm).*

BEHAVIOR: *Occurs primarily around islands, associating with areas of finger coral* (Porites *species). There are regional differences in populations, with the most brightly colored examples originating from the vicinity of the Hawaiian Islands.*

▼ FLAME ANGEL (CENTROPYGE LORICULA)

Diet: *Feeds naturally on marine algae and is easier to acclimatize if these are growing in the aquarium. Substitutes such as* Spirulina *can be offered too, as well as meaty foods.*

Aquarium: *May attack corals and the mantle of clams in a reef tank.*

Compatibility: *Often aggressive toward its own kind, but may settle better in trios of one male and two females in a large aquarium with plenty of retreats. Likely to bully unrelated fish, especially any with similar coloration.*

RESPLENDENT ANGELFISH

Centropyge resplendens

The distinctive coloring of this pygmy angelfish allows it to be identified easily. It is mainly purplish blue, with yellow extending from the lips over the top of the head. There is also yellow evident on the caudal fin.

Distribution: *Restricted to the area around Ascension Island in the southeast Atlantic. Considered Vulnerable by the IUCN (World Conservation Union), although it appears to be quite common here in its localized range.*

Size: *2.5 in (6 cm).*

Behavior: *Seen around its native island in areas of rubble in particular, ranging from about 49 ft (15 m) down to 130 ft (40 m) or more.*

▲ *Keyhole Angelfish (Centropyge tibicen)*

Diet: *Meaty food and algae. Living rock should also be provided, allowing the fish to browse for food.*

Aquarium: *Will attack corals in a reef aquarium.*

Compatibility: *Aggressive toward its own kind and other active fish in aquarium surroundings and will sometimes even pick on species that are much larger.*

KEYHOLE ANGELFISH

Centropyge tibicen

Although relatively dull in color, there is a clear difference between the sexes in this species. Both display a striking white blotch on each side of the body, but while males are a dark blackish blue overall, females are much blacker in color, with the fins on the underside of the body being predominantly yellow.

Distribution: *From Christmas Island, which lies to the south of Sumatra, down to Lord Howe Island off the east coast of Australia, where the largest specimens occur, and north as far as southern Japan. Its easterly range extends to Fiji.*

Size: *7.5 in (19 cm).*

BEHAVIOR: *If the body color starts to lighten under aquarium conditions, this is a sign that there is a problem. This species is shy by nature, so it is important to provide adequate hiding places, which will help the fish settle into its new surroundings.*

DIET: *Offer a varied diet that includes algae, which are essential to the well-being of this angelfish.*

AQUARIUM: *Often proves quite disruptive in a reef tank, since it preys especially on hard corals.*

COMPATIBILITY: *As the largest member of the genus, it can be aggressive.*

PEARL-SCALE ANGELFISH

(HALF-BLACK ANGELFISH)

Centropyge vrolikii

The front half of the body of the pearl-scale angelfish is whitish, broken by blackish brown markings, and the rear part is black. The pectoral fins may have an orange base, with the edge of the caudal, dorsal, and anal fins in adults being bright blue. The appearance of juvenile chocolate surgeonfish (*Acanthurus pyroferus*) is similar to that of this species.

DISTRIBUTION: *Christmas Island in the Indian Ocean, northward via Sumatra to southern Japan and eastward via the Marshall Islands and Vanuatu to Tonga. Its southernmost range extends to Lord Howe Island off the east coast of Australia.*

SIZE: *4.75 in (12 cm).*

BEHAVIOR: *Lives in groups of one male and several females. Just as in other members of the genus, females are able to change into males if necessary. This species may also hybridize with related species, such as the lemonpeel angelfish* (C. flavissimus), *in the Mariana Islands.*

DIET: *Will feed readily on algae, which form an important part of its natural diet, but it can also be persuaded to take a wider range of foods without any great difficulty.*

AQUARIUM: *This species is destructive in a reef tank, particularly toward sponges and corals.*

COMPATIBILITY: *Can be aggressive, especially toward other fish that are introduced after it is established in the aquarium, as well as toward other pygmy angelfish.*

BLUE-SPOTTED ANGELFISH

Chaetodontoplus caeruleopunctatus

As its name suggests, the sides of the body of this angelfish are covered with a series of blue dots. The underlying body color is brownish gray, offset against a bright yellow caudal fin.

DISTRIBUTION: *Restricted to the Philippine Islands and most commonly encountered around the island of Cebu.*

▼ BLUE-SPOTTED ANGELFISH (CHAETODONTOPLUS CAERULEOPUNCTATUS)

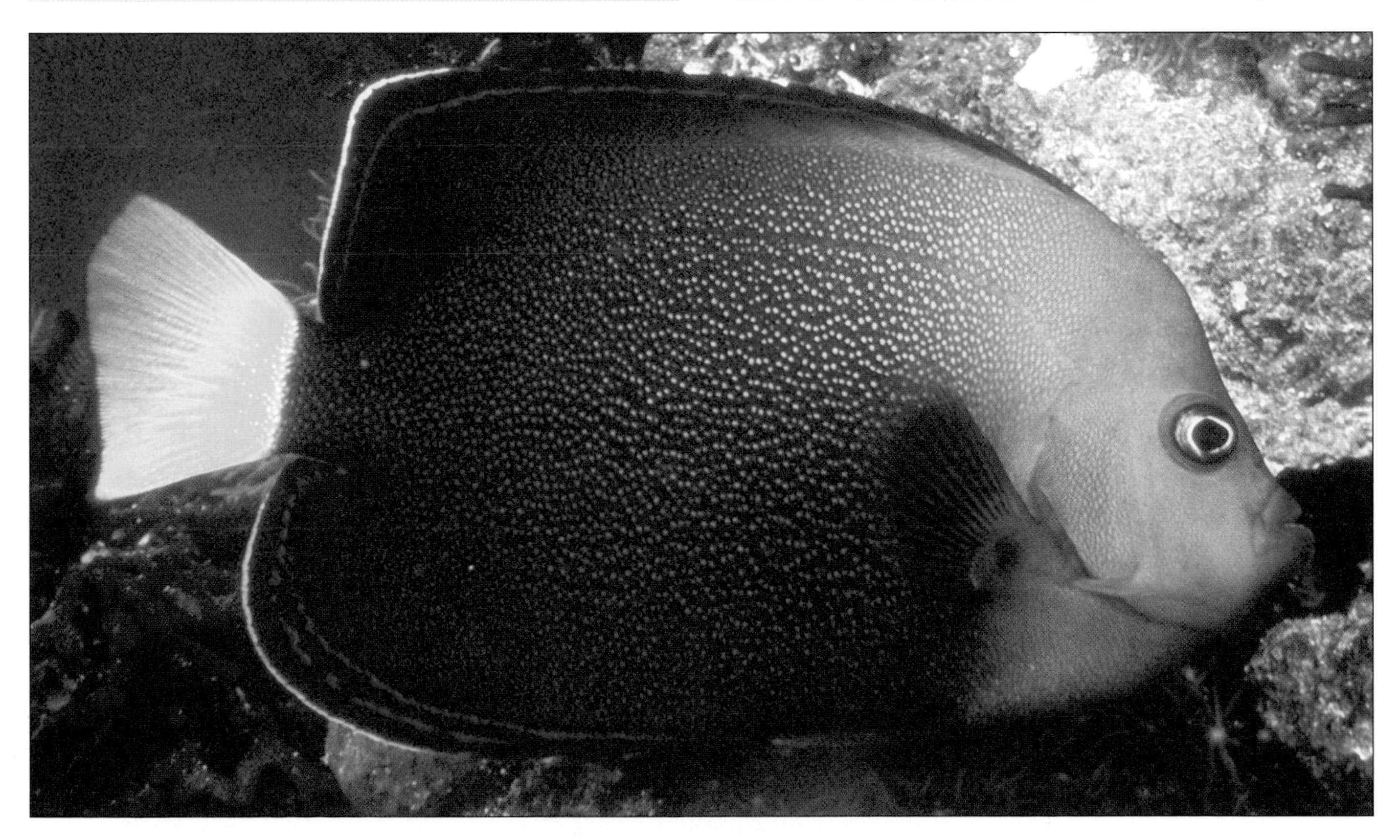

SIZE: *5.5 in (14 cm).*

BEHAVIOR: *Thought to be most common around coastal reefs. Has a relatively bold nature, and tends to be less aggressive than many angelfish, so if you want to house it with similar species, it will need to be introduced to the aquarium first.*

DIET: *A varied range of foods, including marine algae and chopped meaty foods, should be offered.*

AQUARIUM: *Will prey on corals, sponges, and tunicates, so not suitable for a reef tank.*

COMPATIBILITY: *Can be mixed with inoffensive tankmates but, being smaller than related species, it is at risk of being bullied by other angelfish.*

SCRIBBLED ANGELFISH

Chaetodontoplus duboulayi

It is possible to distinguish the two sexes visually in this species. Male scribbled angelfish are a combination of bright blue and yellow, while the females are black and yellow. Adult males are also generally larger and have pointed rather than rounded lobes on the caudal fin. Juveniles, which tend to adapt better than adults to aquarium surroundings, have a much coarser pattern of blue markings on the dark areas of the body. In all cases there is a broad white area behind the eyes.

DISTRIBUTION: *Northwestern Australia and Papua New Guinea via the Great Barrier Reef, extending southward to Lord Howe Island.*

SIZE: *10 in (25 cm).*

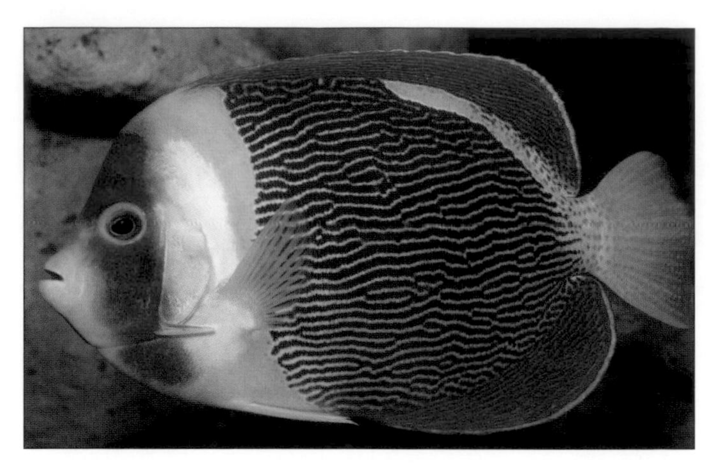

▲ *SCRIBBLED ANGELFISH (CHAETODONTOPLUS DUBOULAYI)*

BEHAVIOR: *Often encountered in relatively open areas of the reef where there is little coral and the water may be quite murky. In this species it is the female rather than the male that soars as part of the display process, and the male then swims under his intended mate. Their floating eggs hatch about a day later.*

DIET: *Meaty foods chopped finely will help substitute for this fish's natural diet of sponges and tunicates. Provide some green food, too.*

AQUARIUM: *Likely to inflict damage in a reef aquarium.*

COMPATIBILITY: *A pair of scribbled angelfish can be housed together in a very spacious aquarium and may even spawn in these surroundings. Likely to be aggressive toward other angelfish, however.*

▼ *BLACK VELVET ANGELFISH (CHAETODONTOPLUS MELANOSOMA)*

BLACK VELVET ANGELFISH

Chaetodontoplus melanosoma

Juveniles differ from adults in their appearance. Their black body has a velvety appearance with yellow markings running down the center of the face as far as the lips. There are also yellow markings behind the head and around the rear of the body. Adults retain the yellow patterning but are paler in color, especially on the head and upperparts of the body. There are regional variations of this, however. Identification of the black velvet angelfish has become more complex because some of the color variants have been classified into distinct species, notably the phantom angelfish (*C. dimidiatus*) and Vanderloos's angelfish (*C. vanderloosi*).

DISTRIBUTION: *Occurs around Indonesia and New Guinea, northward to the Philippines and southern Japan.*

SIZE: *8 in (20 cm).*

BEHAVIOR: *Often occurs in areas of the reef where there are strong currents. Juveniles inhabit deeper water and may be seen individually. Adults can sometimes be observed in small groups.*

DIET: *Algae are recommended, but this fish will also eat finely chopped meaty foods. Sponges and tunicates are important food sources in the wild.*

AQUARIUM: *Likely to cause damage in a reef tank.*

COMPATIBILITY: *This species is not generally aggressive.*

VERMICULATED ANGELFISH
(SINGAPORE ANGELFISH)

Chaetodontoplus mesoleucus

The vermiculated (wavy-line) patterning on the gray rear part of the body of this angelfish is very distinctive. There is a white area in front and a broad black band running down and through the eye. The face is yellowish with blue lips, and yellow may also be present along the top and bottom of the body. There are two different forms—distinguishable by either a gray or a yellow caudal fin—that can occur together in the same area. This angelfish looks a little like a butterflyfish in terms of its body shape but, like all angelfish, it has a distinctive opercular spine pointing backward from the lower part of the gill operculum.

DISTRIBUTION: *Sri Lanka to Indonesia and Papua New Guinea northward to southern Japan.*

SIZE: *6.5 in (17 cm).*

BEHAVIOR: *Often found in coral-rich areas of the reef. Vermiculated angelfish tend to be seen in pairs and sometimes trios consisting of a male and two females. Spawning always takes place at the same site in the territory, with this behavior initiated by the female.*

▲ *VERMICULATED ANGELFISH (CHAETODONTOPLUS MESOLEUCUS)*

DIET: *Eats a variety of marine algae and meaty foods, which should be finely chopped. Algae growing in the aquarium will help in the initial acclimatization process.*

AQUARIUM: *Not to be trusted in a reef tank.*

COMPATIBILITY: *Do not mix with butterflyfish that display similar patterning or with its own kind. However, it can be kept with other angelfish, especially if there are plenty of retreats in the aquarium.*

ORNATE ANGELFISH

Genicanthus bellus

The relative slender body shape of members of this genus, combined with the elongated shape of the caudal fin, has led to them being called swallowtail angelfish. This particular species has black stripes running along its body, interspersed with blue markings, and with red edging along the top of the dorsal fin in

▼ *ORNATE ANGELFISH (GENICANTHUS BELLUS)*

the case of the females. Males are very different in appearance, being grayish brown with yellowish stripes running down the midline and below the dorsal fin. Females can change to males in a period of about a month.

Distribution: *From the Cocos (Keeling) Islands in the Indian Ocean to the Philippines and various islands in the Pacific, including the Marianas, Guam, the Society Islands, and the Marshall Islands. Ranges as far east as Tonga.*

Size: *7 in (18 cm).*

Behavior: *A deepwater species, typically seen below 165 ft (50 m). Because it inhabits such deep waters, our understanding of its full range is limited. Sometimes seen in shoals. Beware of individuals that cannot maintain a level body position. This suggests that their swimbladder has been damaged during collection. Good water movement in the aquarium is essential for maintaining this species.*

Diet: *Feeds naturally on plankton and algae. Wean at first onto brine shrimps, then introduce finely chopped meaty foods.*

Aquarium: *Relatively dim lighting is required to mimic this fish's natural habitat, so do not include in a reef tank that uses high-intensity lighting.*

Compatibility: *Keep in pairs or groups, but note that males housed together are likely to fight.*

BLACK-STRIPED LAMARCK

(LAMARCK'S ANGELFISH)

Genicanthus lamarck

A series of up to six alternating black and blue horizontal stripes running along the body help identify this species. There is also a series of tiny black spots evident on the caudal fin. Females lack the blue spot present at the base of the pectoral fins in males. Juveniles are similar overall, but with coarser spotting.

▼ *Black-striped Lamarck (Genicanthus lamarck)*

▲ *Spot-breast Angelfish (Genicanthus melanospilos)*

Distribution: *Ranges through the Indo-Malaysian region, reaching as far as Vanuatu in the east, and south to the Great Barrier Reef off Australia's east coast. Occurs northward as far as southern Japan.*

Size: *10 in (25 cm).*

Behavior: *Occurs in both shallow and deepwater reef habitats, feeding when the tide is flowing. Replicate these conditions using pumps in the aquarium.*

Diet: *Highly adaptable. Will take a range of plant matter and chopped meaty foods.*

Aquarium: *Ideal for a reef tank, since it is unlikely to harm the occupants, and relatively easy to maintain.*

Compatibility: *Males are likely to be aggressive toward each other. Generally will not interfere with other fish, apart from inoffensive smaller fish such as fairy wrasses.*

SPOT-BREAST ANGELFISH

(JAPANESE SWALLOWTAIL ANGELFISH)

Genicanthus melanospilos

Males have a very distinctive series of black bars on the upper part of the face and around the body, with long streamers on their dorsal, anal, and caudal fins and a black spot on the breast. Females have no bars and are yellowish, becoming more greenish on their underparts. The caudal fin is blue with black edging at the top and bottom. Juveniles resemble adult females but have a prominent black stripe running through their eyes.

Distribution: *From the Malaysian coast northward to the Ryukyu Islands near Japan and south to Rowley Shoals off Australia's northwest coast. In an easterly direction, its range extends to Tonga.*

SIZE: *7 in (18 cm).*

BEHAVIOR: *Most commonly seen in areas where there is active coral growth and where the reef is sloping downward. Encountered both in pairs and in groups consisting of a male and several females.*

DIET: *Can be persuaded to take a variety of foods—both vegetarian and meaty—without much difficulty, and usually proves easy to establish.*

AQUARIUM: *Can adapt well to a reef tank, but may acclimatize better in an aquarium in which the lighting is not as bright.*

COMPATIBILITY: *Avoid mixing males. Active by nature, these angelfish may upset more placid tank occupants, but they are not aggressive and can even be housed with other angelfish.*

BLACK-EDGED ANGELFISH

(WATANABE'S ANGELFISH)

Genicanthus watanabei

The sexes can be distinguished very easily, because only males display narrow black horizontal striping on the sides of their bodies with a blue area above. Females, by contrast, are pinkish blue overall with a dark bar edged with blue running above the eyes and black edging on the dorsal, anal, and caudal fins.

DISTRIBUTION: *Ranges from Taiwan east to the Tuamotu Archipelago and Pitcairn Island, and south as far as New Caledonia and the Austral Islands.*

SIZE: *6 in (15 cm).*

BEHAVIOR: *A deepwater species, usually encountered at depths below 80 ft (25 m) and living in areas where the current is quite strong. A typical group is made up of a single male with two or three females, but large associations may form in some areas.*

DIET: *Will take a variety of vegetable and chopped meaty foods. More willing to feed at first under conditions of subdued lighting .*

AQUARIUM: *Can be kept safely in a reef tank, but bright light may inhibit feeding. Avoid any individuals showing signs of a swimbladder disorder.*

COMPATIBILITY: *Males should be kept separate, but otherwise black-edged angelfish are not aggressive by nature.*

BLUE ANGELFISH

Holacanthus bermudensis

Juvenile blue angelfish have dark bodies with blue vertical stripes and yellowish areas, notably around the jaws and the pectoral fins. A dark band runs through the eyes. The fish lighten in color as they mature—the ends of the dorsal and anal fins become bright yellow, while the sides of the body turn a mottled shade of bluish green with small yellow markings. Although similar to the queen angelfish (*H. ciliaris*), the yellow coloration on the caudal fin is restricted to the tip in the case of the blue angelfish.

DISTRIBUTION: *Western Atlantic, extending from Bermuda to southern Florida and the Bahamas and southward through the Gulf of Mexico*

▼ BLUE ANGELFISH (HOLACANTHUS BERMUDENSIS)

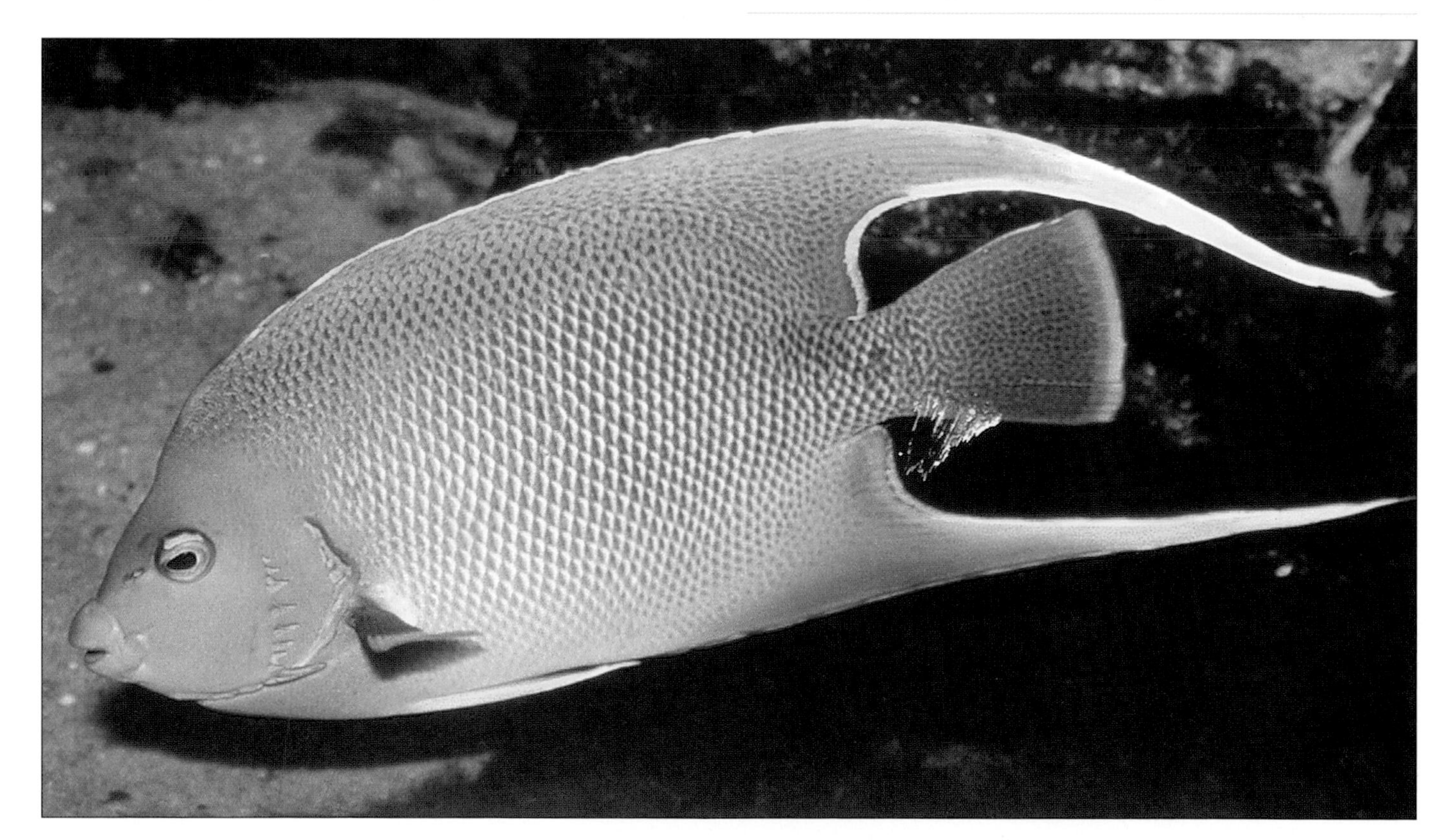

SIZE: *17.75 in (45 cm).*

BEHAVIOR: *Adults tend to wander into deeper water than juveniles, sometimes down to depths of 300 ft (90 m). This angelfish generally favors areas around the mainland rather than around islands.*

DIET: *Juveniles feed primarily on marine algae, with sponges featuring more prominently in their diet as they mature. They usually eat readily in aquarium surroundings and should be offered a variety of foods.*

AQUARIUM: *This fish's large size and its desire to eat sponges creates difficulties as far as housing in a reef tank is concerned.*

COMPATIBILITY: *Will not agree well with other angelfish. It may be possible to keep a juvenile with an adult for a period, but they will require a large aquarium with adequate hiding places.*

CLARION ANGELFISH

Holacanthus clarionensis

This beautiful angelfish is highly variable in appearance. Juveniles are orange with a pattern of blue stripes running vertically down the rear of the body and the face. The stripes tend to become narrower with age. In most cases the lips become blue, although this transformation does not always take place.

DISTRIBUTION: *Eastern Pacific, north as far as the southern tip of Baja California, around the Clarion and Revilla Gigedo Islands, and south to Clipperton Island.*

SIZE: *8 in (20 cm).*

BEHAVIOR: *Some adult clarion angelfish may act as cleaner fish, striking up an unusual relationship with gigantic manta rays* (Manta birostris), *which allow them to pull parasites off their bodies. Males, which can be recognized by being larger than females, may occasionally form single-sex aggregations, but this angelfish is usually solitary.*

DIET: *Feeds on meaty foods.*

AQUARIUM: *Not safe for a reef tank, since it feeds on sponges and tends to prefer a slightly lower water temperature.*

COMPATIBILITY: *Aggressive, so should be kept on its own, but otherwise relatively easy to maintain.*

QUEEN ANGELFISH

Holacanthus ciliaris

The spectacular coloration of queen angelfish can be very variable, and the situation can be made more confusing because this species also hybridizes readily with other *Holacanthus* species occurring in its area of distribution. Juveniles have vertical blue stripes on their bodies, which fade as they mature. The basic body coloration is a series of yellow markings on a bluish green background, with a blue area at the base of the pectoral fins and another on the forehead. The caudal fin is yellow. There are color variants, however, especially from the more isolated areas in which this species occurs, such as around St. Paul's Rocks in the Atlantic Ocean, northeast of Brazil. One of the most common forms has a predominantly blue body with a white caudal fin, while Brazilian specimens may have an orange body offset with black and white markings.

DISTRIBUTION: *Ranges from Florida and the Bahamas south through the Gulf of Mexico and the Caribbean to the coast of South America, extending to Brazil.*

SIZE: *17.75 in (45 cm).*

BEHAVIOR: *Queen angelfish tend to roam around offshore areas of the reef. A single male maintains a large territory that is home to several females and spends time with each of his mates individually during the day, feeding briefly in their company. Spawning takes place just before sunset.*

DIET: *In the wild adults feed almost entirely on sponges, while juveniles eat filamentous algae and remove external parasites from other fish. They can usually be persuaded to eat a wide variety of foods, which must include some greenstuff.*

AQUARIUM: *Not really suitable for a reef aquarium because of its size and preference for feeding on sponges.*

COMPATIBILITY: *Aggressive; attains a large size, so it is not generally possible to keep more than one of these fish together.*

ROCK BEAUTY

Holacanthus tricolor

While juveniles are essentially bright yellow—apart from a black eyespot with a blue surround toward the rear of the body near the base of the dorsal fin—in the adult the rear part of the body is transformed, becoming entirely black. Blue areas above and below the pupils in the eyes also aid identification.

◀ *QUEEN ANGELFISH (HOLACANTHUS CILIARIS)*

DISTRIBUTION: *Ranges from the coast of Georgia east across the Atlantic to Bermuda and south through the northern Gulf of Mexico down to Brazil, reaching Rio de Janeiro.*

SIZE: *14 in (35 cm).*

BEHAVIOR: *Seen not just on reefs but also around jetties, especially in the case of juveniles. Relatively shy by nature, often hiding out of sight in caves and crevices, which explains its common name. Females will change into males and become larger. The face of males turns black during courtship.*

DIET: *Difficult to feed, since juveniles eat the mucus from the bodies of other fish, such as moray eels (family Muraenidae). In order to replicate their natural diet, adults seem to need sponges, without which they are unlikely to thrive, certainly in the longer term. Some specialized angelfish diets include sponges and are recommended where available.*

AQUARIUM: *Not advisable for a brightly lit reef tank, bearing in mind their natural behavior, unless there are plenty of retreats. May also attack coral.*

COMPATIBILITY: *Juveniles are likely to chase other fish; adults are territorial.*

BLUE-RING ANGELFISH

Pomacanthus annularis

Juveniles are black with a series of curved white and blue stripes on the sides of the body. The adults, in complete contrast, are yellowish with curved diagonal stripes running down the sides of their bodies. There is a yellow ocellus (eyespot) surrounded by a blue ring just above the operculum on each side of the body.

DISTRIBUTION: *Extends from East Africa southward to Madagascar and northward via Sri Lanka to southern Japan, reaching as far east in the Pacific as the Solomon Islands.*

▼ *BLUE-RING ANGELFISH (POMACANTHUS ANNULARIS)*

▲ *ARABIAN ANGELFISH (POMACANTHUS ASFUR)*

SIZE: *17.75 in (45 cm).*

BEHAVIOR: *Frequently encountered on coastal reefs and around other sites, such as shipwrecks, that offer retreats. It is important to include similar hiding places in an aquarium housing this angelfish. Having been moved to new surroundings, it will be very shy at first and must be able to conceal itself to feel secure.*

DIET: *This fish usually feeds well once established, and it can often be encouraged to feed initially by the introduction of live brine shrmps. Juveniles especially will feed on algae, and chopped meaty foods should also be offered.*

AQUARIUM: *Not to be trusted in a reef aquarium, since it attacks coral, sponges, and tunicates in these surroundings.*

COMPATIBILITY: *Has a justifiable reputation for being somewhat aggressive, especially toward other angelfish.*

ARABIAN ANGELFISH

Pomacanthus asfur

Matt-black coloration with a broad yellow stripe extending from the dorsal fin and tapering off toward the ventral surface of the body helps identify this angelfish, as well as its bright yellow tail with blue edging at the tip. Occasionally some individuals have white rather than yellow markings. Juveniles have a similar basic patterning to adults, but they have blue vertical stripes running through the darker bluish black areas of the body.

DISTRIBUTION: *Restricted to the western Indian Ocean, occurring from the Gulf of Aden and the Red Sea south as far as the island of Zanzibar off the Tanzanian coast in East Africa.*

SIZE: *15.75 in (40 cm).*

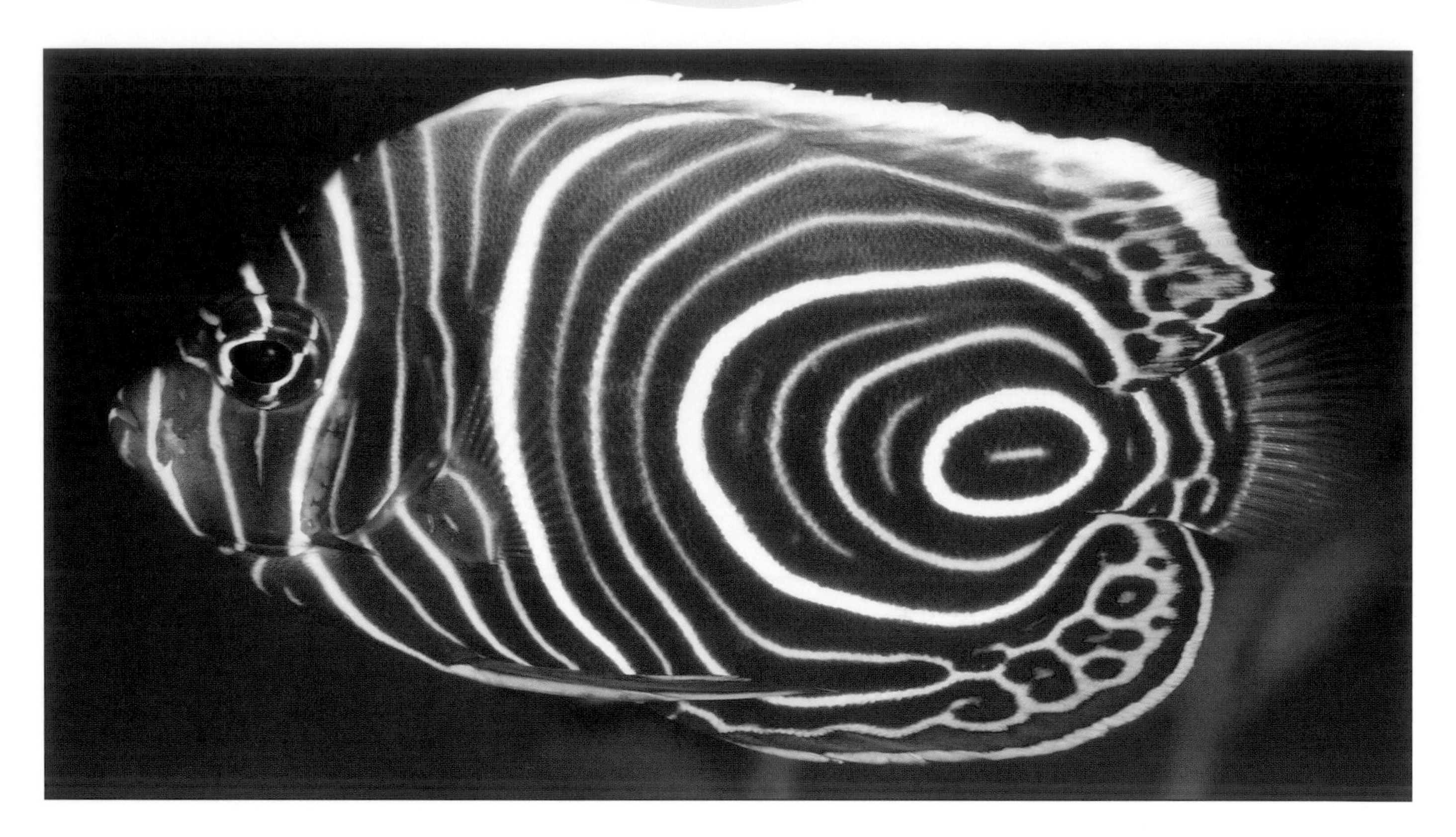

BEHAVIOR: *Often lives in fairly turbid waters and is not conspicuous on the reef because it naturally tends to hide away. It frequents areas where there is active coral growth, and occurs in relatively shallow water.*

DIET: *Feeds on sponges and tunicates. Slow to adjust to alternative foods, especially without adequate retreats accessible within the aquarium.*

AQUARIUM: *May cause little damage alongside stony coral, but cannot be entirely trusted in a reef tank, especially if its natural foods are included.*

COMPATIBILITY: *Quarrelsome both with its own kind and with other angelfish, so must be housed individually.*

EMPEROR ANGELFISH

Pomacanthus imperator

The emperor angelfish is one of the most variable angelfish in terms of its patterning (particularly in the juvenile phase), reflecting in part its wide distribution. Generally, young specimens have concentric white and blue circles curling around in the direction of the caudal fin, with the lines nearer the head being more vertical. The basic color pattern changes as they grow older, and they develop a relatively horizontal pattern of yellow and blue stripes. The area of the body in front of the pectoral fins is dark, linking on the underside of the body with a dark band running through the eyes. Individuals from the Indian Ocean tend to have white faces, whereas in emperor angelfish from the Red Sea and the Pacific Ocean that area is blue.

SYNONYM: *P. nicobariensis.*

▲ *EMPEROR ANGELFISH* (POMACANTHUS IMPERATOR)

DISTRIBUTION: *Occurs in the Red Sea and East Africa and ranges in the Pacific as far north as southern Japan and the Ogasawara Islands. Extends south to the Great Barrier Reef and the Austral Islands. To the east it reaches the Hawaiian, Line, and Tuamotu Islands.*

SIZE: *15.75 in (40 cm).*

BEHAVIOR: *The color change typically occurs when juveniles reach just over 3 in (8 cm), but it sometimes takes place later in aquarium surroundings. The resulting appearance is never as vivid as in wild adults, possibly because of dietary differences. Unfortunately, they tend not to settle as well in aquarium surroundings when adult.*

DIET: *Juveniles may feed on the mucus of other fish, and sponges feature prominently in the diet of adults. Offer a variety of items, including special marine angelfish food and chopped meaty items.*

AQUARIUM: *Likely to cause some damage in a reef aquarium.*

COMPATIBILITY: *An adult and a juvenile may agree temporarily in a large aquarium if the juvenile is introduced first, but the smaller fish will be harassed once it starts to change color.*

BLUE-GIRDLED ANGELFISH

Pomacanthus navarchus

The juvenile coloration differs markedly from that of the adult in this species. It is predominantly black with pale blue stripes running vertically down the entire body from mouth to tail. With age the dorsal fin and central area of the body become yellowish, and the stripes fade. Eventually this area of the body turns yellow with blue spots. The caudal fin and lower part of the face and throat are also yellow, while the rest of the body is blue.

DISTRIBUTION: *From Malaysia and Indonesia south to Rowley Shoals off Australia's northwest coast, and eastward to Papua New Guinea and the Great Barrier Reef. Reaches as far north as the Philippines.*

SIZE: *10 in (25 cm).*

BEHAVIOR: *Juveniles spend much of their time hiding on the reef. Adults are more likely to be seen in the open, usually on their own. Suitable retreats must be provided in the aquarium. The size at which juveniles transform into adults is variable—sometimes not until they are about 8 in (20 cm) long.*

DIET: *Delicate at first, this angelfish may need livefood such as brine shrimps to persuade it to eat. Feeds naturally on sponges, tunicates, and possibly the waste expelled by anemones.*

AQUARIUM: *Risky in a reef tank because of its feeding habits.*

COMPATIBILITY: *In spite of its shy nature, it is frequently aggressive toward others of its own kind, especially once established.*

FRENCH ANGELFISH

Pomacanthus paru

Juveniles of this species are much more vividly colored than adults. They have yellow striping on their bodies, encircling the caudal fin, and a yellow stripe running down between the eyes and continuing around the jaws. The stripes disappear with age, but first the scales on the body develop yellow edging at the rear, with a more pronounced spot present at the base of each pectoral fin. The underlying body color turns more grayish, and the area around the lips becomes white.

DISTRIBUTION: *Ranges from the coast of Florida and the Bahamas in the Atlantic down through the Gulf of Mexico and the Caribbean as far as Brazil. Also occurs farther east, around Ascension Island and St. Paul's Rocks.*

▼ BLUE-GIRDLED ANGELFISH (POMACANTHUS NAVARCHUS)

▲ FRENCH ANGELFISH (POMACANTHUS PARU)

SIZE: *16 in (41 cm).*

BEHAVIOR: *French angelfish live in pairs and are highly territorial, driving away others of their own kind. They may, however, spend more than an hour a day attracting cleaner shrimps and fish to groom them. Unlike some angelfish, no discernible courtship occurs prior to spawning, which takes place at sunset.*

DIET: *Omnivorous, although sponges tend to predominate in its natural diet, with marine algae being important, too. Can be weaned onto a diet based on meaty foods and greenstuff.*

AQUARIUM: *Not recommended for a reef tank because of its feeding habits.*

COMPATIBILITY: *Not as aggressive as some larger angelfish. Juveniles may pester other occupants, however, since they act as cleaner fish themselves at this stage.*

SEMICIRCLE ANGELFISH
(KORAN ANGELFISH)

Pomacanthus semicirculatus

A bluish black background color, broken by relatively broad white stripes on the body and blue stripes especially around the edge, help separate juveniles of this species from other *Pomacanthus* species. The tail is also stripy. Adult semicircle angelfish develop a greenish body color with blue speckling, and there is more evident blue coloration on the head.

DISTRIBUTION: *From the Red Sea and East Africa northward as far as southern Japan and south down to New South Wales and Lord Howe Island off the east coast of Australia. Its easterly distribution extends as far as Samoa.*

SIZE: *15.75 in (40 cm).*

BEHAVIOR: *Juveniles are found in shallow waters, to a depth of about 16 ft (5 m), but adults are found in much deeper water, down to 160 ft (40 m). They often seek cover in caves or artificial sites such as shipwrecks on the reef. The characteristic color change may start in juveniles at any stage from 3 to 6 in (8–16 cm) long.*

DIET: *Offer marine algae and a variety of other foods, especially chopped meaty foods. This fish may also be persuaded to take flake.*

AQUARIUM: *Cannot be trusted in a reef tank because of its omnivorous feeding habits.*

COMPATIBILITY: *Even juveniles are aggressive and must not be mixed with their own kind or other angelfish. Companions need to be chosen carefully.*

SIX-BAR ANGELFISH
(SIX-BANDED ANGELFISH)

Pomacanthus sexstriatus

This angelfish typically has six incomplete brownish bands that start on each side of the body behind the pectoral fins and continue to the base of the caudal fin. The face is brownish, broken by a white stripe behind the eyes. The pale yellow body is also covered in a mesh of small bluish spots. The juvenile appears very different, however, displaying a series of fairly straight white vertical bands down the sides of the body on a blackish background. Some blue markings are evident as well.

▼ *SIX-BAR ANGELFISH (POMACANTHUS SEXSTRIATUS)*

DISTRIBUTION: *From Malaysia and Indonesia northward to the Ryukyu Islands near Japan and south to Rowley Shoals off the northwestern coast of Australia. Extends eastward as far as the Solomon Islands, which lie to the east of New Guinea.*

SIZE: *18 in (46 cm).*

BEHAVIOR: *Pairs occupy a large territory, so these angelfish do not appear very common, although in certain parts of their range, such as the Great Barrier Reef, they are evidently more numerous. They are said to make a grunting sound if threatened.*

DIET: *A typically varied diet of greenstuff and meaty foods is required.*

AQUARIUM: *The feeding habits and size of this angelfish preclude it from being kept in a reef tank.*

COMPATIBILITY: *Both juveniles and adults are combative by nature. They are therefore better kept singly, especially in view of their potential adult size.*

YELLOW-FACE ANGELFISH
(BLUE-FACE ANGELFISH)

Pomacanthus xanthometopon

The confusing common names for this species are both correct in their own way. There is a bright yellow area on the face encircling the eyes and also a more extensive blue area below, encompassing the jaws. There is also a blue spot on the dorsal fin above the caudal peduncle. The tail is yellow, as are the pectoral fins. The body coloration is a rich blue, and the scales have yellow edging. Juveniles display almost vertical alternating blue and white stripes on the sides of their bodies and they have distinctive yellow areas on the dorsal and caudal fins. Their pectoral fins are completely yellow, even at this stage.

DISTRIBUTION: *Ranges from the Maldives in the Indian Ocean northward as far the Ryukyu Islands near Japan, and extends southward to the Great Barrier Reef and across to Vanuatu.*

SIZE: *15 in (38 cm).*

BEHAVIOR: *Usually sighted on a reef in areas where coral is common and there are caves that can serve as retreats. Juveniles venture quite close inshore, seeking shelter in caves, and are believed to feed on algae.*

DIET: *Juveniles adapt more easily to substitute diets than adult fish, which will feed mainly on sponges and tunicates.*

AQUARIUM: *Likely to be disruptive in a reef tank.*

COMPATIBILITY: *Less aggressive than some related species, but its size means that it can easily be a bully, especially in a relatively small aquarium.*

CORTEZ ANGELFISH

Pomacanthus zonipectus

In its juvenile color phase the Cortez angelfish is a yellow-banded species. It has a black body, and narrower blue stripes separate the curved yellow markings directed toward the rear of the body. Although dark in color, adults are very different in appearance. There is a whitish yellow stripe behind the eyes, and a more definite yellow area below the chin. The pectoral fins are patterned with yellow and black markings. Just behind this area is a lemon-colored crescent that extends vertically on the body with a diffuse and broader white area behind. Vermiculations (wavy-line markings) extend over the rest of the body, and the caudal fin is suffused with white.

DISTRIBUTION: *Eastern Pacific, ranging south from Magdalena Bay, Baja, through the Gulf of California as far as Peru in South America. Also occurs around the Galápagos Islands west of Ecuador.*

SIZE: *19 in (48 cm).*

BEHAVIOR: *Tends to be seen in relatively shallow waters above 49 ft (15 m). Sometimes encountered in the open, over sandy areas as well as among coral. This species is less territorial than many of its relatives, and can occasionally be found in groups.*

DIET: *Algae are very important in the diet of both juveniles and adults, and sponges feature prominently in the diet of mature individuals. Ensure that alga is growing in the aquarium, and supplement it as necessary. Also offer chopped meaty foods.*

AQUARIUM: *Not safe for a reef tank.*

COMPATIBILITY: *Do not mix juveniles, particularly with other species displaying similar coloration. Adults are also best housed individually.*

▲ *ROYAL ANGELFISH (PYGOPLITES DIACANTHUS)*

ROYAL ANGELFISH
(REGAL ANGELFISH)

Pygoplites diacanthus

Juveniles of this species have a distinctive ocellus (eyespot) on each side of the body near the rear of the dorsal fin. They are orange in color and have a series of bluish white stripes with narrow black edging extending down the sides of their bodies. The ocellus disappears in adults and is replaced by a prominent blue area. A number of different color variants have been described through the extensive range of this species, with the base color varying from yellow through to orange. The patterning can vary widely, from almost nonexistent in individual fish from the Red Sea to circular in some specimens from the Pacific. In the latter the facial coloring is more bluish gray overall.

DISTRIBUTION: *Ranges from the Red Sea and East Africa through the Indo-Pacific, reaching northward to the Ryukyu Islands and southward to the Great Barrier Reef. Extends eastward to the Tuamotu Archipelago.*

SIZE: *10 in (25 cm).*

BEHAVIOR: *Found in areas of the reef where coral is plentiful and where there are suitable retreats nearby. Juveniles are particularly shy and solitary but adults associate in small groups of a male and two or three females.*

DIET: *Known to eat sponges and tunicates, but often hard to persuade it to take a substitute diet in the aquarium. May initially take mysid shrimps, however, and will also eat algae, so prepare the aquarium in advance. Those from the Red Sea are often easier to acclimatize than fish from the Pacific.*

AQUARIUM: *May nibble at corals, but the choice of food here, including the opportunity to forage on living rock, may induce this shy fish to start feeding.*

COMPATIBILITY: *Must be kept on its own in an aquarium with plenty of hiding places. Often prefers relatively subdued lighting. Can be aggressive toward its own kind.*

HAWKFISH
FAMILY CIRRHITIDAE

The small fish that make up the family Cirrhitidae are known as hawkfish. This is because of their habit of adopting a suitable perch on a prominent piece of rock or coral and swooping down on prey that emerges below them, in a similar way to a bird of prey. They are not generally difficult to maintain. Their natural distribution is centered on the Indo-Pacific region. There are some 32 species, divided among nine genera. Although some are popular aquarium occupants, most are rarely kept.

SPOTTED HAWKFISH

(BLOTCHED HAWKFISH; THREAD-FIN HAWKFISH)

Cirrhitichthys aprinus

Reddish coloration on the sides, becoming paler toward the underparts, help identify this fish. There is a series of whitish markings there, too, creating something of a marbling effect that helps break up the body shape. When the fish is viewed from above, small bristlelike projections, called cirri, can be seen along the back. They are brownish red in color and their function is also to disrupt the fish's outline. There is a characteristic spot on the operculum covering the gills.

DISTRIBUTION: *The exact range of this fish is not well documented, but it is known to include the area around the Maldives off the southwest coast of India, extending from there into the western Pacific Ocean.*

SIZE: *5 in (13 cm).*

BEHAVIOR: *Tends to be found in coastal waters in areas of the reef where coral is present and also in rocky stretches, both of which provide vantage points for this predatory fish. These hawkfish associate in groups made up of a single male and several females, with the male being recognizable as the largest fish. This is not necessarily a reliable way of sexing them, however, because there can often be a considerable variation in size among hawkfish from different groups. Obviously the age of a fish also has a bearing on its size.*

DIET: *Will feed readily on a range of meaty foods, including shrimps and lancefish, and may even take flake food.*

AQUARIUM: *Generally suitable, although likely to prey on crustaceans.*

COMPATIBILITY: *May be bullied by other, larger species of hawkfish and may also prey on smaller companions, which should therefore be excluded from its aquarium. Generally not a problem to maintain, however, and its relatively sedentary nature means that it will not disturb nervous fish.*

▼ *SPOTTED HAWKFISH* (CIRRHITICHTHYS APRINUS)

DWARF HAWKFISH

(FALCO'S HAWKFISH)

Cirrhitichthys falco

The dwarf hawkfish displays the typical red and white coloration that is often a feature of members of this family. It also has some red patches on the body, some of which form more definite horizontal stripes. The red coloration in this case is relatively pale, bordering on pink, and there are some brownish markings extending from the area behind the eyes up along the base of the dorsal fin. As in the spotted hawkfish (*C. aprinus*), there are prominent bristlelike cirri, which are yellow at the tips, on the rays of the first part of the dorsal fin. There is also dark red spotting present on the dorsal and caudal fins.

SYNONYM: *C. serratus.*

DISTRIBUTION: *Extends through the Indo-Pacific region from the Maldive Islands off India's southwest coast, northward to the Ryukyu Islands, and southward as far as the southern part of Australia's Great Barrier Reef. Its most easterly distribution is New Caledonia.*

SIZE: *3 in (7 cm).*

BEHAVIOR: *This hawkfish has been found in waters as deep as 150 ft (45 m), but it is probably more commonly encountered on reefs close to shore, in areas where coral predominates. Its relatively small size means that it can use the coral heads as hunting platforms, although it may often choose to spend more time on the aquarium substrate than most other hawkfish. Although this species can occasionally be seen in pairs, it is more usual for a male to associate with several females. Spawning occurs under cover of darkness. Like other hawkfish, it has extremely sharp eyesight, and its torpedolike body shape helps it swim very fast to seize passing quarry.*

DIET: *Will prey on crustaceans and also smaller fish. Feed a meaty diet, offering a range of suitable foods rather than just one type. Converts very easily to artificial diets, especially if food is dropped within range of its chosen hunting promontory in the aquarium.*

AQUARIUM: *Not to be trusted alongside crustaceans in a reef tank, but will not harm other occupants.*

COMPATIBILITY: *Small companions such as dartfish are at risk from this predatory species, but larger fish will be ignored.*

▲ *DWARF HAWKFISH (CIRRHITICHTHYS FALCO)*

CORAL HAWKFISH

(PIXY HAWKFISH)

Cirrhitichthys oxycephalus

The color of this species is particularly variable, which can make identification difficult. Some individuals are predominantly pink with whitish marbling on the sides of the body. At the other extreme, some have a blotched appearance with evident grayish, blackish, and red markings on their bodies, with the underparts being mainly white rather than reddish. The cirri, or small bristles, at the tips of the dorsal fin are made up of white and red bands. Colored markings extend into the rear area of the dorsal fin and are also likely to be conspicuous in the caudal fin. One relatively consistent feature is the presence of brown speckling above and behind the eyes. The pectoral fins are large, as in other hawkfish, and are used to help anchor the fish onto the rocks.

SYNONYM: *C. corallicola.*

DISTRIBUTION: *Ranges from the Red Sea down the east coast of Africa to the vicinity of East London, extending across the Indo-Pacific region as far as the Marquesan Islands. Its northerly distribution extends to the Marianas, while in the south it is found down to New Caledonia. Also occurs from the Gulf of California south to the Pacific coast of Colombia, and is present around the Galápagos Islands in the eastern area of the Pacific.*

SIZE: *4 in (10 cm).*

BEHAVIOR: *As its name suggests, this hawkfish is found in close association with coral, which provides a convenient area from which to watch for prey and also plenty of retreats in which it can conceal itself from would-be predators. This fish is not an especially strong swimmer in open water in any event, because it lacks a swimbladder, which would normally help it maintain its buoyancy. In common with related species, males defend their territories*

▼ *CORAL HAWKFISH (CIRRHITICHTHYS OXYCEPHALUS)*

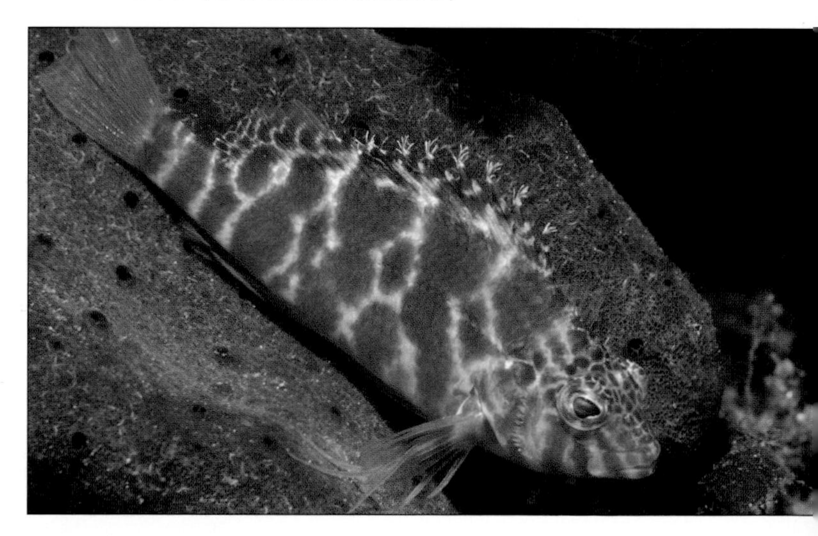

fiercely, proving surprisingly ferocious in spite of their small size. A male will live with a group of females rather than as a member of a pair.

DIET: *Preys on shrimps and similar crustaceans, so provide meaty foods, which can include chopped shrimps and some fish.*

AQUARIUM: *Can be safely included in a reef tank provided no crustaceans are present.*

COMPATIBILITY: *Cannot be trusted with smaller companions, which it may attack. Males will be aggressive toward potential rivals. A single individual can, however, be housed successfully in an aquarium with larger, nonaggressive companions.*

STOCKY HAWKFISH

(FRECKLED HAWKFISH)

Cirrhitus pinnulatus

The coloration of this hawkfish is more subdued than that of other species. Although its markings can be highly variable, they often consist of brownish bands along the sides of the body, alternating with a white background. Some individuals are much darker in color than others, and their white areas may be reduced to spots, appearing dusky as well in some cases. On the head the pattern may be one of lines rather than blotches.

▲ *STOCKY HAWKFISH (CIRRHITUS PINNULATUS)*

SYNONYMS: *C. alternatus; C. maculatus; C. spilotoceps.*

DISTRIBUTION: *From the Red Sea and East Africa extending through the Indo-Pacific region northward as far as southern Japan. Ranges southward as far as Kermadec and Rapa Islands and eastward through the Pacific via Micronesia and the Hawaiian Islands. A separate Atlantic population is also present off the coast of South Africa.*

SIZE: *12 in (30 cm).*

BEHAVIOR: *This hawkfish is restricted to the shallows and only recorded at depths down to 10 ft (3 m). It is found in areas where there can be quite strong currents. Its large size (compared to that of some of its relatives) may help prevent it from being swept away.*

DIET: *Crabs are the major prey items of this species, but it will also prey on other invertebrates such as brittlestars and sea urchins. Meaty foods should therefore be offered.*

AQUARIUM: *Not suitable for a reef tank in which there are crustaceans or other likely prey items.*

COMPATIBILITY: *Hunts smaller fish as well, so companions need to be chosen with care—they should not be aggressive themselves.*

SWALLOWTAIL HAWKFISH

(LYRE-TAILED HAWKFISH)

Cyprinocirrhites polyactis

The overall impression often created by the coloration of the swallowtail hawkfish is that it is orange. However, when seen close up, there are often paler areas on the body, creating a blotched effect. Young fish may be a brighter shade than adults, whose color can vary from a yellowish shade through to orange-brown, with brownish markings likely to be most evident close to the base of the dorsal fin. The caudal fin is elongated, resulting in a highly distinctive lyre-tailed appearance, while the cirri on the first part of the dorsal fin are small and relatively inconspicuous. The pectoral fins are orange and white and have a barred appearance.

SYNONYMS: *C. stigma; C. ui.*

DISTRIBUTION: *East Africa down to the southeastern coast of South Africa, ranging through the Indo-Pacific northward as far as southern Japan. Extends south as far as the northern part of New Zealand and ranges eastward through the Pacific to Tonga.*

SIZE: *6 in (15 cm).*

BEHAVIOR: *Occurring in relatively deep water—down to 430 ft (132 m)—this particular hawkfish hunts for food off the bottom, seizing crustaceans and similar creatures. It is often found in areas of the reef where the sides fall away steeply and also in areas where the current is strong. In aquarium surroundings, however, the fish often spends long periods resting on the decor and only hovers in characteristic fashion (like the butterfly after which it is named) if there is a strong current. It has sometimes been noted associating with larger anemones in the aquarium, apparently seeking protection by lurking under their tentacles.*

▼ *SWALLOWTAIL HAWKFISH* (CYPRINOCIRRHITES POLYACTIS)

▲ *FLAME HAWKFISH* (NEOCIRRHITES ARMATUS)

DIET: *Predatory, so requires meaty foods such as chopped shrimps and fish.*

AQUARIUM: *Not to be trusted with crustaceans and similar invertebrates in a reef tank.*

COMPATIBILITY: *Relatively social with its own kind, but a newcomer that is introduced to the aquarium at a later date alongside established individuals is likely to be persecuted. May also prey on smaller companions.*

FLAME HAWKFISH

Neocirrhites armatus

This species is one of the most brightly colored of all hawkfish. It is generally fiery red, although this color may sometimes fade and become more orange-red in aquarium surroundings. There are black bands encircling the eyes, with a prominent black streak extending along the base of the dorsal fin, the rays of which are red.

DISTRIBUTION: *Centered on western and central parts of the Pacific, ranging north as far the Ryukyu Archipelago, south through Micronesia down to Australia's Great Barrier Reef, and east as far as the Austral Islands, which lie to the south of Tahiti.*

SIZE: *3.5 in (9 cm).*

BEHAVIOR: *Occurs in relatively shallow water, down to about 33 ft (10 m) below the surface. Closely associated with live corals, hiding among the branches and seeking shelter there if danger threatens. It may encourage circulation of water within the coral, wafting in food and providing nitrogen through its waste, which the coral can use for growth. Lives in pairs rather than forming harems.*

DIET: *Feeds on a variety of invertebrates, including crabs and mollusks. Similar meaty foods need to be provided in the aquarium. Shrimps may help retain the natural vibrancy of the color of this hawkfish.*

AQUARIUM: *Not to be trusted in a reef tank, because it is adept at catching crustaceans and can even pull snails from the safety of their shells.*

COMPATIBILITY: *Likely to prove territorial and aggressive toward other fish living close to the aquarium substrate, but not especially troublesome toward companions that swim higher up in the water.*

LONG-NOSE HAWKFISH

Oxycirrhites typus

The long narrow mouthparts of this hawkfish enable it to be identified relatively easily, especially in view of its distinctive patterning. Its basic body coloration is silvery, broken by a series of red vertical and horizontal bands that result in a highly individual pattern of conjoined rectangles on the body. Banding is also present in the dorsal fin. Overall, this hawkfish has a fairly slender body shape.

DISTRIBUTION: *Ranges from the Red Sea south to South Africa and across the Indo-Pacific, reaching southern Japan to the north, and extending south as far as New Caledonia. Found eastward to the Hawaiian Islands and is also present from the Gulf of California down to the coast off northern Colombia. Also encountered around the Galápagos Islands.*

SIZE: *5 in (13 cm).*

BEHAVIOR: *Occurs primarily on steep slopes on the edge of reefs down to depths of 330 ft (100 m). It associates mainly with black corals and gorgonians, using them as platforms from which to look for passing prey. It is a highly territorial species by nature. Long-nose hawkfish are also surprisingly agile and are not well suited to being kept in an open-top aquarium, since they can jump out of the water and end up on the floor.*

DIET: *Feed on a variety of meaty foods, which should include shrimps. Food should be chopped, although this hawkfish can swallow quite large pieces, in spite of the shape of its snout.*

AQUARIUM: *Do not mix with crustaceans or small fish, which may also be preyed on.*

COMPATIBILITY: *It is possible to keep true pairs together, provided they are introduced to the aquarium at the same time. Two males, however, will fight savagely, and if the fish do not appear to be agreeing, they must be separated before one is badly injured. Smaller companions may be eaten but other, nonaggressive fish (especially those that do not have a similar elongated body shape) will be left alone.*

BLACK-SIDE HAWKFISH
(FRECKLED HAWKFISH; FORSTER'S HAWKFISH)

Paracirrhites forsteri

This hawkfish varies greatly in appearance, depending on its age and also on its place of origin. This has given rise to taxonomic confusion in the past. One of the most attractive forms, originally classified as *P. typee*, has a maroon body with a yellow caudal fin. Another attractive color phase has a white lower half to its body. The upper part is dull red, becoming significantly darker toward the tail, with black coloration there. There is also a very narrow yellow band running down each side of the dorsal fin. Around the head is a pattern of red spots, becoming black in the red striped area. This color scheme is most evident in individuals that are found close to major landmasses. However, in black-side hawkfish originating from Oceania the dark red area may be replaced by green with contrasting greenish yellow underparts and dark spotted patterning on the face.

DISTRIBUTION: *Red Sea and East Africa eastward through the Indo-Pacific (but absent from the Persian Gulf and the Gulf of Oman), reaching as far north as southern Japan. Extends southward to New Caledonia and the Austral Islands, which lie south of Tahiti. Also present throughout Micronesia and the Society Islands.*

SIZE: *8.75 in (22 cm).*

BEHAVIOR: *Tends to rest on exposed branches of* Acropora *species coral and similar large corals, using them as vantage points for hunting purposes. Typically found in waters down to 100 ft (30 m), but has been recorded in much deeper waters in some localities.*

DIET: *Feeds on fish and crustaceans, so requires similar meaty foods in aquarium surroundings.*

AQUARIUM: *Will prey readily on crustaceans, although will not harm sessile invertebrates.*

COMPATIBILITY: *Can be mixed safely only with much larger fish because of its predatory feeding habits. Will also tend to bully other fish, so only much bigger and robust companions, such as triggerfish, are suitable companions. Avoid keeping these hawkfish together as well, unless they are in true pairs, because fighting will break out in an aquarium. Both members of the pair must be introduced to the aquarium at the same time to minimize the risk of any conflict.*

◀ BLACK-SIDE HAWKFISH (PARACIRRHITES FORSTERI)

FAIRY BASSLETS AND GRAMMAS FAMILY GRAMMATIDAE

There are only three species in this attractive family—*Gramma loreto, G. melacara*, and *G. linki*—all of which occur in the Caribbean Sea. They are striking in terms of their coloration, but they tend to be shy by nature.

ROYAL GRAMMA

Gramma loreto

This is a particularly attractive fish. The front of its body is purple with a black spot at the front of the dorsal fin and black streaking extending through the eyes. The rear part of the body is a contrasting shade of bright yellow, although its exact appearance varies depending on the individual. In some cases yellow coloration is replaced with orange.

DISTRIBUTION: *Ranges from Bermuda and the Bahamas southward across the Caribbean and down the coast of Central America as far as northern South America.*

SIZE: *3.25 in (8 cm).*

BEHAVIOR: *The royal gramma is encountered on the reef near caves into which it can retreat at any hint of danger. It may sometimes swim upside down under ledges, and may do this in aquarium surroundings, but it is not a cause for concern. It is a shy fish and rarely strays far from cover. It is thought that males are significantly larger than females. The royal gramma feeds naturally on the external parasites of other fish. Spawning occurs in pits excavated in the substrate and carefully lined with algae.*

DIET: *Provide a range of meaty food. Will also feed readily on brine shrimps.*

AQUARIUM: *Ideally suited to a reef aquarium that includes suitable retreats in the rockwork to make this small fish feel secure. It should then prove less nervous and spend longer in the open.*

COMPATIBILITY: *Can be kept in groups and will even spawn in aquarium surroundings. It is likely to prove territorial in defending its hiding places, so it is not a good idea to mix this fish with other cave-dwelling fish, since this can lead to conflicts. Otherwise, this species will agree well with nonaggressive species of a similar size.*

▼ *ROYAL GRAMMA (GRAMMA LORETO)*

ANEMONEFISH AND DAMSELFISH FAMILY POMACENTRIDAE

The Pomacentridae is divided into two broad groups: the anemonefish (subfamily Amphiprioninae), also known as clownfish, and the damselfish and chromids (subfamilies Pomacentrinae and Chrominae). Both groups are extremely popular with marine aquarists. As their name suggests, anemonefish associate with sea anemones in their natural habitat. They are often shades of orange or red with white markings, which explains their alternative common name of clownfish—their appearance is reminiscent of the bright makeup usually worn by circus clowns.

Over recent years, successful spawning of clownfish has become commonplace, and tank-bred stock is now widely available. Availability has been helped in part by the fact that these fish are hermaphrodite, so it is easy to obtain breeding pairs. Interestingly, however, having been reared in aquarium surroundings without exposure to predators, these captive-bred clownfish display little or no desire to associate with sea anemones. This has simplified their care, since it is much easier to keep the fish on their own than having to maintain them alongside sea anemones.

Damselfish differ markedly in color from clownfish, often having a bluish or yellow appearance. They prefer to seek protection among coral. They are highly valued because they can aid the maturation process of a recently established marine aquarium. Unfortunately, some species can be aggressive and territorial, which can create difficulties when trying to introduce more delicate fish alongside them in due course. Choosing the right species at the outset is therefore vital.

Both groups are easy to cater for in terms of their feeding requirements, since they browse not just on foods but also on algae that develop in the tank. They will sample dry foods as well as fresh items, and some can benefit from being given food with natural coloring agents from time to time.

▼ *YELLOW-TAIL CLOWNFISH (AMPHIPRION CLARKII)*

▲ *TWO-BAR ANEMONEFISH (AMPHIPRION ALLARDI)*

TWO-BAR ANEMONEFISH

Amphiprion allardi Subfamily Amphiprioninae

In this species the white band behind the eyes is much broader than the second band, which occurs approximately midway along the length of the dorsal fin. The sides of the body are relatively dark, and the underside of the body and the fins are yellow, with the exception of the caudal fin, which is typically white. Although it is one of the less commonly available species, the two-bar anemonefish presents no particular problems in terms of its care in the aquarium.

DISTRIBUTION: *Restricted to the vicinity of Africa and is found down the east coast from Kenya southward, extending as far as the coast of Durban in South Africa.*

SIZE: *6 in (15 cm).*

BEHAVIOR: *A resident species (like other anemonefish), found in association with various anemones, such as* Heteractis aurora *and* Stichodactyla mertensii.

DIET: *Provide a range of fresh and prepared items, including zooplankton and crustacean-based foods.*

AQUARIUM: *Water temperature should be around 77 to 79°F (25–26°C), with a recommended typical SG reading of 1.020–1.024.*

COMPATIBILITY: *Take care not to overcrowd these fish in the aquarium. They are best kept in pairs.*

TWO-BAND ANEMONEFISH

Amphiprion bicinctus Subfamily Amphiprioninae

The body coloration of the two-band anemonefish is dark, especially in young fish. The fins are generally orangey, but there may be a dark mark on the caudal fin and at the back of the dorsal fin that disappear with age. The white band extending over the top of the head is broader than the second band. The latter begins on the upper part of the body where the spiny part of the dorsal fin ends.

DISTRIBUTION: *From the Red Sea southeast to the Chagos Archipelago in the Indian Ocean.*

SIZE: *5.5 in (14 cm).*

BEHAVIOR: *Found on coral reefs, often in the vicinity of* Heteractis *species of anemone, but may also form a symbiotic relationship with either* Stichodactyla gigantea *or* Entacmaea quadricolor.

DIET: *Provide a range of fresh and prepared items, including zooplankton and crustacean-based foods.*

AQUARIUM: *A very suitable species for the reef aquarium.*

COMPATIBILITY: *A relatively peaceful species; a small group may live together without serious conflict, especially in a largish aquarium.*

◄ *SADDLE ANEMONEFISH (AMPHIPRION EPHIPPIUM)*

AQUARIUM: *A relatively hardy and easily managed species that provides a good introduction to the group. Suitable for a reef aquarium, but it may bully less assertive fish, such as seahorses, in these surroundings and can sometimes prey on small invertebrates, especially crustaceans.*

COMPATIBILITY: *Trying to introduce adult fish can be difficult, since they can prove quarrelsome. It is better to start out with two young fish, which should form a pair.*

YELLOW-TAIL CLOWNFISH

Amphiprion clarkii Subfamily Amphiprioninae

As might be expected from its extensive distribution, the yellow-tail clownfish displays considerable variation in appearance throughout its wide range. This variation has led to it being classified previously under a series of different names. Typically, it has two relatively broad white bands, one of which is located behind the eyes, and the other just in front of the caudal peduncle. In young fish there is usually a third band present on the caudal peduncle, but this disappears with age. In spite of this clownfish's common name, the caudal fin may sometimes be white rather than yellowish. The body coloration itself is relatively dark, tending to be lighter on the underside of the body in the vicinity of the head.

SYNONYMS: *A. boholensis; A. japonicus; A. melanostolus; A. papuensis; A. snyderi; A. xanthurus.*

DISTRIBUTION: *One of the most wide-ranging species, extending from the Persian Gulf eastward to Micronesia and Melanesia and northward as far as southern Japan and the Ryukyu Islands.*

SIZE: *6 in (15 cm).*

BEHAVIOR: *Individual populations have developed in isolation, which accounts for the variability in appearance, although this is a relatively adaptable species. As expected, a number of different anemones may be colonized by this clownfish, particularly those belonging to the genera* Heteractis *and* Stichodactyla.

DIET: *Provide a range of fresh and prepared items, including zooplankton and crustacean-based foods.*

SADDLE ANEMONEFISH

(RED SADDLEBACK CLOWN)

Amphiprion ephippium Subfamily Amphiprioninae

A particularly beautiful species, the saddle anemonefish is so called because of the black saddlelike marking on the sides of its body. It can extend from below the dorsal fin on each side to near the base of the tail, although this area of black can vary noticeably from individual to individual. In the case of juveniles, it is entirely absent—they have a white streak just behind the head. The rest of the body is a fiery orange color.

SYNONYM: *A. calliops.*

DISTRIBUTION: *Extends through the eastern part of the Indian Ocean from the Andaman and Nicobar Islands south to Sumatra and Java, and northward to Thailand and Malaysia.*

SIZE: *5.5 in (14 cm).*

BEHAVIOR: *Unlike many other members of the group, the saddle anemonefish can often be encountered close to the shore, in what can be fairly murky waters. They typically live in pairs and may associate with the anemones* Heteractis crispa *and* Entacmaea quadricolor.

DIET: *Provide a range of fresh and prepared items, including zooplankton and crustacean-based foods.*

AQUARIUM: *Can be kept with invertebrates in a reef tank.*

COMPATIBILITY: *The bright coloration of this anemonefish signals a somewhat aggressive nature. For this reason, it is not a good idea to keep them in groups or with weaker companions in the aquarium.*

TOMATO CLOWNFISH

(BRIDLED CLOWNFISH)

Amphiprion frenatus Subfamily Amphiprioninae

A feature of this species is a white stripe edged with black that extends around the head behind the eyes and narrows toward the throat. There is also a black patch on each flank and a dark area around the eyes. Variations occur in some populations, however—in some individuals a second white band highlighted by a narrow black border is evident around the middle of the body. The black markings on the body are not consistent

▲ *TOMATO CLOWNFISH (AMPHIPRION FRENATUS)*

throughout the life of the fish. Older tomato clownfish in particular tend to be more melanistic, turning almost completely black in some cases.

SYNONYMS: *A. macrostoma; A. polylepis.*

DISTRIBUTION: *Ranges from the Gulf of Thailand eastward to Palau, reaching Java in the south and extending as far as southern Japan to the north.*

SIZE: *5.5 in (14 cm).*

BEHAVIOR: *Usually encountered near populations of the anemone* Entacmaea quadricolor, *in which it hides.*

DIET: *Provide a range of fresh and prepared items, including zooplankton and crustacean-based foods.*

AQUARIUM: *Hardy and easy to keep, this anemonefish is well suited to a reef aquarium.*

COMPATIBILITY: *Aggressive, especially when mature, so keep singly or in pairs rather than in larger groups.*

WIDE-BAND ANEMONEFISH

Amphiprion latezonatus

This particular anemonefish differs greatly in appearance from those that are more commonly seen in aquaria, since it is predominantly black in color. There is a relatively narrow pale band behind the eyes and another small white band at the base of the tail, but its name derives from the wide grayish patch bordered by white that is seen around the center of the body.

DISTRIBUTION: *A relatively small area of the western Pacific, extending from Australia's Great Barrier Reef eastward to New Caledonia.*

SIZE: *5.5 in (14 cm).*

BEHAVIOR: *It naturally associates with* Heteractis crispa. *It occurs not only on coral reefs but can also be found on rocky reefs throughout its area of distribution. This species tends not to be found as close to the surface as other anemonefish.*

DIET: *Provide a range of fresh and prepared items, including zooplankton and crustacean-based foods.*

AQUARIUM: *A typical reef aquarium will suit this anemonefish. The provision of a suitable anemone will help wild-caught fish settle into aquarium surroundings more readily.*

COMPATIBILITY: *Beware of mixing this and other anemonefish with small crustaceans.*

FIRE CLOWNFISH

(CINNAMON CLOWNFISH; RED-AND-BLACK CLOWNFISH)

Amphiprion melanopus Subfamily Amphiprioninae

The fire clownfish gets its name from the juveniles, which are red with distinctive white stripes around the broadest part of the body and around the caudal peduncle. Adults, however, are duller in appearance, with relatively restricted orange coloration merging with black on the sides of the body, especially toward the tail. The ventral fins on the underside of the body are also black, and there is a relatively broad white stripe behind the eyes.

▼ *FIRE CLOWNFISH (AMPHIPRION MELANOPUS)*

▲ *MALDIVE ANEMONEFISH (AMPHIPRION NIGRIPES)*

SYNONYMS: *A. arion; A. verweyi.*

DISTRIBUTION: *Found throughout much of the Pacific region eastward from Bali to the Caroline Islands and north to the southern Philippines.*

SIZE: *4.75 in (12 cm).*

BEHAVIOR: *Occurs in the vicinity of coral reefs, down to depths of about 60 ft (18 m). It occurs most commonly in association with the anemone* Entacmaea quadricolor *and much less frequently with* Heteractis *species.*

DIET: *Provide a range of fresh and prepared items, including zooplankton and crustacean-based foods. This anemonefish also eats microscopic planktonic creatures.*

AQUARIUM: *A large aquarium lessens the risk of aggressive behavior in this amenomefish. The volume of the tank should be at least 30 gallons (114 l) if this species is to be included, since it is potentially one of the larger anemonefish.*

COMPATIBILITY: *Much more easily achieved by starting with young tank-bred individuals. Introducing two adult members of the species can be very difficult, especially without introducing them first on neutral territory. This species will also bully weaker companions.*

MALDIVE ANEMONEFISH

(BLACK-FINNED ANEMONEFISH; BLACK-FOOTED CLOWNFISH)

Amphiprion nigripes Subfamily Amphiprioninae

The Maldive anemonefish is yellowish, often with a pinkish tinge evident on the top part of its body. There is usually a white stripe extending around the head behind the eyes, with a narrow blackish border. The pelvic fins are invariably black in color, although the anal fin may be yellowish, like the other fins. With a limited distribution, the Maldive anemonefish is often less commonly available than a number of other members of the genus. It is much easier to settle in captive-bred rather than wild-caught stock of this relatively shy species.

DISTRIBUTION: *Occurs in the Indian Ocean around the Maldive Islands and Sri Lanka.*

SIZE: *4.25 in (11 cm).*

BEHAVIOR: *Lives on coral reefs, being observed down to a depth of 80 ft (25 m), and forms a symbiotic relationship with the anemone* Heteractis magnifica.

DIET: *Provide a range of fresh and prepared items, including zooplankton and crustacean-based foods.*

AQUARIUM: *Suitable for an invertebrate tank. Including an appropriate anemone will make wild-caught individuals easier to establish.*

COMPATIBILITY: *Generally does not get on well in groups. Can be more tolerant of other species of anemonefish, particularly those displaying different patterning, especially if the tank is large.*

CLOWN ANEMONEFISH

Amphiprion ocellaris Subfamily Amphiprioninae

The most popular and widely kept member of the genus, the orange clownfish is also one of the most brightly colored. It is a brilliant shade of orange, with pure white bands encircling the body and prominent black edging on the fins. Individuals can usually be recognized by slight differences in their markings. Tank-bred specimens are often more yellow overall than those seen in the wild, but feeding them a diet rich in natural coloring agents can make them more orange.

SYNONYMS: *A. bicolor; A. melanurus.*

DISTRIBUTION: *Eastern part of the Indian Ocean across to the northwest coast of Australia; also northward from the Asian coast as far as the Ryukyu Islands in the western Pacific.*

SIZE: *4.25 in (11 cm).*

BEHAVIOR: *Lives on reefs, seeking shelter among the tentacles of* Heteractis *and* Stichodactyla *anemones. Those bred in aquaria are unlikely to display a close affinity with these invertebrates.*

DIET: *Provide a range of fresh and prepared items, including zooplankton and crustacean-based foods.*

AQUARIUM: *Can be housed satisfactorily in a 20-gallon (76-l) tank, but avoid overcrowding.*

▼ CLOWN ANEMONEFISH (AMPHIPRION OCELLARIS)

▲ ORANGE CLOWNFISH (AMPHIPRION PERCULA)

COMPATIBILITY: *These anemonefish tend to be less aggressive toward each other than toward related species, particularly if obtained as a young group at the outset.*

ORANGE CLOWNFISH

Amphiprion percula Subfamily Amphiprioninae

The orange clownfish has very similar coloration to that of the clown anemonefish (*A. ocellaris*), but it can be differentiated by the thicker black lines separating the orange and white areas on the body. The dorsal fin is also usually blacker in this species. The appearance of the three white bands that are present—behind the head, around the middle of the body, and on the caudal peduncle—is extremely variable and enables individuals to be distinguished easily from one another.

SYNONYM: *A. tunicatus.*

DISTRIBUTION: *From waters around New Guinea down to Australia's Great Barrier Reef off the coast of Queensland in the western Pacific and to the Solomon Islands and Vanuatu.*

SIZE: *4.25 in (11 cm).*

BEHAVIOR: *Occurs in relatively shallow water (compared with* A. ocellaris*). Encountered in groups consisting of a breeding pair and as many as four immatures. Typical host anemones for the orange clownfish belong to the* Heteractis *and* Stichodactyla *genera.*

DIET: *Provide a range of fresh and prepared items, including zooplankton and crustacean-based foods.*

AQUARIUM: *Like most of its relatives, the orange clownfish is a bold feeder, emerging readily even if sheltering in the tentacles of an anemone when food is offered.*

COMPATIBILITY: *Can be territorial by nature.*

PINK ANEMONEFISH

Amphiprion perideraion Subfamily Amphiprioninae

A white stripe extends from above the snout and runs along the length of the back to the base of the caudal fin in this species. A white band also extends down each side of the body at the back of the head, running across the gill covers. This feature separates it at a glance from *A. akallopisos*, which is otherwise similar. The body coloration is pinkish orange, often appearing more yellowish on the lower parts.

SYNONYMS: *A. amamiensis; A. rosenbergi.*

DISTRIBUTION: *The western Pacific ocean from the Gulf of Thailand north to the Ryukyu Islands near Japan, south as far as Australia's Great Barrier Reef, and east to New Caledonia.*

SIZE: *4 in (10 cm).*

BEHAVIOR: *Has been encountered down to depths approaching 130 ft (40 m). Typically associates in its reef environment with sea anemones of the genus* Heteractis, *especially* H. magnifica; *may sometimes be found alongside other anemonefish in these surroundings.*

DIET: *Provide a range of fresh and prepared items, including zooplankton and crustacean-based foods.*

AQUARIUM: *Recommended for a reef aquarium in which there is an anemone to provide cover. It is unusual in that pairs can be recognized with certainty: males have orange edging on the upper and lower surface of the caudal fin and running along the soft (rear) part of the dorsal fin.*

COMPATIBILITY: *One of the shyer species and therefore likely to be bullied by more assertive clownfish.*

SADDLE-BACK CLOWNFISH

(WHITE-SADDLED CLOWNFISH)

Amphiprion polymnus Subfamily Amphiprioninae

This relatively dark-colored species is similar to the sebae anemonefish (*A. sebae*)—both have a white saddlelike area midway along the back. However, the saddle-back clownfish can be distinguished by the white edging around the caudal fin. There is some variability in appearance among individuals from different parts of its wide range. Those found in the vicinity of Papua New Guinea, for example, have a deeper orange shade on their underparts than those occurring elsewhere.

▼ PINK ANEMONEFISH (AMPHIPRION PERIDERAION)

▲ *Saddle-back Clownfish (Amphiprion polymnus)*

▶ *Yellow Clownfish (Amphiprion sandaracinos)*

Synonyms: *A. intermedius; A. laticlavius; A. polynemus; A. trifasciatus.*

Distribution: *Western Pacific from southeast Asia and northern Australia north to the Ryukyu Islands and east to the Solomon Islands.*

Size: *5 in (13 cm).*

Behavior: *Lives in association with* Heteractis crispa *and* Stichodactyla haddoni *sea anemones. May occur at slightly greater depths than related species, being recorded around reefs down to 100 ft (30 m).*

Diet: *Provide a range of fresh and prepared items, including zooplankton and crustacean-based foods.*

Aquarium: *This clownfish has gained a reputation for being one of the more difficult species to establish, but it usually feeds readily.*

Compatibility: *A territorial clownfish.*

YELLOW CLOWNFISH

Amphiprion sandaracinos Subfamily Amphiprioninae

In contrast to many anemonefish, which have white bands encircling the body, this species has a white stripe running up from the snout and extending along the midline of the back as far as the caudal peduncle.

Distribution: *Ranges in the western Pacific, extending from western Australia northward as far as Taiwan and the Ryukyu Islands close to Japan. Distributed eastward via New Guinea to New Britain and the Solomon Islands.*

Size: *5.5 in (14 cm).*

Behavior: *Lives in close association with reefs and has been recorded down to a depth of about 65 ft (20 m). Its favored host anemones are* Stichodactyla mertensii *and* Heteractis crispa. *Tends to be observed in pairs or small numbers rather than in large groups.*

▲ *SEBAE ANEMONEFISH (AMPHIPRION SEBAE)*

DIET: *Provide a range of fresh and prepared items, including zooplankton and crustacean-based foods.*

AQUARIUM: *Suitable for a reef aquarium.*

COMPATIBILITY: *Dominant females can prove aggressive toward their own kind and toward other species of anemonefish.*

SEBAE ANEMONEFISH

Amphiprion sebae — Subfamily Amphiprioninae

Sebae anemonefish vary markedly in appearance across their wide range. Their body color is often predominantly yellow, but sometimes individuals display just a yellow band on the body. Those that are found in the waters around the Indonesian island of Bali are predominantly black.

DISTRIBUTION: *From the Arabian Peninsula eastward via India and Sri Lanka as far as Sumatra and Java.*

SIZE: *A relatively large species, growing up to 6.5 in (16 cm).*

BEHAVIOR: *Sebae anemonefish are found in coastal areas on reefs down to 82 ft (25 m), usually in association with the sea anemone* Stichodactyla haddoni.

DIET: *Provide a range of fresh and prepared items, including zooplankton and crustacean-based foods.*

AQUARIUM: *Can be kept in a reef aquarium.*

COMPATIBILITY: *It is usually safe to keep the sebae anemonefish in a reef aquarium but, as with other related species, it may prey on small crustaceans known as featherdusters. Offer food two or three times a day in order to avoid the fish foraging too extensively for its own food in these surroundings.*

SCISSOR-TAIL SERGEANT

Abudefduf sexfasciatus — Subfamily Pomacentrinae

These damselfish can be identified by the presence of six black stripes on their bodies. Four of them are vertical, with the first

▼ *SCISSOR-TAIL SERGEANT (ABUDEFDUF SEXFASCIATUS)*

◄ GOLDEN DAMSELFISH (AMBLYGLYPHIDODON AUREUS)

one running down to the pectoral fin, but it is the patterning on the caudal fin that is distinctive—two black stripes run above and below the fork in the tail. Like other commonly kept members of the genus, its underlying body coloration is bluish green suffused with a slight silvery yellow.

DISTRIBUTION: *From the Red Sea down the coast of eastern Africa as far as Mozambique. In the western Pacific, it reaches southern Japan to the north and Lord Howe Island off the east coast of Australia to the south. Ranges across the Pacific to the Tuamotu Archipelago, but is not present in the Hawaiian Islands.*

SIZE: *6.25 in (16 cm).*

BEHAVIOR: *Can be found in relatively shallow water, often quite close to the shore, particularly where soft corals are present.*

DIET: *Will eat prepared foods of all types.*

AQUARIUM: *Normally unlikely to be disruptive in a reef aquarium, provided there are no other fish present, which it may bully.*

COMPATIBILITY: *Not well disposed to its own kind, and it becomes increasingly territorial and aggressive with maturity.*

INDO-PACIFIC SERGEANT

Abudefduf vaigiensis Subfamily Pomacentrinae

This species displays black barring on the sides of its body, contrasting with an underlying greenish blue body color. There are five well-defined black bars, the first of which extends to the pectoral fins, but they do not encircle the body. The caudal fin itself is deeply forked. This fish is similar in appearance to the closely related Atlantic sergeant major (*A. saxatilis*).

DISTRIBUTION: *From the Red Sea and the coast of East Africa eastward to the Line and Tuamotu Islands in the Pacific. Occurs as far north as southern Japan and reaches Australia in the south.*

SIZE: *8 in (20 cm).*

BEHAVIOR: *Lives on reefs. Spawning appears to be a coordinated occurrence, with large numbers congregating at specific localities for the purpose. The event is synchronized with the tides to ensure subsequent widespread dispersal of the young.*

DIET: *Will eat prepared foods of all types.*

AQUARIUM: *Usually safe to keep in a reef aquarium.*

COMPATIBILITY: *Increasingly intolerant of its own kind as it matures. This damselfish also grows to a relatively large size, so companions need to be chosen with care.*

GOLDEN DAMSELFISH

Amblyglyphidodon aureus Subfamily Pomacentrinae

This beautiful damselfish is completely golden yellow in color, apart from some slight bluish purple spotting in the facial area, especially around the eyes. Similar faint markings are evident along the underside of the fish's body as well.

DISTRIBUTION: *From the Andaman Islands off India's east coast south to Rowley Shoals off northwestern Australia, through the Pacific eastward to Tonga. Its northerly range reaches the Ryukyu Islands.*

SIZE: *5 in (13 cm).*

BEHAVIOR: *The golden damselfish frequents areas of the reef that have a relatively strong current. When spawning, it seeks out sea-whip coral and gorgonians, which it uses as breeding sites. The males watch over the eggs until they hatch. These damselfish often congregate to feed on zooplankton wafted on the currents.*

DIET: *Will eat prepared foods of all types.*

AQUARIUM: *This fish is suitable for a reef aquarium and will not usually harm other occupants.*

COMPATIBILITY: *Relatively peaceful by nature, especially if a group can be introduced to the aquarium at the same time.*

GREEN CHROMIS

Chromis caerulea Subfamily Chrominae

An overall pale green coloration with a turquoise-blue suffusion that is especially evident on the fins, combined with a distinctive lyre-tailed appearance, help identify this species.

Distribution: *Ranges over a wide area from the Red Sea and East Africa across the Pacific Ocean as far east as the Line Islands and the Tuamotu Archipelago.*

Size: *2.5 in (6.5 cm).*

Behavior: *Occurs in groups across the entire reef, although most often congregates in areas where live coral predominates.*

Diet: *Will eat prepared foods of all types.*

Aquarium: *Quite suitable for a reef aquarium.*

Compatibility: *Since it is tolerant by nature, the green chromis thrives in groups of six or so. Odd individuals will not thrive on their own. They usually appear somewhat nervous and become reluctant to feed well if kept in isolation.*

DAMSELFISH
(BLUE DAMSELFISH)

Chromis chromis Subfamily Chrominae

The coloration of these damselfish changes significantly as they mature. While the young are bluish violet, adults assume a golden-brown color with dark edging around the individual scales running down the sides of the body.

Synonyms: *C. castanea; C. mediterranea.*

◀ *Green Chromis (Chromis caerulea)*

Distribution: *Occurs on the eastern side of the Atlantic from the coast of Portugal into the Mediterranean and down the west coast of Africa as far as Angola. Commonly sighted around the Canary and Azores Islands.*

Size: *8 in (20 cm).*

Behavior: *Lives in shoals, often close to the shore, especially where the coast is rocky. May often be encountered in areas where seagrass (Posidonia species) is established. Males become territorial during spawning, cleaning rocks where females are enticed to lay their eggs.*

Diet: *Will eat prepared foods of all types.*

Aquarium: *Originating from more temperate waters than most marine aquarium species, this damselfish can thrive at a water temperature of just 70° F (21° C). An SG (specific gravity) reading approaching 1.025 is recommended, which corresponds to that of their natural habitat. Both these factors mean that this fish is better suited to a species-only setup, although they can be kept successfully with invertebrates that have similar requirements.*

Compatibility: *Lively by nature but, like most other* Chromis *damselfish, it is not especially aggressive.*

SCALY CHROMIS

Chromis lepidolepis Subfamily Chrominae

The scaly chromis has rather subdued coloration, being olive-brown overall. However, there are black areas present along the top of the dorsal fin and there is also a black spot evident at both the top and the bottom of the caudal fin. There is also a black area on the anal fin.

▼ *Scaly Chromis (Chromis lepidolepis)*

▲ *BICOLOR CHROMIS (CHROMIS MARGARITIFER)*

DISTRIBUTION: *East Africa across the Indo-Pacific region north to southern Japan and south to Australia, extending to New Caledonia.*

SIZE: *3.5 in (9 cm).*

BEHAVIOR: *A rather shy species, it does not venture far from hiding places on the reef into which it can dart back if it feels threatened.*

DIET: *Will eat prepared foods of all types.*

AQUARIUM: *An ideal occupant for a reef tank, since it will not disturb any invertebrates present.*

COMPATIBILITY: *Relatively tolerant toward its own kind, but introducing a new individual to an existing group is likely to prove problematic.*

BICOLOR CHROMIS

Chromis margaritifer Subfamily Chrominae

This damselfish's dark body contrasts markedly with its white caudal peduncle and caudal fin. The latter extends into two filaments at the top and bottom. The spiny area at the top of the dorsal fin is blue.

DISTRIBUTION: *From the vicinity of Christmas Island off Java's west coast down to the northwest coast of Australia and across the Pacific to the Line Islands and the Tuamotu Archipelago.*

SIZE: *3.5 in (9 cm).*

BEHAVIOR: *Often encountered on the outer area of the reef, sometimes on its own or in a small group.*

DIET: *Will eat prepared foods of all types.*

AQUARIUM: *Ideally suited to a reef aquarium.*

COMPATIBILITY: *Rather intolerant toward other fish, including its own kind, especially as it matures.*

BLUE-GREEN DAMSELFISH

Chromis viridis Subfamily Chrominae

The color of this damselfish varies, with some individuals appearing a lighter shade of blue than others, which may be greener overall. A striking difference between the sexes becomes

apparent when they are ready to spawn—at this time males can be recognized by their predominantly bright yellow color, which becomes blackish toward the tail. An absence of black markings at the base of the pectoral fins distinguishes this species from the very similar black axil chromis (*C. atripectoralis*).

DISTRIBUTION: *Occurs over a wide area from the Red Sea right across the Pacific region to the Tuamotu Archipelago and the Marquesan Islands. It reaches the Ryukyu Islands in the north and New Caledonia in the south.*

SIZE: *3 in (8 cm).*

BEHAVIOR: *This fish is found in association with* Acropora *species corals in relatively calm areas of the reef. When breeding, the male is polygamous, preparing the nest site and encouraging females to spawn there. He watches over the eggs after they have been laid and fans them with movements of his caudal fins to keep them oxygenated. The young hatch about three days later.*

DIET: *Will eat prepared foods of all types.*

AQUARIUM: *Very suitable for a reef aquarium in which* Acropora *coral is present. The young will prove territorial.*

COMPATIBILITY: *It is usually possible to keep these damselfish in small groups without problems even once they are mature, especially in a spacious aquarium. They often tend to prefer the upper area of the aquarium, swimming above the coral just as they do on the reef.*

▲ *BLUE DEVIL (CHRYSIPTERA CYANEA)*

BLUE DEVIL

(SAPPHIRE DEVIL; BLUE DAMSELFISH)

Chrysiptera cyanea Subfamily Pomacentrinae

The blue devil's coloration is very variable across its wide area of distribution. Bright blue predominates, with black streaking often apparent on the sides of the head and tiny yellow spots on the sides of the body. The area around the lips is yellow, as is the caudal peduncle, while the tail fin is orange with a bluish border. However, blue devils from the vicinity of Indonesia can be recognized by their blue tails, offset with a darker border.

SYNONYMS: *C. assimilis; C. gaimardi; C. punctatoperculare; C. uniocellata.*

DISTRIBUTION: *Ranges from northwestern Australia and eastern parts of the Indian Ocean northward via the Philippines and Taiwan to the Ryukyu Islands and across the Pacific via New Guinea to the Caroline Islands (southeast of Hawaii).*

SIZE: *3.5 in (8.5 cm).*

BEHAVIOR: *In terms of its water chemistry requirements, the blue devil is relatively hardy by nature. It is an ideal choice for adding to a marine tank that is maturing. Nevertheless, keep a watch for stress-related ailments such as white spot (ick), which may be less apparent than in related species because of the spotted appearance of this fish.*

DIET: *Will eat prepared foods of all types.*

AQUARIUM: *A good choice for a reef aquarium, but choose companions with care. Males especially can be very aggressive, not just toward their own kind but also toward gentler tank occupants, such as gobies.*

COMPATIBILITY: *Avoid keeping more than one pair in an aquarium. Blue devils can be sexed easily, however, since the caudal fin of females is clear rather than colored. Successful spawning in these surroundings may be feasible.*

AZURE DEMOISELLE

Chrysiptera hemicyanea Subfamily Pomacentrinae

Although displaying the typical blue and yellow coloration that is characteristic of a number of damselfish, the azure demoiselle can be identified easily by the yellow coloration that runs along the entire underside of its body. In addition, the blue coloration on its body has an even hue.

DISTRIBUTION: *Occurs around Indonesia southward to Scott Reef and Rowley Shoals off the northwestern coast of Australia.*

SIZE: *2.75 in (7 cm).*

BEHAVIOR: *Typically lives in small groups in the vicinity of* Acropora *species corals, which provide hiding places from predators.*

DIET: *Will eat prepared foods of all types.*

AQUARIUM: *Suitable for a reef aquarium, since it will not interfere with the other occupants. Usually feeds on plankton. Include coral if possible.*

COMPATIBILITY: *Not especially aggressive, but not particularly tolerant of its own kind, especially once mature. Hiding places will help reduce the risk of bullying.*

▼ AZURE DEMOISELLE (CHRYSIPTERA HEMICYANEA)

▲ GOLD-TAIL DEMOISELLE (CHRYSIPTERA PARASEMA)

GOLD-TAIL DEMOISELLE

Chrysiptera parasema Subfamily Pomacentrinae

As its name suggests, this beautiful species is characterized by its conspicuous golden-yellow caudal peduncle and fin. The rest of its body is blue. The gold-tail demoiselle is one of the damselfish that can often be spawned successfully in the aquarium.

SYNONYM: *Abudefduf parasema.*

DISTRIBUTION: *Restricted to western areas of the Pacific Ocean, from Papua New Guinea and the Solomon Islands northward via the Philippines to the Ryukyu Islands south of Japan.*

SIZE: *2.75 in (7 cm).*

BEHAVIOR: *Tends to be found in calmer areas of coral reefs, most often in the vicinity of patches of* Acropora *species coral.*

DIET: *Will eat prepared foods of all types.*

AQUARIUM: *An ideal choice for a reef aquarium, especially in view of its relatively tolerant nature. Unlikely to harm any invertebrates kept in these surroundings.*

COMPATIBILITY: *It may be possible to keep the gold-tail demoiselle in a group, since it is less aggressive when adult than other damselfish. The aquarium must be spacious and should include a number of retreats, to minimize the risk of any territorial disputes. Unfortunately, this fish is likely to be bullied by other damselfish, and should not therefore be mixed with them.*

WHITE-TAIL DASCYLLUS

(HUMBUG DASCYLLUS)

Dascyllus aruanus Subfamily Chrominae

This is another damselfish with black and white barring on its body. In the case of the white-tail dascyllus, the foremost of the three black stripes extends across the front of the face, while the third one ends at the base of the caudal peduncle, which is white.

Distribution: *Extends over a wide area from the coast of East Africa and the Red Sea across the Pacific, reaching southern Japan and the vicinity of Sydney on Australia's east coast to the south. Its easterly range extends via the Line Islands as far as the Tuamotu Archipelago and the Marquesan Islands.*

Size: *4 in (10 cm).*

Behavior: *Where there are large areas of* Acropora *species coral on the reef, this species is often encountered in groups, although where this habitat is not available, shoals are smaller. The coral provides them with cover if they are threatened.*

Diet: *Will eat prepared foods of all types.*

Aquarium: *This popular damselfish will thrive in reef tanks where suitable coral is included. It will feed on algae there and is unlikely to harm other occupants, although other fish may be persecuted.*

Compatibility: *Keep adults singly or in pairs. White-tail dascyllus have been bred successfully in aquarium surroundings, and are devoted parents. The male constructs a nest in which the female lays her eggs, and then defends the site until the young hatch between three and five days later. They feed on plankton that drifts through the water.*

MARGINATE DASCYLLUS

Dascyllus marginatus Subfamily Chrominae

The marginate dascyllus can be identified by the pale lemon coloration of its body, and its scales have a slight pale bluish tinge on the edges. The fins are generally dark in color, with the exception of the softer rear part of the dorsal fin and the entire caudal fin, which are yellowish white. There is a dark vertical band extending across each eye.

Distribution: *Restricted to the western part of the Indian Ocean, extending from the Red Sea to the Gulf of Oman.*

Size: *2.5 in (6 cm).*

Behavior: *This damselfish is often to be found in the vicinity of various corals, including members of the genera* Stylophora *and* Acropora.

Diet: *Will eat prepared foods of all types.*

Aquarium: *Suitable for a reef aquarium, in which appropriate corals should be included to provide natural retreats.*

▼ *Marginate Dascyllus (Dascyllus marginatus)*

COMPATIBILITY: *Displays typical damselfish behavior, with the fish becoming increasingly intolerant of each other as they mature. They are relatively lively by nature, swimming throughout the aquarium, which should have a minimum volume of 18.5 gallons (70 l) for an adult.*

RETICULATE DASCYLLUS

Dascyllus reticulatus Subfamily Chrominae

These damselfish are so called because of the reticulated patterning on their bodies, with the dark edges highlighting their scales. The front part of the body is brownish, but the rest of the body is lemon-white in color—apart from the caudal peduncle, which is bluish. As with other members of the group, it has a stub-nosed appearance.

SYNONYM: *D. reticulata.*

DISTRIBUTION: *From the eastern Indian Ocean, extending from the Cocos (Keeling) Islands south to Lord Howe Island on Australia's east coast and north to southern Japan. Occurs across the Pacific as far as Samoa and the Line Islands.*

SIZE: *3.5 in (9 cm).*

BEHAVIOR: *Often found in close proximity to the heads of branching coral, darting in for protection if danger threatens. Do not be surprised if the young fish hide among rocks in the aquarium—this behavior is quite usual.*

▲ *RETICULATE DASCYLLUS (DASCYLLUS RETICULATUS)*

DIET: *Will eat prepared foods of all types.*

AQUARIUM: *Settles well in these surroundings, feeding readily, and will not be disruptive in a reef tank, provided that there are no other fish for it to bully.*

COMPATIBILITY: *Do not purchase a young group of these damselfish, because they will inevitably become highly territorial and aggressive toward each other as they mature. At that stage they must then be housed singly or in pairs.*

DOMINO DAMSELFISH
(THREE-SPOT DAMSELFISH)

Dascyllus trimaculatus Subfamily Chrominae

There can be a considerable variation in appearance among individuals of this species, depending on their age and area of origin. Adults are predominantly black in color, with a small white blotch present on each side of the upper part of the body below the dorsal fin. This white area tends to be larger in young fish, and they also display another white spot on the forehead, which explains why they are known as domino damselfish. Adaptable by nature, this species is often recommended for starting the conditioning process in a new saltwater tank.

SYNONYMS: *D. trimaculatum; D. unicolor.*

Distribution: *Extends widely in the Indo-Pacific region from the Red Sea and East Africa north to the southerly waters off Japan and south to the vicinity of Sydney on Australia's east coast. Its range extends to the Pitcairn Islands and reaches the Line Islands, but not Hawaii.*

Size: *5.5 in (14 cm).*

Behavior: *Often occurs in groups, with younger individuals seeking protection among the tentacles of sea anemones and the spines of sea urchins. When adults approach spawning condition, they turn reddish gray. The eggs are deposited on rockwork or coral and are guarded by one of the adult fish—probably the male—until they hatch.*

Diet: *Will eat prepared foods of all types.*

Aquarium: *Suitable for a reef aquarium; will help check algal growth.*

Compatibility: *May be kept in groups when young, but they become more territorial as they grow older, and bullied individuals appear paler. They will also harass less assertive companions. However, they will usually agree well with more dominant species such as angelfish or pufferfish if they are housed together in a fish-only setup.*

BLACK DAMSELFISH

(BOW-TIE DAMSELFISH)

Neoglyphidodon melas Subfamily Pomacentrinae

Although all immature damselfish in the genus *Neoglyphidodon* are brightly colored, they are unfortunately soon transformed into very drab adults. The young of this particular species are especially striking in their appearance. They have brilliant yellow upperparts and pale blue on the sides of the body, and there are black areas restricted to the fins. Once they reach about 2 inches (5 cm) long, however, they start to change color, ultimately becoming completely bluish black.

Distribution: *Found over a very wide area from the Red Sea and East Africa through the Indian Ocean to the Pacific, extending eastward as far as the Solomon islands and Vanuatu. Its northward distribution extends via the Philippines and Taiwan to the Ryukyu Islands, while to the south it extends as far as northern Australia.*

Size: *7 in (18 cm).*

Behavior: *The black dameselfish is another relatively solitary species, encountered on its own or in pairs. Associates quite commonly with* Tridacna *clams on the reef.*

Diet: *Will eat prepared foods of all types.*

Aquarium: *Can be a problem in a reef aquarium, because it will eat soft coral. The young will benefit from the provision of* Acropora *corals to give them shelter.*

Compatibility: *Difficult to accommodate with other fish when older, because of its solitary nature. The size of this damselfish means that it can prove a formidable bully.*

▼ *Black Damselfish* (Neoglyphidodon melas)

▲ BEHN'S DAMSELFISH (NEOGLYPHIDODON NIGRORIS)

BEHN'S DAMSELFISH

(BLACK AND GOLD CHROMIS)

Neoglyphidodon nigroris Subfamily Pomacentrinae

It is only the young of this species that display significant yellow coloration, and the black areas at this stage are restricted to two horizontal stripes running along their bodies. One extends down the midline, while the other commences above the eye. The caudal fin is particularly long and forked, and the dorsal and anal fins are also attenuated. In the case of the westerly population, which extends as far as Bali, adult fish turn black in color, while those from farther east become brownish and the yellow coloration is restricted to the area around the tail.

SYNONYM: *Paraglyphidodon behnii.*

DISTRIBUTION: *Extends from the Andaman Islands off India's east coast across Indonesia and the Philippines, north to Taiwan and the Ryukyu Islands, and via New Guinea to the Solomon Islands and Vanuatu.*

SIZE: *5 in (13 cm).*

BEHAVIOR: *A rather solitary species, usually occurring in areas of active coral growth on the reef.*

DIET: *Will eat prepared foods of all types.*

AQUARIUM: *Suitable for a reef tank, but not recommended as a companion for other fish.*

COMPATIBILITY: *Short-tempered and aggressive as adults, these fish are among the least tolerant damselfish. They also grow relatively large*

CAERULEAN DAMSEL

Pomacentrus caeruleus Subfamily Pomacentrinae

As its name suggests, the caerulean damsel is generally a stunning shade of deep blue, especially when young. It often has black markings on the face as well. There are regional differences in coloration, and some individuals tend to develop yellow underparts as they become older, the yellow areas extending across the caudal peduncle, which can make identification difficult. However, the caerulean damselfish tends to be a deeper shade of blue than other damselfish with similar coloration, such as the gold-belly damselfish (*P. auriventris*), and this can serve as a point of distinction.

SYNONYM: *P. pulcherrimus.*

DISTRIBUTION: *From the shores of East Africa across the Indian Ocean to the Maldives off the southwestern tip of India.*

◀ LEMON DAMSEL (POMACENTRUS MOLUCCENSIS)

SIZE: *4 in (10cm).*

BEHAVIOR: *A lively and rather territorial species.*

DIET: *Will eat prepared foods of all types.*

AQUARIUM: *Another very popular and widely kept damselfish. It is highly valued as the first occupant of a new tank, since it can help create suitable conditions for species that are more sensitive in terms of their water chemistry requirements. Also suitable for inclusion in a reef aquarium.*

COMPATIBILITY: *Like other damselfish, it is a little quarrelsome, especially once mature.*

LEMON DAMSEL
(MOLUCCA DAMSEL)

Pomacentrus moluccensis Subfamily Pomacentrinae

As its name suggests, this damselfish is lemon-yellow in color, although it sometimes displays faint blue speckling on its fins and possibly the forehead as well, depending on its area of origin.

SYNONYMS: *P. popei; P. sufflavus.*

DISTRIBUTION: *Found around the Andaman Islands off the eastern coast of India, extending southward to Rowley Shoals off the northwestern coast of Australia and across the Pacific to Tonga, reaching as far north as the Ryukyu Islands near Japan.*

SIZE: *3.5 in (9 cm).*

BEHAVIOR: *Often occurs on reefs in among branching corals that offer some protection against predators.*

DIET: *Will eat prepared foods of all types.*

AQUARIUM: *An ideal occupant for a reef tank, helping keep unwanted algal growth under control. Presents no danger to other occupants.*

COMPATIBILITY: *Unfortunately, these damselfish are not tolerant of the company of their own kind in aquarium surroundings. A range of retreats should help if you are trying to keep several together. Apart from problems of aggression, which are especially common among adults, the lemon damsel is an easy fish to maintain.*

PHILIPPINE DAMSEL

Pomacentrus philippinus Subfamily Pomacentrinae

In spite of its name, this species extends over a wide area, rather than occurring only in the vicinity of the Philippines. Adult Philippine damsels are a fairly dull shade of blue, with the rear of the scales being darker in color, bordering on black. Immature individuals are more colorful, however, thanks to the yellow coloration that predominates on their fins. Their body is a richer shade of blue.

DISTRIBUTION: *Ranges over a wide area from the Maldives in the Indian Ocean south to Rowley Shoals off the northwest coast of Australia. It extends northward to the Ryukyu Islands and eastward through the Pacific as far as New Caledonia and Tonga.*

SIZE: *4 in (10 cm).*

BEHAVIOR: *Tends to be sighted along vertical areas of reef, where it often seeks to hide under overhangs.*

DIET: *Will eat prepared foods of all types.*

AQUARIUM: *Suitable for inclusion in a reef aquarium.*

COMPATIBILITY: *May be slightly more tolerant of the company of its own kind than many damselfish, but trying to introduce a newcomer alongside an established individual can lead to conflict and bullying. The best hope of success is to obtain two juveniles and place them together in the aquarium at the same time.*

▼ PHILIPPINE DAMSEL (POMACENTRUS PHILIPPINUS)

▲ OCELLATE DAMSELFISH (POMACENTRUS VAIULI)

OCELLATE DAMSELFISH
(PRINCESS DAMSELFISH)

Pomacentrus vaiuli Subfamily Pomacentrinae

The prominent blue-edged black spot above the caudal peduncle, on the dorsal fin, helps explain the common name of this species—it resembles an eye. This may confuse predators and cause them to strike at this point rather than at the head of the fish. The background coloration is variable, but the upper part of the body and the head tend to be yellowish or tan. The rest of the body is bluish, sometimes with a decidedly purplish hue. The base of the pectoral fins is typically yellow.

DISTRIBUTION: *Occurs in the Pacific region from the Moluccas (islands of Indonesia), extending eastward to Samoa and south as far as Rowley Shoals off Australia's northwestern coast. The Izu Islands mark the most northerly part of its range.*

SIZE: *4 in (10 cm).*

BEHAVIOR: *Can be found in the relative shallows of a reef down to depths of more than 130 ft (40 m).*

DIET: *Will eat prepared foods of all types.*

AQUARIUM: *Swims throughout the tank and appreciates the presence of retreats in which it can hide. In an invertebrate tank it favors coral for this purpose. Will browse on filamentous algae.*

COMPATIBILITY: *Ocellate damselfish are not tolerant of others of their own kind, including other species of damselfish.*

BEAUGREGORY

Stegastes leucostictus Subfamily Pomacentrinae

The upper part of the body of this damselfish is bluish with some blackish spotting. There is also a bluish hue evident in the caudal fin. The lower half of the body is a contrasting shade of yellow.

DISTRIBUTION: *This damselfish is found in Atlantic rather than Pacific waters, extending from the coast around southern Florida east to Bermuda and southward from the Gulf of Mexico to Brazil.*

SIZE: *4 in (10 cm).*

BEHAVIOR: *The beaugregory often lives close inshore, and it can sometimes even be caught in the shallows. It can be found in a range of habitats, from mangrove areas to sandy beaches and reefs.*

DIET: *Will eat prepared foods of all types.*

AQUARIUM: *Not entirely to be trusted in a reef aquarium, since it may eat polychaetes and gastropods, but will otherwise settle well in these surroundings.*

COMPATIBILITY: *Usually displays typical damselfish behavior, so that attempting to keep a group of fish together is unlikely to be feasible once they start to mature.*

DRAGONETS AND MANDARINFISH FAMILY CALLIONYMIDAE

The common names of the fish in the small family Callionymidae are dragonets and mandarinfish. These bottom-dwelling fish are found in all the oceans of the world, both temperate and tropical. They often display what can be described as psychedelic patterning on their bodies. It may serve as a warning of danger to potential predators, because the mucus that covers their bodies is thought to be toxic. Although they appear extremely vivid against a drab background, in their natural habitat of mixed algae, these fish are extremely well camouflaged.

Mandarinfish breed in the water column. A pair may mate every evening for many months. The female produces eggs that float in layers of plankton. They hatch out and, once developed, the young swim back down to the seabed.

FINGER DRAGONET

Dactylopus dactylopus

This species gets its name from the fingerlike structure at the front of each pelvic fin, created by the fusion of the pelvic ray and the associated spine. It also has a tall dorsal fin, the rays of which are serrated at the tips. The anal fins display blue spotted markings, while the body color is a variable shade of brown with white spots and blotches. Sexing is straightforward, since males have filaments on the tips of the first dorsal fin.

DISTRIBUTION: *Extends from northwestern parts of Australia via Indonesia northward to southern Japan.*

SIZE: *12 in (30 cm).*

BEHAVIOR: *Occurs in muddy or sandy habitats and is sometimes found in estuaries as well. Its mottled coloration helps this slow-swimming fish blend*

▼ *FINGER DRAGONET (DACTYLOPUS DACTYLOPUS)*

▲ MANDARINFISH (SYNCHIROPUS SPLENDIDUS)

in against the background, concealing its presence. It is not confined to shallow areas, however, and has been encountered down to depths of 180 ft (55 m). Buries itself if threatened and also at night.

DIET: *Can be difficult to feed if not kept in an aquarium with live sand, which will provide edible items, such as worms, alongside living rock. It is important to try to place food within easy reach of the fish without disturbing it. Clean tweezers can be used to drop pieces of finely chopped shrimps and similar foods within its reach, while a small pipette is useful for releasing brine shrimps.*

AQUARIUM: *Ideal for a reef aquarium, provided there is a good expanse of sand to form the substrate.*

COMPATIBILITY: *Keep in individual pairs, since males are likely to fight. Choose inoffensive companions and avoid fish, such as puffers, that are likely to nip their elaborate fins.*

MANDARINFISH
(MANDARIN DRAGONET)

Synchiropus splendidus

It is virtually impossible to confuse the highly patterned mandarinfish with any other fish that is commonly kept in a reef aquarium. Its scaleless body is a blue or green color transversed by orange, blue, or green wavy lines. (The colors resemble the bright robes of a Chinese mandarin, which explains the fish's common name.) The tail is bright red with blue edging. It has a short cylindrical body, while its head is roughly triangular in shape. Its fins, especially the pectorals, are large. The male mandarinfish is usually larger than the female and has a large pointed dorsal fin that is not always evident.

SYNONYMS: *Callionymus splendidus; Pterosynchiropus splendidus; Neosynchiropus splendidus.*

DISTRIBUTION: *Extends through the western Pacific, from the Ryukyu Islands near Japan, south to Australia.*

SIZE: *2.5 in (6 cm).*

BEHAVIOR: *Inhabits shallow protected lagoons and inshore reefs, where coral and rubble predominate. It is constantly browsing for small invertebrates among the algae covering dead coral. Mandarinfish are usually found in small groups and are generally peaceful and shy by nature.*

DIET: *This fish is omnivorous. It will feed on small livefoods, preferably crustaceans—such as copepods, amphipods, and isopods—as well as some frozen foods. Newly hatched live brine shrimps are ideal to start with. Ensure that plenty of algae are available.*

AQUARIUM: *Well suited to an invertebrate tank. It is important to have plenty of rocks, among which it can hide, and a sandy substrate.*

COMPATIBILITY: *This mandarinfish is intolerant of its own species. Males are likely to fight, so it is best to keep a single male and female together. If conditions are suitable, they may mate regularly. May be kept alongside other similarly sized fish, provided they are not aggressive.*

Wrasses Family Labridae

The family Labridae is large and diverse. Generally only the smaller and more colorful species are kept in aquaria. In many cases the appearance of young fish differs from that of adults. These fish can often undergo a sex change, with the dominant female becoming a male when there is no male present. Wrasses are quite active by nature and they rely on the propulsive power of their pectoral fins rather than on their caudal fin when swimming. Some species will burrow into the substrate—this is quite normal behavior. Just as they would do on the reef, they may also act as cleaner fish, seeking to remove parasites from other types of fish that share their quarters.

RED-TAIL TAMARIN WRASSE

(PSYCHEDELIC WRASSE)

Anampses chrysocephalus

Females of this species can be identified by a series of fine white spots on the head and along the back, which become larger on the sides of the body. There is a broad white stripe at the base of the caudal fin, and the remainder of the tail is dark red, apart from its faint blue edging. In contrast, with a bright yellowish orange head transversed with irregular blue streaks and spots, the male lives up to the species' alternative common name of psychedelic wrasse. The remainder of the male's body is dusky red with blue markings on the scales, while the tail is a dull blue and darker blue at the tip.

▼ *Red-tail Tamarin Wrasse (Anampses chrysocephalus)*

▲ LYRETAIL HOGFISH (BODIANUS ANTHIOIDES)

DISTRIBUTION: *Restricted to the areas around the Hawaiian and Midway Islands in the Pacific Ocean.*

SIZE: *6.75 in (17 cm).*

BEHAVIOR: *It is usual to see a single male in the company of a number of females over the reef. They are lively fish and cannot be kept in an open tank, or they may jump out.*

DIET: *Difficult to feed, particularly at first, but most likely to respond to being offered livefood such as brine shrimps. May also forage for food among living rock, which should be included in the tank at first. Can be gradually weaned onto inert, finely chopped meaty foods, but needs to be offered small amounts several times daily.*

AQUARIUM: *Can be problematic in a reef tank, since it will eat various invertebrates there.*

COMPATIBILITY: *Females agree well together. In the absence of a male, a female may transform into a male. Do not mix with aggressive companions.*

LYRETAIL HOGFISH

Bodianus anthioides

The lyre-tailed appearance of these wrasses is emphasized by the darker markings that may be present running down the caudal fin. Their appearance is highly individual, however, with the exception of an orange area that extends from just in front of the eye back to the level of the start of the dorsal fin, which has a black spot at the front. The remainder of the body is whitish with variable spotting extending down to the tail.

DISTRIBUTION: *Extends from the Red Sea southward down the coast to South Africa and across the Indian Ocean. Ranges north to southern Japan and south to New Caledonia and the Austral Islands. Its easterly range reaches the Line Islands and the Tuamotu Archipelago.*

SIZE: *9.5 in (24 cm).*

BEHAVIOR: *Occurs (usually singly) over areas of the reef where sessile invertebrates are prevalent. It may blow away the sand with its mouth to find prey such as worms hidden beneath the surface. Young fish often act as cleaner fish on the reef, removing parasites from other fish, and behave in a similar way in aquarium surroundings.*

DIET: *Requires meaty foods, especially shellfish—a mainstay of its diet in the wild. Food may have to be chopped up into smaller pieces.*

AQUARIUM: *Cannot be trusted alongside crustaceans in a reef tank.*

COMPATIBILITY: *Active but not usually aggressive, although it may prey on significantly smaller companions, so select tankmates carefully.*

AXIL-SPOT HOGFISH

Bodianus axillaris

Sexing these wrasses is straightforward, because females are predominantly black in color with large white spots on their body. Males have a dark area on the front part of the body, with a silvery tinge on the sides. There is a black spot at the base of the pectoral fins, and others on the dorsal and anal fins. The iris is vivid red in color. Young fish are similar to the females.

DISTRIBUTION: *Extends down the eastern coast of Africa from the Red Sea to South Africa and across the Indian Ocean northward as far as Japan. Also ranges through the Pacific to the Marshall and Marquesan Islands, reaching the Tuamotu Archipelago.*

SIZE: *8 in (20 cm).*

BEHAVIOR: *Juveniles of this species act as cleaners, emerging from their hiding places under ledges to pluck off parasites from other fish on the reef. They are solitary by nature.*

▼ AXIL-SPOT HOGFISH (BODIANUS AXILLARIS)

DIET: *Requires meaty food, such as shrimps, which should be chopped up as necessary.*

AQUARIUM: *Not to be trusted in a reef tank containing crustaceans and mollusks, which form its natural prey.*

COMPATIBILITY: *Can prove to be something of a bully, so choose companions carefully. Avoid mixing with smaller fish. Should be kept singly rather than in a group of its own kind.*

DIANA'S HOGFISH

Bodianus diana

Orange coloration, paler on the flanks, tends to predominate in this fish. Like other members of the genus, it has a fairly pronounced snout, which explains why they are commonly known as hogfish. A series of white spots runs alongside the base of the dorsal fin and there are variable black markings on the fins, too.

DISTRIBUTION: *Ranges from the Red Sea southward as far as South Africa and across the Indian Ocean northward to Japan. Extends through the Pacific as far as the Marshall Islands and Samoa.*

SIZE: *10 in (25 cm).*

BEHAVIOR: *Solitary by nature, even when young. This hogfish hides away on the reef and is quite shy, although it frequents areas on the reef where it will groom other fish, removing their parasites. It can be very aggressive, however, particularly as it grows older.*

DIET: *Prefers meaty foods, and usually feeds readily on a wide range of items. Defrost frozen foods thoroughly before offering them, and chop them up into small pieces as necessary.*

AQUARIUM: *Not recommended for a reef tank because it will feed on various invertebrates that are likely to be found there, including crustaceans and mollusks. Will also eat small fish.*

▲ *DIANA'S HOGFISH (BODIANUS DIANA)*

COMPATIBILITY: *Antisocial by nature and not easily intimidated by bigger companions, so it can be difficult to incorporate this fish successfully into a communal tank.*

SPOT-FIN HOGFISH
(CUBAN HOGFISH)

Bodianus pulchellus

A very attractively colored member of its genus, the spot-fin hogfish is so called because of the black spot evident at the front of the dorsal fin. The front part of the body is reddish. There are traces of two black stripes extending back from the eyes and a broader yellow stripe running along the lower area of the body, below the eyes. The rear of the body, including the caudal fin, is bright yellow. There is also what is sometimes described as an intermediate phase, in which the body coloring can appear blackish rather than red. Juveniles are entirely yellow, apart from the black area on their dorsal fin.

DISTRIBUTION: *Found in the western part of the Atlantic, ranging as far north as the coast of South Carolina. Extends east to Bermuda through the Gulf of Mexico and the Caribbean via the Antilles to the northern coast of South America.*

SIZE: *9 in (23 cm).*

BEHAVIOR: *A very lively fish that tends to occur in water above a depth of 80 ft (25 m). It has a bold nature and is not unduly disturbed by the presence of divers. It tends to be more common in the waters of Florida than in the Caribbean. Juveniles may form shoals with blue-head wrasses* (Thalassoma bifasciatum) *and, like others of their genus, they may act as cleaner fish, sometimes removing parasites from much larger species.*

◀ *SPOT-FIN HOGFISH (BODIANUS PULCHELLUS)*

DIET: *Requires meaty foods, chopped as necessary. Usually proves to have a healthy appetite, in common with related species, and can be persuaded to take flake food in due course.*

AQUARIUM: *Avoid mixing with potential prey species such as crustaceans and mollusks in a reef tank.*

COMPATIBILITY: *Reasonably tolerant when young, but should be housed only with other fairly assertive species. Remain alert for any signs of bullying, particularly as the fish grow older.*

SPANISH HOGFISH

Bodianus rufus

The basic underlying coloration of this hogfish is yellow, with much of the area from the level of the eye extending along the upper part of the back being purplish, often with a reddish tinge. In some individuals, however, this coloration may be more evident and can even reach the underparts. The pelvic and anal fins as well as the front part of the dorsal fin may be similarly colored. Juvenile fish display more extensive purple areas on their bodies than the adults.

DISTRIBUTION: *From the southern coast of Florida east to Bermuda, extending via the Bahamas and the Caribbean Sea to the Gulf of Mexico down to southern Brazil in South America.*

▼ *SPANISH HOGFISH (BODIANUS RUFUS)*

SIZE: *15.75 cm (40 in).*

BEHAVIOR: *Proves to be a relatively conspicuous reef inhabitant. Juveniles will seek fish to clean in these surroundings. This hogfish will hunt a variety of invertebrate prey and will even attack sea urchins. Like some other marine fish, they are hermaphrodite, with females being able to change into fertile males. This particular hogfish has been known to hybridize with the spot-fin hogfish* (B. pulchellus) *in the wild.*

DIET: *Predatory by nature, so requires a diet based on meaty foods such as shrimps, which may need to be chopped up. May be persuaded to take flake food, too. Always try to offer a varied diet.*

AQUARIUM: *Unsuitable for a reef aquarium, since it preys on various occupants, ranging from brittlestars to clams..*

COMPATIBILITY: *Although they are initially shy in the aquarium, these hogfish can soon develop into bullies. They should be housed only with equally assertive species in a spacious aquarium that provides them with adequate swimming space.*

HARLEQUIN TUSKFISH

Choerodon fasciatus

The harlequin tuskfish is characterized by unusual vertical-striped patterning. Its coppery-colored banding is edged with violet and black, and a particularly broad stripe encircles the pectoral fins. The intervening areas are a whitish shade, and the rear of the body is blackish, adjoining a broad white area at the base of the caudal fin. Harlequin tuskfish vary somewhat in their depth of coloration depending on their distribution. Specimens from Australia's Great Barrier Reef are regarded as the most colorful.

DISTRIBUTION: *Occurs in the western Pacific, ranging from Taiwan north to the Ryukyu Islands near Japan. There is another, apparently separate, population lying to the south, extending along the Great Barrier Reef from the coast of Queensland southward and eastward to New Caledonia.*

SIZE: *12 in (30 cm).*

BEHAVIOR: *Solitary by nature, the harlequin tuskfish lives on the seaward area of the reef, seeking out invertebrates that form the basis of its diet. It is territorial by nature and, having powerful teeth in its jaws, it can prove to be a fearsome opponent.*

DIET: *Meaty foods of various types are required. Chop them up if necessary. Offer a range of foods to ensure a varied diet.*

AQUARIUM: *Certainly not to be trusted in the confines of a reef tank, since it will prey on many of the invertebrates present, from echinoderms to crustaceans. Also likely to eat smaller fish.*

COMPATIBILITY: *Must be housed singly but can be mixed with other fish of similar size that are not likely to be intimidated.*

CLOWN CORIS
(TWIN-SPOT CORIS)

Coris aygula

This potentially large wrasse undergoes a dramatic change in appearance as it grows. Juveniles, which are most commonly seen, display the distinctive black ocelli, or eyespots, on the dorsal fin. The spots are highlighted with a white surround. The body itself is whitish overall with brown spots over the head and with fine black cross-hatching becoming more prominent toward the rear. There are usually two yellowish areas on the body below the ocelli. The fins are a dark greenish color with black edging and spotting. Adults differ markedly, becoming dark green in color. At this stage it is possible to sex the fish, because the males have much longer pelvic fins and a truncated rather than a rounded caudal fin. They also develop a more bulbous head shape.

▼ CLOWN CORIS (*CORIS AYGULA*)

DISTRIBUTION: *Occurs from the Red Sea down the coast of East Africa and through the Indian Ocean. Extends northward as far as southern Japan and reaches southward to Lord Howe Island off the east coast of Australia. Its range through the Pacific extends to the Line and Ducie Islands.*

SIZE: *47.25 in (120 cm).*

BEHAVIOR: *Young juveniles are often encountered in shallow pools on the reef. They are typically found in sandy areas of the reef, which allows them to burrow into the substrate, helping to conceal their presence.*

DIET: *Needs a meaty diet with pieces of crustacean and similar foods, cut up as necessary.*

AQUARIUM: *The clown coris is not suitable for a reef tank, since its predatory nature means that it is liable to eat many of the invertebrate occupants of the tank.*

COMPATIBILITY: *Should be housed only with assertive companions of similar size and will need a very spacious aquarium as it grows larger. Decor must be carefully supported, since this fish is adept at digging and can move relatively large pieces of rock.*

QUEEN CORIS
(FORMOSAN CORIS)

Coris formosa

Young fish of this species are orange in color with a variable series of white blotches extending over their bodies. The blotches have black edging, and the most prominent of them extends down from the front part of the dorsal fin. As these fish mature, their coloration changes dramatically . Adult queen coris have a yellowish head and they develop a light blue diagonal stripe behind each eye. The body is a greenish blue color with a random patterning of black spots superimposed on it. The edge of the dorsal fin is reddish, as is the base of the tail, which ends in a broad bluish margin.

SYNONYM: *C. frerei.*

DISTRIBUTION: *Restricted to the Indian Ocean, from the southern part of the Red Sea southward down the African coast to the South African province of Natal and east as far as Sri Lanka off the east coast of India.*

SIZE: *23.75 in (60 cm).*

BEHAVIOR: *Adults are found in deeper water than the young of this species, which not infrequently occur in tidal pools. Here they will burrow into the sandy substrate to conceal themselves. Large adults use their curved*

▲ QUEEN CORIS (CORIS FORMOSA)

elongated snout to search for food, flipping over rocks that may be hiding places for invertebrates such as crabs and other crustaceans. They will also feed on sea urchins and various mollusks.

DIET: *Meaty foods of various types, especially crustaceans. May also be persuaded to eat some flake food.*

AQUARIUM: *Not suitable for a reef tank, particularly as it grows older, since it is likely to prey on many of the invertebrates there.*

COMPATIBILITY: *Less aggressive than various other members of the genus, but avoid mixing this species with its own kind. Generally will not interfere, however, with other fish of a similar size.*

YELLOW-TAIL CORIS

(RED CORIS)

Coris gaimard

The young of this species are especially striking. They display a bright orange body color with a series of white blotches of variable size extending down the back onto the sides of the body. The blotches are highlighted with black borders, although these bands are less pronounced than in the young of other members of the genus. Adults are completely different in appearance to juveniles, with sky-blue bands on the sides of the face, which is largely pale orange in color. The rest of the body is of a dark greenish shade with a variable light blue spotted patterning that often becomes more intense toward the tail. The tail is yellow in females. The narrow dorsal and ventral fins have a series of horizontal orange and blue stripes.

DISTRIBUTION: *Ranges from Christmas and Cocos (Keeling) Islands in the eastern part of the Indian Ocean northward as far as southern Japan and southward to Australia's Great Barrier Reef. Extends through the Pacific to the Hawaiian, Society, and Tuamotu Islands.*

SIZE: *15.75 in (40 cm).*

BEHAVIOR: *This active reef fish will hunt a variety of invertebrate prey. It is often seen among rubble in which crabs and similar creatures may be lurking, as well as in more open sandy areas. It is here that the yellow-tail coris burrows at night, concealing its presence. The aquarium should therefore contain about 4 inches (10 cm) of sand into which young fish can burrow for this purpose.*

DIET: *A range of meaty foods. Young fish in particular have very healthy appetites and need frequent feeding. It may be better to choose slightly*

larger individuals at the outset, since they often seem to settle better in aquarium surroundings.

Aquarium: *Cannot be trusted in a reef tank, because many of the invertebrates there constitute its natural prey.*

Compatibility: *The yellow-tail coris is not especially aggressive toward other aquarium occupants, although it may bully smaller fish. This wrasse, as with others of the genus, can grow to a large size, however, and needs suitably spacious accommodation.*

BIRD-MOUTH WRASSE

Gomphosus caeruleus

The body coloration of mature males of this species is a dark shade of bluish black. The central area of the caudal fin is greenish blue. The dorsal fin is also bluish and there is a white edge to the pectoral fins. Females and young males, however, are brown in color. The characteristic snout of the bird-mouth wrasse is significantly longer in adults than in juveniles, which enables them to be distinguished easily.

Distribution: *Ranges from East Africa south to the coast of Natal in South Africa. Also found in an easterly direction across the Indian Ocean as far as the Andaman Sea, which lies to the east of India.*

Size: *12 in (30 cm).*

Behavior: *These lively fish are quite prominent over the reef. The young serve as cleaner fish, seeking to remove parasites from other reef fish. Adults use their long mouthparts to probe for small prey in among rocks. They also feed on worms.*

Diet: *Requires a meaty diet that should include small items such as vitamin-enriched brine shrimps and mysid shrimps.*

▲ *Bird-mouth Wrasse* (Gomphosus caeruleus)

Aquarium: *Can help rid a reef tank of unwanted pests, such as fireworms, that may have been introduced with living rock. Unfortunately, it will also prey on crustaceans and other desirable reef inhabitants.*

Compatibility: *Do not mix males of this species or house this lively fish with sensitive or significantly smaller companions, on which it may prey. Because it can jump well, the bird-mouth wrasse needs to be housed in a covered aquarium .*

BIRD WRASSE

Gomphosus varius

The characteristic elongated beak of this species, combined with its body shape, make it reminiscent of a miniature dolphin in appearance. Males are greenish in color—often appearing bluer on their underparts—whereas females are brown. The young of both sexes resemble females but they lack the pronounced snout seen in adults.

▼ *Bird Wrasse* (Gomphosus varius)

DISTRIBUTION: *Extends from the Cocos (Keeling) Islands in the Indian Ocean south to Rowley Shoals off the northwestern coast of Australia. Ranges eastward to southern Japan and south in the Pacific to Lord Howe and Rapa Islands. Its easterly distribution extends as far as the Hawaiian and Marquesan Islands and the Tuamotu Archipelago.*

SIZE: *12 in (30 cm).*

BEHAVIOR: *Hunts for small invertebrates, probing for prey on the reef with its snout. May also catch small fish on occasion, battering them into pieces that can be swallowed easily. This wrasse is very active by nature, but it retreats at night into caves rather than burrowing into the substrate.*

DIET: *Meaty foods that should include vitamin-enriched brine shrimps and mysid shrimps.*

AQUARIUM: *Not trustworthy in surroundings in which there are smaller fish. May also prey on various invertebrates, including featherstars.*

COMPATIBILITY: *Avoid mixing with smaller companions and keep males apart from each other. Can otherwise be accommodated in a tank with nonaggressive species.*

GOLDEN WRASSE

(CANARY WRASSE)

Halichoeres chrysus

Individuals of this species range in color from a canary yellow to a golden shade, and the coloration is often more intense on the head than elsewhere on the body. Pale yellow-green bands may also be evident on the head. Males display a prominent black spot edged with white near the front of the dorsal fin. Females have a second spot, which has yellow rather than white edging, while young individuals may have a third black spot toward the rear of the dorsal fin.

DISTRIBUTION: *Found in the eastern Indian Ocean, extending from Christmas Island, which lies to the south of Sumatra, southward to Rowley Shoals off Australia's northwestern coast. From here it extends around the coast to the southern part of the Great Barrier Reef. It ranges northward as far as southern Japan and through the Pacific via the Solomon Islands to Tonga.*

SIZE: *4.75 in (12 cm).*

BEHAVIOR: *Occurs around the edge of the reef, often in sandy areas. This fish will burrow into the sand at night or if it is threatened. Golden wrasses serve as cleaner fish for other fish on the reef and will often display this behavior in aquarium surroundings as well.*

DIET: *Feeds on invertebrates such as brine and mysid shrimps. Provide other meaty foods as well. May even take flake food.*

AQUARIUM: *Not trustworthy in a reef tank, since it is likely to prey on a variety of the occupants, such as shrimps and fanworms, although it will also consume troublesome invertebrates, such as flatworms. There may be a case for transferring these fish to an aquarium to control such pests, in which case it is advisable to move vulnerable species elsewhere on a temporary basis. Requires a covered aquarium, since it is able to jump well.*

▲ *CHECKERBOARD WRASSE* (HALICHOERES HORTULANUS)

COMPATIBILITY: *Can be kept successfully in groups, creating an attractive display. May also be mixed with relatively small placid species such as firefish (family Microdesmidae).*

CHECKERBOARD WRASSE

Halichoeres hortulanus

The young of this species are black and white with a yellow hue in the vicinity of the caudal peduncle. Adults are much more colorful, with a dark checkerboard patterning on the side of their bodies behind the pectoral fins. There are a number of prominent pinkish stripes on the sides of the head, offset against a bluish green background color. The rear of the body, including the dorsal and caudal fins, is yellow and there are often one or two yellow saddlelike markings on the back. A black spot may also be evident on the caudal peduncle, although the precise patterning of individual checkerboard wrasses can vary depending on their area of origin.

DISTRIBUTION: *Ranges widely from the Red Sea southward as far as Sodwana Bay off the South African coast and extending through the Indian Ocean. Its northerly range reaches as far as southern Japan, and southward it reaches Australia's Great Barrier Reef. Found through the Pacific as far as the vicinity of the Line and Marquesan Islands as well as around the Tuamotu Archipelago.*

SIZE: *10.75 in (27 cm).*

BEHAVIOR: *Juveniles frequent surge channels on the reef, while adults range more widely over sandy areas. They have powerful mouthparts and they seek out well-protected prey, such as sea urchins and mollusks.*

DIET: *Offer meaty foods, allowing frozen items to thaw properly first and cutting them into suitable-sized pieces as necessary.*

AQUARIUM: *Will prey on various occupants of the reef tank.*

COMPATIBILITY: *Its size means that this wrasse needs suitably spacious aquarium surroundings, and it must not be mixed with smaller species that could become its prey.*

BLACK-EYE THICK-LIP WRASSE

(BLACK-EDGE THICK-LIP WRASSE; HALF-AND-HALF THICK-LIP)

Hemigymnus melapterus

It is very easy to distinguish juveniles from adults in this species. In the young the front part of the body is white and the rear is sharply demarcated in areas of green and black. They also have pink and blue streaking around the eyes, which expands to cover a wider area of the head as they mature. Conversely, the yellow coloration evident on the caudal fin disappears with age, and the wrasses develop the characteristic thick lips. The appearance of adults tends to be bluish with cross-hatching evident on the rear of the body and blue streaking on the tail.

DISTRIBUTION: *Ranges from the Red Sea and East Africa through the Indian Ocean via Micronesia in the Pacific to Polynesia and Samoa.*

SIZE: *35.5 in (90 cm).*

BEHAVIOR: *Young occur in the shallows, seeking the relative protection of coral, while adults are found in deeper water down to depths of about 100 ft (30 m). The diet of this fish also changes significantly with age. While the youngsters feed on planktonic crustaceans, adult fish hunt significantly larger invertebrate prey.*

DIET: *The young can be difficult to feed, but live brine shrimps will help establish them in aquarium surroundings. The aquarium should also include live rock, because creatures present in the rock can provide an additional source of food for the wrasse. Frequent feeding is necessary. It may be possible to wean this fish onto inert meaty foods, but it is not an easy species to manage.*

AQUARIUM: *Not to be trusted in the company of invertebrates in a reef tank otherwise many, including fanworms and brittlestars, are likely to fall victim to this wrasse.*

COMPATIBILITY: *Keep separated from others of its own kind. Bear in mind that this wrasse can attain a very large size, which can make choosing suitable companions very difficult.*

BICOLOR CLEANER WRASSE

Labroides bicolor

The front half of the body of this fish is dark in color, except for what appears to be a pale yellowish line running down the body in some instances, just below the dorsal fin. Sometimes there is also a second lighter area running through the pectoral fin. These lines broaden out to create a variable pale yellow area on the rear of the body, although the exact depth of coloring as well as its extent varies depending on the individual.

▼ *BICOLOR CLEANER WRASSE (LABROIDES BICOLOR)*

▲ *FOUR-LINE WRASSE (LARABICUS QUADRILINEATUS)*

DISTRIBUTION: *Ranges from East Africa across the Indian Ocean north as far as southern Japan and south as far as Lord Howe Island off the east coast of Australia. Extends across the Pacific to the Line, Marquesan, and Society Islands.*

SIZE: *6 in (15 cm).*

BEHAVIOR: *Young of this species are shy by nature and solitary, but adults are conspicuous, particularly in the vicinity of the coral heads that serve to demarcate their cleaning stations. The resident wrasse draws other fish to its station for grooming and will nibble at their bodies to remove parasites. It will also take the opportunity to consume some of the protective mucus covering the bodies of the fish, and will defend its station against incursions by other bicolor cleaner wrasses.*

DIET: *Offer a range of meaty foods. Bear in mind these wrasses must be kept with a large group of other fish that they can clean to obtain mucus.*

AQUARIUM: *Suited to a reef tank, since it is unlikely to cause harm to the other occupants.*

COMPATIBILITY: *Must be kept in a community of fish—preferably not with members of its own species, given its territorial nature. Not always easy to maintain because of its requirement for mucus as part of its diet.*

FOUR-LINE WRASSE
(RED SEA CLEANER WRASSE)

Larabicus quadrilineatus

Stunning purplish blue coloration helps distinguish this wrasse. It also has a pair of lighter blue stripes running along its body. The higher one passes above the eyes to the snout, whereas the lower stripe ends at the lower part of the eye. There is an area of black evident on the caudal fin, and the overall body shape of this wrasse is decidedly elongate.

DISTRIBUTION: *Confined to the Red Sea and the Gulf of Aden in the western Indian Ocean.*

SIZE: *4.5 in (11.5 cm).*

BEHAVIOR: *The young of this species are active cleaners of other fish on the reef, and the adults are found in areas where coral predominates. As they mature, their diet progresses from invertebrates (parasites) to the polyps of stony corals.*

DIET: *A difficult species to maintain. Adults may be persuaded to eat live brine shrimps and are likely to find edible items by browsing over live rock. Finely chopped meaty foods may also be eaten in due course, but at first it may be necessary to sacrifice some coral to meet the needs of this wrasse.*

AQUARIUM: *Stony corals will be attacked by adults, although juveniles can be housed safely in these surroundings.*

COMPATIBILITY: *Keep individually, but provide companions for juveniles to clean. Choose the tankmates carefully, in case they decide to prey on the wrasse.*

STRIATED WRASSE
(SECRETIVE WRASSE)

Pseudocheilinus evanidus

A small and somewhat shy species of wrasse, recognizable by its attractive, predominantly orange-red or red coloration. There is a series of approximately 25 very fine white horizontal lines running down the sides of the body. A broad whitish stripe, often with pale blue edging, is evident below each eye, running in the direction of the jaws.

DISTRIBUTION: *Ranges from the Red Sea down the east coast of Africa to South Africa and across the Indian Ocean into the Pacific. It is found as far north as the Izu Islands and extends in an easterly direction to the Hawaiian Islands and the Tuamotu Archipelago.*

SIZE: *3.25 in (8 cm).*

BEHAVIOR: *Tends to occur at depths below 66 ft (20 m), usually in areas of the reef where there is either branching coral or rubble that will provide a range of suitable retreats for this rather solitary fish.*

DIET: *Brine shrimps, preferably vitamin enriched, are usually taken readily, as are live grass shrimps. These foods are especially useful to encourage newly acquired specimens, whose diet in the wild consists of small invertebrates, to eat well from the outset in aquarium surroundings.*

AQUARIUM: *Relatively safe to include in a reef tank, although adult fish may prey on some crustaceans. Will thrive in an environment with living rock.*

▼ *STRIATED WRASSE (PSEUDOCHEILINUS EVANIDUS)*

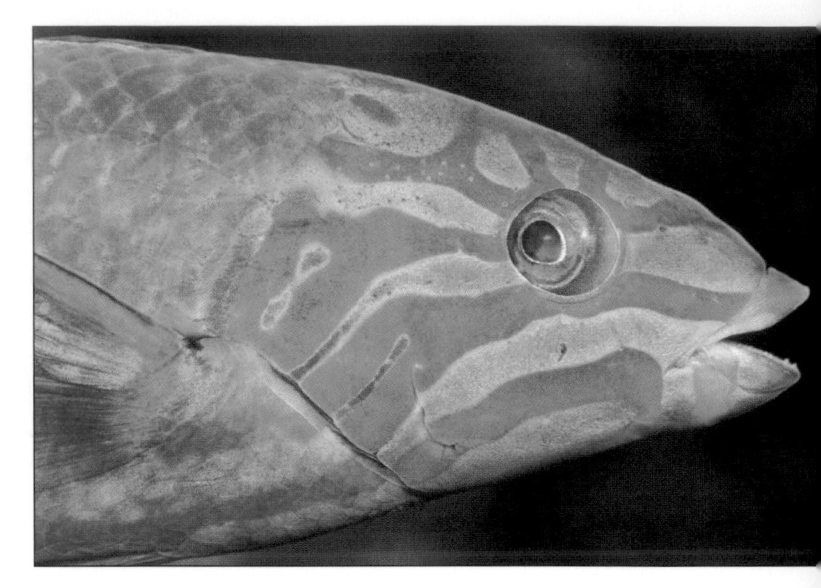

▲ *MOON WRASSE (THALASSOMA LUNARE)*

COMPATIBILITY: *Somewhat territorial and should be kept separate from its own kind. Can be relatively aggressive for its size, and is better housed with larger species that will not attack it.*

MOON WRASSE

Thalassoma lunare

In this genus adults are always more colorful than juveniles. The moon wrasse displays an irregular series of broad pink to pale violet markings on the head. Its body is greenish, becoming bluer on the underparts, with a black spot breaking up the narrow pink edging at the base of the dorsal fin. The caudal fin is yellow, and also edged above and below with pink stripes. Males develop blue heads as they grow bigger, and females are able to transform themselves into males.

DISTRIBUTION: *Ranges from the Red Sea and East Africa across the Indian Ocean northward as far as southern Japan, while extending south to Lord Howe Island off the east coast of Australia and to northern New Zealand. Its easterly range across the Pacific reaches the Line Islands.*

SIZE: *10 in (25 cm).*

BEHAVIOR: *Tends to occur in relatively shallow waters to a maximum depth of about 65 ft (20 m), and is sometimes found in estuaries.*

DIET: *Prefers a meaty diet ranging from brine shrimps to thawed prepared foods. May even take marine flake in due course.*

AQUARIUM: *Will prey on a number of the occupants of a reef tank, including crustaceans and mollusks.*

COMPATIBILITY: *Do not mix with smaller companions, which are likely to be harassed or even preyed on. A pair can be housed together successfully in a large aquarium, however, and this wrasse can be kept with bigger fish.*

BLENNIES FAMILY BLENNIDAE

These lively small fish can be identified by the presence of bristlelike projections, called cirri, above the eyes. They also have a relatively long, tall, dorsal fin running along the length of the body. Male blennies are usually larger than females. Some of them may change color during the spawning period.

HI-FIN BLENNY

Atrosalarias fuscus

Juveniles of this species are typically much more colorful than the adults. They are predominantly golden yellow with faint whitish markings that create a marbling effect on the head and tiny black spots on the body. Adults, in contrast, are very dark in color, bordering on black, although some populations have a reddish caudal fin.

DISTRIBUTION: *Restricted to western parts of the Indian Ocean, ranging from the Red Sea eastward as far as the coast of Pakistan.*

SIZE: *4 in (10 cm).*

BEHAVIOR: *Lives in shallow waters, and is sometimes found in estuaries. Often encountered resting on branching corals, occasionally feeding on coral polyps. Like other blennies, it is not a particularly active fish.*

DIET: *Microalgae are the natural foods that predominate in the diet of this blenny in the wild, and it will require access to a similar food source in the aquarium. It needs a herbivorous diet, which can include prepared foods based on* Spirulina *algae, for example, but this should be supplemented by algae growing in the aquarium.*

AQUARIUM: *May occasionally cause slight damage to coral polyps. Otherwise it will be ideal for a reef tank and will help keep algal growth in check.*

COMPATIBILITY: *Avoid mixing with larger predatory species. Also, keep hi-fin blennies separate from each other, unless you can obtain a compatible pair.*

▼ *HI-FIN BLENNY (ATROSALARIAS FUSCUS)*

BICOLOR BLENNY

Ecsenius bicolor

Bicolor blennies can display wide variation in their coloring. Some have a dark blackish front part to the body, with the rear being yellow or even a dull shade of orange. In other individuals the body may be black on the upperparts with a white area below, and it can also have a yellow tail in some cases. The coloration of both sexes is transformed during the spawning period. At first, males are red with white bars, but immediate ly after spawning they become dark blue with lighter patches on the sides of the body. When a female is ready to spawn, she turns light brown and yellowish orange in color.

DISTRIBUTION: *Ranges from the Maldive Islands in the Indian Ocean north to the Ryukyu Islands near Japan. Extends southward to Australia's Great Barrier Reef. Found through Micronesia in the Pacific, reaching as far as the Phoenix Islands.*

SIZE: *4 in (10 cm).*

BEHAVIOR: *Since this small fish often hides away out of sight on the reef for long periods, its more colorful hindparts are well disguised. It frequents areas of the reef where there is algal growth. It is therefore important to ensure that algae are growing well in the aquarium before introducing the fish to this environment.*

DIET: *In the aquarium this blenny will browse on algae, which should be supplemented with a range of herbivorous foods.*

AQUARIUM: *Sometimes nips at the coral polyps and the mantles of clams, but is generally not a serious threat to invertebrates in a reef tank. Will thrive in these surroundings.*

COMPATIBILITY: *Likely to be aggressive toward other bicolor blennies unless you have a true pair. Agrees well with most other fish, but may not prove compatible with others, such as gobies, that also spend much of their time close to the substrate in the aquarium.*

▼ *BICOLOR BLENNY (ECSENIUS BICOLOR)*

▲ *STRIPED FANG BLENNY (MEIACANTHUS GRAMMISTES)*

STRIPED FANG BLENNY

Meiacanthus grammistes

Brown and pale yellowish horizontal stripes running from the long dorsal fin all along the body help identify this species. At the rear of the body there may be a more evident pale blue suffusion in the yellow striped areas. Here, too, the brown stripes tend to break up, giving rise to spots instead.

DISTRIBUTION: *Ranges from the Ryukyu Islands near Japan southward through Indochina and Papua New Guinea down to the northwestern coast of Australia and south to the Great Barrier Reef on the east coast.*

SIZE: *4.5 in (11 cm).*

BEHAVIOR: *This blenny is usually seen on its own, although occasionally it will join with others to form small groups. As its name suggests, it is unusual in having venomous fangs in its mouth. Other potentially predatory fish on the reef instinctively seem to recognize this, and give the blenny a wide berth. Even if a striped fang blenny is ingested into the mouth of another fish, it will use its fangs to bite, and the predator is likely to react by spitting out its intended meal.*

DIET: *Both meaty foods, chopped as required, and herbivorous items should be offered to this omnivorous species.*

AQUARIUM: *Will thrive in aquarium surroundings without upsetting any of the occupants. Should be housed in an aquarium alongside live rock, even if not being kept as part of a reef tank, since this will provide valuable additional feeding opportunities.*

COMPATIBILITY: *Be careful when choosing companions for this species. They must not interfere with the blenny, otherwise they are likely to receive a painful bite.*

GOBIES
FAMILY GOBIIDAE

The Gobiidae is the largest of the families of marine fish, consisting of about 1,900 species separated into more than 210 genera. Some representatives of the group are found in brackish and freshwater environments as well. All gobies are small fish that can usually be found close to the substrate. Although they are not strong swimmers, gobies have evolved an interesting way of being able to anchor themselves in place in a current. Their pelvic fins have fused together and developed into a cuplike sucker with which they can attach themselves to rocks. The sucker may even be used to hold onto larger fish that are groomed by certain members of the family.

In the saltwater aquarium gobies are easy to cater for, frequently settling well as part of a community setup. They are usually suitable for a reef tank, in which they may display close links with various invertebrates, notably alpheid shrimps and some sponges. A few species have even been bred successfully in this environment.

OLD GLORY
(RAINFORD'S GOBY)

Amblygobius rainfordi

This is one of the more colorful members of its genus, with a series of yellow or orange horizontal stripes running down the length of the body. The stripes have fine black edging and they are separated by wider grayish bands. The overall shape of this goby is relatively narrow and elongate. There may be odd blue spots down the back with evident eyespots, known as ocelli, present at the base of the dorsal fin toward the rear and in the vicinity of the upper part of the caudal peduncle.

DISTRIBUTION: *Occurs from Rowley Shoals off the northwestern coast of Australia northward via Indonesia as far as the Philippines. Extends around the Australian coast to the southern part of the Great Barrier Reef, ranging eastward across the Pacific to Fiji and Tonga.*

SIZE: *2.5 in (6.5 cm).*

BEHAVIOR: *This is not a burrowing species. Instead, it prefers to conceal itself among coral and rocks on the reef. It frequents coastal stretches of reef, where the water is turbid, staying close to the sandy or muddy base.*

▼ OLD GLORY (*AMBLYGOBIUS RAINFORDI*)

In an aquarium it will behave in a very similar fashion. This is quite normal and not a sign of ill health.

DIET: *Will thrive in an aquarium in which filamentous algae are already established as well as live rock, among which it will find food by itself. Needs to be offered tiny crustaceans, including vitamin-enriched brine shrimps, very finely chopped meaty foods, and herbivorous items.*

AQUARIUM: *Ideal for a reef tank but avoid introducing occupants such as dottybacks, which may bully it.*

COMPATIBILITY: *Young will often settle in a large aquarium together, but adults are more quarrelsome and territorial by nature.*

LEMON GOBY

(POISON GOBY)

Gobiodon citrinus

The lemon goby is a very colorful species. Its bright appearance probably acts as a warning to would-be predators. Its overall color is bright yellow, but it often takes on a more orange hue in the vicinity of the gills. There is a pale blue stripe running down the base of the dorsal fin on each side and two short stripes running up from the level of the mouth to the eyes, crossing through the pupil. There is a stripe extending down the back of the head and another one behind, extending from the pectoral fins as far as the midline of the body.

DISTRIBUTION: *Occurs from the Red Sea southward down the African coast as far as Delagoa Bay, Mozambique, and across the Indian Ocean. Reaches north as far as southern Japan and extends to the southern end of the Great Barrier Reef off Australia's southeast coast. Present through the Pacific as far east as Samoa.*

SIZE: *2.5 in (6.5 cm).*

BEHAVIOR: *This goby is very closely associated on the reef with colonies of* Acropora *coral, sheltering and feeding within its branches. It produces a toxic mucus that makes it unpalatable to predatory species of fish. It relies on this feature to avoid predation, since it is not a powerful swimmer. In the aquarium it will often adopt rocks as well as coral and use them as retreats.*

▼ *LEMON GOBY (GOBIODON CITRINUS)*

DIET: *Live brine shrimps can be valuable in helping this goby settle, although once established it can be given thawed foods such as mysid shrimps. It may find other edible items by foraging alongside live rock in a reef tank.*

AQUARIUM: *Very suitable for a reef aquarium, although may occasionally nip at coral polyps without causing serious harm. Should be provided with coral or a coral substitute for it to find hiding places, mimicking the environment in which it occurs in the wild.*

COMPATIBILITY: *If there is sufficient space and coral in the aquarium, lemon gobies can be kept in a small group without squabbling. Pairs have even spawned in such surroundings.*

YELLOW GOBY

(OKINAWA GOBY)

Gobiodon okinawae

The entire body of the yellow goby is a stunning shade of pure yellow, usually with a very slight hint of blue in the iris around the pupil in the eye. Some individuals are a brighter shade of yellow than others, however, and may have slightly orange areas on the body. Unfortunately, these differences do not provide a reliable means of sexing the fish. They probably reflect slight variations that are features of particular populations. If water conditions in the aquarium are poor, these gobies tend to turn a brownish shade.

DISTRIBUTION: *Ranges from Indonesia and the Philippines northward to southern Japan and southward as far as Rowley Shoals off the northwestern coast of Australia. Extends around the Australian coast to the Great Barrier Reef and across the Pacific to Palau and the Marshall Islands, which form part of Micronesia.*

SIZE: *1.5 in (3.5 cm).*

◀ *YELLOW GOBY (GOBIODON OKINAWAE)*

BEHAVIOR: *This goby is closely associated with areas of* Acropora *coral, whose branching structure provides it with plenty of vantage points. Not only does it swim among the branches, it also rests on them. Yellow gobies can be found in groups numbering up to 15 individuals, although they maintain their own territories within an area.*

DIET: *Small meaty foods, such as vitamin-enriched brine shrimps, and other similar foods cut into tiny pieces as required. Live brine shrimps are favored, but the yellow goby will also seek out small prey among living rock.*

AQUARIUM: *Ideal for a reef tank, since it does not generally cause any harm to invertebrates.*

COMPATIBILITY: *It may be possible to introduce two of these gobies together into the aquarium, especially one of each sex. However, they tend to be territorial, and any newcomer will be relentlessly persecuted. This goby is inoffensive toward other fish. Although protected by a toxic mucus covering the body, it should not be mixed with predatory species or others, such as hawkfish, that may harass it.*

NEON GOBY

Gobiosoma oceanops

A narrow elongated body is characteristic of this popular species. It has a neon-blue horizontal stripe running down the center of its body, bordered by black stripes above and below, with a paler area on the underparts below the jaws. When seen from above, the patterning on the snout is highly individual. The stripes in that area are separated by black markings that extend to the tip of the snout. Although it is not generally possible to sex these gobies, the blue stripe of the male darkens significantly prior to spawning.

SYNONYM: *Elacatinus oceanops.*

DISTRIBUTION: *Occurs in the western Atlantic, ranging from southern Florida and the coast of Texas across the Gulf of Mexico as far as Belize in Central America. Not known from the Bahamas or other oceanic Caribbean islands.*

SIZE: *2 in (5 cm).*

BEHAVIOR: *This small fish can be found lurking in association with coral, but it is also a prominent member of the reef community. It is often left unmolested by potential predators because it serves as a cleaner fish. Neon gobies form groups for this purpose, occurring over certain parts of the reef to which other fish come to have parasites pulled off their bodies. It is even possible to attract a wild neon goby to investigate a finger underwater on the reef, with a view to cleaning it. This species has been bred relatively frequently in aquarium surroundings. Spawning takes place in a small cave or a similar retreat—sometimes even a large shell—and the adults guard the eggs determinedly. Hatching takes at least one week. The fry require rotifers as a first food, after which they can be reared on live brine shrimps. The main drawback in aquarium surroundings is the fact that neon gobies are short-lived—their life expectancy is typically no more than 12 to 24 months.*

DIET: *Small meaty foods, such as vitamin-enriched brine shrimps and mysid shrimps. Will also forage for food wherever live rock is included. Can help control the white spot parasite by biting off the cysts before they develop.*

AQUARIUM: *An ideal choice for a reef tank. May even breed successfully in these surroundings.*

COMPATIBILITY: *Pairs are compatible, but in a small aquarium fighting is likely if they are housed in a group. In view of their small size, it is generally safer to avoid mixing these gobies with larger aggressive companions.*

BLUE-BANDED GOBY
(CATALINA GOBY)

Lythrypnus dalli

This is one of the most beautiful of all gobies, partly because of its underlying vivid red coloration. A purplish blue area extends between and around the eyes, and a variable number of similar bands run vertically down the sides of the body. These bands become progressively narrower as they get nearer to the tail, although they have usually disappeared by the middle of the body. Each of these gobies, however, displays highly individual patterning. Males can be recognized by the spines on the dorsal fin, which are longer than those of the females.

▼ *NEON GOBY (GOBIOSOMA OCEANOPS)*

▼ *BLUE-BANDED GOBY (LYTHRYPNUS DALLI)*

DISTRIBUTION: *Occurs in the eastern Pacific Ocean, ranging from the vicinity of Morro Bay in the central area of the state of California southward as far as Guadalupe Island, which lies off the northern central part of Baja California, Mexico.*

SIZE: *2.5 in (6.5 cm).*

BEHAVIOR: *The blue-banded goby is conspicuous. It usually rests in the open on a rock or similar surface because, like many other gobies, it is not a strong swimmer. If danger threatens, it will dart back to a nearby retreat, not infrequently taking advantage of the spines of sea urchins for this purpose. Females lay their somewhat oblong-shaped eggs in suitable empty shells, where they are guarded by the males. In aquarium surroundings the resulting fry can then be reared on rotifers followed by brine shrimps. Unfortunately, blue-banded gobies appear to have a naturally short life span, sometimes no more than 12 months.*

DIET: *Small meaty foods, especially crustaceans such as mysid shrimps and vitamin-enriched brine shrimps. Livefoods can help establish this fish. it will also seek edible items in an aquarium that contains living rock.*

AQUARIUM: *Ideally suited to a reef tank, which will provide it with both cover and food as well as spawning opportunities.*

COMPATIBILITY: *Avoid mixing this goby with larger, potentially predatory, companions. The blue-banded goby is relatively territorial by nature, but it can be possible to keep several of them in a spacious aquarium provided there are suitable retreats. Alternatively, concentrate on keeping just a single breeding pair.*

BLACK-FIN DARTFISH
(SCISSOR-TAIL GOBY)

Ptereleotris evides

Light bluish coloration is evident on the front half of the body of this goby. There is a blackish area on the snout and a much darker, slightly purplish, area on the rear. Its most distinctive feature, however, is the broad second part of the blackish dorsal fin, which is located directly above the similiarly colored anal fin on the underside of the body. The first part of the dorsal fin is low and inconspicuous, while the caudal fin has a slightly scissor-tailed appearance. Young fish of this species can be identified by the presence of a relatively large oval black spot located at the base of the caudal fin.

▼ BLACK-FIN DARTFISH (PTERELEOTRIS EVIDES)

DISTRIBUTION: *From the Red Sea and East Africa through the Indian Ocean. Reaches as far north as the Ryukyu and Ogasawara Islands near Japan. Occurs southward to the eastern coast of New South Wales and Lord Howe Island, and ranges across the Pacific as far as the Line and Society Islands.*

SIZE: *5.5 in (14 cm).*

BEHAVIOR: *Found in the more tranquil open areas of reefs where there is a sandy base into which it can burrow. Juveniles may be encountered in groups, but adults tend to occur in pairs. They feed off the bottom, darting away if danger threatens rather than seeking out the protection of their burrows—although they retire there as soon as darkness falls. The black-fin dartfish can prove shy at first in aquarium surroundings, but will settle down well and soon prove to be quite conspicuous.*

DIET: *This goby feeds on zooplankton in the water and will therefore be attracted by live brine shrimps. It requires a meaty diet, which should be provided in the form of fine particles of food.*

AQUARIUM: *Very suitable for a reef aquarium that has a fine sandy substrate into which it can burrow.*

COMPATIBILITY: *Not aggressive and can usually be kept in small groups. It can also be mixed with other members of its genus and with other peaceful species of similar size.*

ZEBRA DARTFISH
(ZEBRA DART GOBY)

Ptereleotris zebra

The zebra dartfish can be identified by its distinctive coloration. Its background color is a variable yellowish green. Overlaying this is a prominent series of approximately 20 orange stripes with narrow blue edging. The stripes are vertical but they do not extend up to the dorsal fin. There is a broader but shorter red stripe, and two adjacent blue lines are evident at the base of the

▼ ZEBRA DARTFISH (PTERELEOTRIS ZEBRA)

▲ *Maiden Goby (Valenciennea puellaris)*

pectoral fins. On the forehead is a bluish area. There is a dark area below the eye, and diagonal blue bands can be seen running over the operculum.

DISTRIBUTION: *Ranges from the Red Sea across the Indian Ocean north to the Ryukyu Islands close to Japan. Ranges via the Mariana and the Marshall Islands in the western Pacific and extends eastward through the Pacific as far as the Marquesan and the Line Islands.*

SIZE: *4.5 in (11 cm).*

BEHAVIOR: *This particular goby is a shoaling fish. Even when adult, it is seen in shoals that will quickly retreat to suitable hiding places together if they are threatened. It occurs on the reef in relatively open areas of shallow water, swimming in search of zooplankton above the substrate. It may spawn in the aquarium, in which case the female will watch over the eggs. Rotifers should be provided initially as the rearing food for the fry, followed by brine shrimps.*

DIET: *Provide finely chopped meaty foods, although brine shrimps are especially favored. Feeding is not particularly difficult if there is a group of these fish, because once one starts to feed then all its companions will join in the feast.*

AQUARIUM: *Ideal for a reef tank. Will sometimes bury itself into the sandy substrate in an aquarium.*

COMPATIBILITY: *This species should be kept in small groups in a covered aquarium because it can jump well, especially if it feels threatened. It can also be mixed with other nonagressive species that will not harass it.*

MAIDEN GOBY
(ORANGE-SPOTTED SLEEPER GOBY)

Valenciennea puellaris

A yellowish to pale brown background color is a feature of this goby. It has a variable pattern of elongated blotches arranged roughly in lines running down the length of the body. The appearance of maiden gobies can vary markedly through their range. There are generally sky-blue markings on the cheeks and a more clearly defined orange stripe with a blue border on the lower part of the body toward the caudal fin. The iris surrounding the pupil is also orange, while the fins tend to be clear.

DISTRIBUTION: *Ranges from the Red Sea through the Indian Ocean, extending northward up to southern Japan and south to the Great Barrier Reef off Australia's east coast. Extends through the Pacific via New Caledonia as far east as Samoa.*

SIZE: *6.75 in (17 cm).*

BEHAVIOR: *Confined to sandy areas on the reef, where rubble is also present. This goby will bury itself into the sand near rocks, creating shallow burrows that are occupied by individual pairs. The maiden goby has relatively large mouthparts to help it move the sand, but it is not always easily maintained, because it tends to hunt instinctively for small worms and similar prey in the substrate. Weaning it onto alternative foods is not always straightforward.*

DIET: *You will need to offer a wide range of foods in order to find items that it will eat readily. Mysid shrimps may be taken as well as other finely chopped meaty food. It may ultimately eat marine flake food, too.*

AQUARIUM: *Suitable for a reef tank with a fine sandy substrate, although its burrowing activity may disturb some sessile invertebrates. Likely to eat small bristleworms in these surroundings.*

COMPATIBILITY: *Keep separately or in pairs, rather than in groups, since this goby is quarrelsome by nature. Can be mixed with other small, non-aggressive fish, however, without causing fighting.*

BLUE-BAND GOBY

(YELLOW-HEADED SLEEPER GOBY; BLUE-CHEEK GOBY)

Valenciennea strigata

The most characteristic feature of the blue-band goby is a prominent bluish streak, bordered above and below by a thin black line that runs from just in front of each eye back to just above the base of the pectoral fins. There is also usually a blue spot behind the eye and other blue markings on the throat. The area from the jaws to the pectoral fins is otherwise yellow, and the rest of the body is yellowish brown on the upperparts, becoming paler on the sides. The dorsal fin is divided into two parts and, like the other fins, is clear.

DISTRIBUTION: *Ranges from East Africa across the Indian Ocean, extending north to the Ryukyu Islands near Japan. Reaches southward to the south-eastern part of the Australian coast, and also occurs around Lord Howe Island. Extends across the Pacific as far as the Tuamotu Archipelago.*

SIZE: *7 in (18 cm).*

BEHAVIOR: *A burrowing species that often feeds by excavating mouthfuls of sand in search of edible items, ranging from invertebrates to fish eggs. This can very be beneficial in the aquarium if you are operating an*

▲ BLUE-BAND GOBY (VALENCIENNEA STRIGATA)

undergravel filter, because the digging activity helps prevent the sand from becoming compacted, which would reduce the water flow and therefore the filter's efficiency. This goby is likely to remain close to its burrow, even when there is no obvious danger, and will dart back inside if there is any threat. It usually becomes most active under cover of darkness.

DIET: *Offer a variety of meaty food, especially small crustaceans such as brine shrimps and mysid shrimps. May also be persuaded to eat marine flake food once established.*

AQUARIUM: *Suitable for a reef tank, although its continual digging may upset some sessile invertebrates in these surroundings and may also undermine tank decor.*

COMPATIBILITY: *An ideal member of a community aquarium of small nonaggressive fish. Makes an ideal introduction to the gobies, since it is easy to maintain, either in pairs or larger groups, in a suitably spacious aquarium.*

DIAGONALLY BANDED SHRIMP GOBY

Amblyeleotris diagonalis

As its name suggests, this small goby has a series of broad diagonal stripes running down its body. The stripes are dark reddish brown in color and they encircle the body, starting just in front of the pectoral fin and extending down the body to the base of the caudal fin. The area from the upper jaw to the eyes is similarly marked and has a narrower broken band behind.

DISTRIBUTION: *Ranges from the Red Sea and east Africa through the Indian Ocean via Madagascar and the Persian Gulf eastward to Born Island in the Andaman Sea and to Sri Lanka and Indonesia. Extends south to Australia's Great Barrier Reef and through the Pacific as far as the Solomon Islands, which lie to the east of New Guinea.*

SIZE: *3.5 in (9 cm).*

BEHAVIOR: *This species and other prawn gobies are so called because of their very close association with shrimps of the genus* Alpheus, *with which they form a symbiotic relationship, meaning that both species benefit. The shrimps are responsible for digging the burrows that they occupy with the gobies, and they repair them as necessary. For their part, the gobies warn the shrimps of danger by darting back to the burrow. They also provide food for them by bringing it to the burrow, where the shrimps scavenge on the leftovers.*

DIET: *Requires small meaty foods such as brine shrimps and other items that will need to be finely chopped.*

AQUARIUM: *Very suitable for a reef tank, especially if it can be kept in the company of an* Alpheus *species of shrimp. The tunneling activities of the shrimps are unlikely to cause any major effects in the aquarium, although always check that rockwork is firmly positioned.*

COMPATIBILITY: *Usually placid toward other nonaggressive tankmates, but males are likely to fight, anchoring onto each other with their jaws. A true pair, however, should prove quite compatible.*

ORANGE-SPOTTED PRAWN GOBY

Amblyeleotris guttata

As its name suggests, the random orange patterning over the pale brownish body of this shrimp goby aids its identification. There may be darker brown bands on its body, most evident on the underside in the vicinity of the pectoral fin. The eyes are located high on the head, with the pupils displaying alternate brown and white markings.

DISTRIBUTION: *Confined to the western part of the Pacific, extending from the Philippines to the Ryukyu Islands near Japan and southward to Scott Reef and the Great Barrier Reef off Australia's east coast.*

SIZE: *3.5 in (9 cm).*

BEHAVIOR: *This prawn goby occurs in sandy areas of the reef, living in close association with the shrimp* Alpheus ochrostriatus. *It inhabits the shrimp's burrows, into which it retreats at the slightest hint of danger. If caught in the open, it may jump. It is advisable to bear this in mind if you are keeping the fish in an aquarium, especially if it is not with alpheid shrimps.*

DIET: *Feeds on small meaty food including brine shrimps. Thaw and chop frozen foods into small pieces before offering them to the fish.*

AQUARIUM: *Very suitable for a reef tank when in the company of alpheid shrimps, but liable to prey on any significantly smaller crustaceans.*

▲ ORANGE-SPOTTED PRAWN GOBY (AMBLYELEOTRIS GUTTATA)

COMPATIBILITY: *Pairs will agree well together and will thrive in the company of other small nonaggressive species. Males are likely to fight, however, and will have to be kept separate.*

BROWN STREAK PRAWN GOBY

(MASKED SHRIMP GOBY)

Amblyeleotris gymnocephala

Although the brown streak prawn goby is similar to the diagonally banded shrimp goby, its banded pattern tends not to slope backward on the body. Instead, it encircles the body. Furthermore, there is a characteristic narrow horizontal dark line that runs in the midline from the back of the eye through the first brownish band. The body may also have some dark spots among the white areas that separate the individual bands. The color of the spots corresponds to the color of the bands.

▼ BROWN STREAK PRAWN GOBY (AMBLYELEOTRIS GYMNOCEPHALA)

DISTRIBUTION: *Ranges through the Indian Ocean into the western part of the Pacific, but its precise range remains unclear at present.*

SIZE: *5.5 in (14 cm).*

BEHAVIOR: *This shrimp goby is often encountered close to the shore, occurring in brackish water in mangrove areas. It is also found in sandy areas in bays. It is not easy to spot, however, since it will retreat out of sight to the relative safety of the shrimp's burrow at any sign of danger.*

DIET: *Although it is one of the larger members of its genus, it requires a diet similar to that of other prawn gobies, based on small meaty foods that may have to chopped up into small pieces.*

AQUARIUM: *Suitable for a reef aquarium, but should not be housed with much smaller crustaceans that may be preyed on (alpheid shrimps being the notable exception).*

COMPATIBILITY: *Males are territorial and will fight if housed together, but this species is otherwise peaceful and can be housed in a reef tank with other similar occupants.*

SLENDER SHRIMP GOBY
(PERIOPHTHALMA PRAWN GOBY)

Amblyeleotris periophthalma

Banded patterning is a feature of this species. On close examination, however, a series of small orange spots is evident behind the eyes, running from the white band into the adjacent brown band. The banding itself may have an orange hue, and mottling is often apparent lower down the body in the otherwise white areas. Brown spots are also evident in the transparent dorsal fin.

DISTRIBUTION: *Extends from the Red Sea across the Indian Ocean via Micronesia as far as the island of Samoa in the Pacific, although its exact distribution is not currently known.*

SIZE: *3.5 in (9 cm).*

BEHAVIOR: *Found in burrows excavated by alpheid shrimps, and its tendency to dart back there at any hint of danger means that studying this fish can be difficult. The shrimps themselves have very poor eyesight but have learned to rely on the fish's rapid movement to alert them to any approaching threat in their immediate environment. The high position of the eyes of the slender shrimp goby helps ensure that it has a good view of its surroundings.*

DIET: *Offer finely chopped meaty foods. Will also feed on other similar items such as vitamin-enriched brine shrimps.*

AQUARIUM: *Ideal for a reef tank when housed in the company of alpheid shrimps, although this fish will live on its own as well without any problem.*

COMPATIBILITY: *Males are territorial and will fight if housed together, particularly in a relatively small aquarium. Otherwise, this species is inoffensive by nature and can be mixed with a variety of nonaggressive fish that are suitable for inclusion in a reef tank.*

▼ *SLENDER SHRIMP GOBY (AMBLYELEOTRIS PERIOPHTHALMA)*

▲ STEINITZ'S PRAWN GOBY (AMBLYELEOTRIS STEINITZI)

STEINITZ'S PRAWN GOBY

Amblyeleotris steinitzi

The banded appearance of this prawn goby tends to be less distinctive than that of many related species. The brown banding is generally faint, while the whitish areas display very pale bluish wavy lines extending right down the body rather than just in the immediate vicinity of the head. There is a fine pattern of small pale yellow spots on the dorsal fin, which is split into two parts. There is no patterning on the other fins.

DISTRIBUTION: *Ranges from the Red Sea across the Indian Ocean, extending northward to the Yaeyama Islands near Japan and southward as far as the Great Barrier Reef off the eastern coast of Australia. Extends via Micronesia through the Pacific to Samoa.*

SIZE: *3.25 in (8 cm).*

BEHAVIOR: *Steinitz's prawn goby occurs in sandy parts of the reef in association with alpheid shrimps. Its coloration is similar to that of the shrimps, which may make it easier for the shrimps to identify with them. It possibly also gives the fish greater protection from predators, which consider them to be poorly sighted shrimps rather than fish.*

DIET: *Offer a range of meaty foods chopped into small pieces as required. Will readily feed on brine shrimps.*

AQUARIUM: *Suitable for a reef tank, where it can be accommodated with alpheid shrimps. The burrowing activities of the shrimps are unlikely to cause a serious disturbance in these surroundings.*

COMPATIBILITY: *Males are quarrelsome toward each other and should not be housed together, but otherwise it is a peaceful species that can be housed with other fish of similar size and temperament.*

MAGNUS'S PRAWN GOBY

Amblyeleotris sungami

The appearance of these prawn gobies is quite variable. The basic pattern is a series of brown and whitish bands encircling the body. The first stripe may extend below the eyes down to the jaws, but this is not a consistent feature. There is usually a brown area on the top of the head, however, between the eyes. The brown bands vary considerably in width from individual to individual. They are a relatively light shade of brown, sometimes darker at the edges. The fins are clear.

DISTRIBUTION: *Ranges from the Red Sea southward as far as Seychelles in the Indian Ocean and eastward into the western part of the Pacific, although its precise range is currently unclear.*

SIZE: *4 in (10 cm).*

BEHAVIOR: *This prawn goby occurs on sandy areas of the reef. Its banded patterning helps conceal it in this environment. There are regional variations in patterning, based on the background of a particular habitat. In some areas, for example, they are whiter than in others. As with other members of the genus, they live in association with alpheid shrimps, taking advantage of the crustaceans' burrows, into which they can retreat if danger threatens. This makes them relatively inconspicuous on the reef.*

DIET: *Brine shrimps are favorite food items, but this prawn goby will eat other meaty foods, which need to be finely chopped.*

AQUARIUM: *An ideal inhabitant for a reef tank, especially if housed in the company of its partner shrimps, so they can display their fascinating pattern of behavior together.*

COMPATIBILITY: *Males are likely to be territorial but otherwise this small fish is inoffensive and will not disturb other aquarium occupants.*

▼ MAGNUS'S PRAWN GOBY (AMBLYELEOTRIS SUNGAMI)

GORGEOUS PRAWN GOBY

Amblyeleotris wheeleri

This is one of the most colorful members of the genus, as its name suggests. However, its appearance can vary markedly among individuals. It displays the typical banded patterning that tends to be a feature of prawn gobies. The bands are relatively wide and can range in color from orange and white to red and yellow. Their appearance seems to depend both on their place of origin and on their current surroundings—these gobies become paler under bright light. The red areas generally have a darker overlay toward the tail, and the body is covered with a series of mainly white spots. Small reddish spotting is also likely to be evident on the dorsal fin.

DISTRIBUTION: *Ranges over a very wide area from East Africa through the Indian Ocean northward to southern Japan and southward as far as Australia's Great Barrier Reef. Ranges eastward through the Pacific Ocean to the area around Fiji.*

SIZE: *3.25 in (8 cm).*

BEHAVIOR: *The gorgeous prawn goby often occurs in shallower areas of the reef than most other species. It is regularly found among coral rubble, although it can also be encountered down to 100 ft (30 m). Its lifestyle is otherwise similar to that of related species, and it occurs in association with alpheid shrimps. It associates most often with* Alpheus ochrostriatus *in the wild but will live in association with other species.*

DIET: *Small meaty foods such as vitamin-enriched brine shrimps should feature in the diet of this fish, and its food should be chopped up into small pieces as necessary.*

AQUARIUM: *This species will thrive in a reef tank, especially in the company of alpheid shrimps, which will dig burrows into which the prawn goby can retreat.*

COMPATIBILITY: *While not aggressive in the company of other fish, males of this species will be quarrelsome toward each other and must be housed apart. Be careful not to mix with larger companions that may prey on them.*

▼ *GORGEOUS PRAWN GOBY (AMBLYELEOTRIS WHEELERI)*

▲ *FLAG-TAIL SHRIMP GOBY (AMBLYELEOTRIS YANOI)*

FLAG-TAIL SHRIMP GOBY

Amblyeleotris yanoi

This is one of the more colorful shrimp gobies. It has whitish underparts and a series of orange and dark patches on its body. The patches are significantly variable to allow individuals to be distinguished from each other. The body shape tapers down its length, and the relatively large caudal fin is often compressed.

DISTRIBUTION: *Restricted to the vicinity of the western Pacific. Extends from the Ryukyu Islands near Japan southward to Indonesia as far as the islands of Bali and Flores.*

SIZE: *2.5 in (6 cm).*

BEHAVIOR: *Like other members of the family, this fish is not a powerful swimmer and remains close to the seabed. It is often found in relatively shallow sandy areas near the coast, but it can also be observed at depths of 100 ft (30 m) on occasion. It forms a close association with the red-banded shrimp (*Alpheus randalli*) through its range.*

DIET: *Offer vitamin-enriched brine shrimps and mysid shrimps as well as other meaty foods. These items may need to be chopped up finely for this small fish.*

AQUARIUM: *Highly recommended for a reef tank, where it will live peacefully alongside other occupants. Keep this goby in the company of alpheid shrimps if possible.*

COMPATIBILITY: *Males are territorial and are likely to disagree with each other. Try to find compatible fish for this species—fish that are being housed together by a stockist, for example—because this should reduce the likelihood of subsequent conflict.*

YELLOW PRAWN GOBY

Cryptocentrus cinctus

In spite of its name, there is a distinctive whitish color phase of the yellow prawn goby. The coloration of the yellow variety is not especially vibrant. The yellow is most evident on the underside of the body, with the remainder of the body often being yellowish green. There is a dark stripe present above the upper jaw and four or five nondescript dark bars running down the sides of the body, although the latter tend to be far less evident in the yellow variety. Whitish or pale bluish speckling is present on the sides of the head, extending to the dorsal fins, while the ventral fins are spotted and streaked with distinctive sky-blue markings.

DISTRIBUTION: *Occurs in the western part of the Pacific, ranging from the Yaeyama Islands near Japan southward via Singapore as far as Australia's Great Barrier Reef. Extends eastward to the Micronesian islands of Palau and Truk.*

SIZE: *3 in (7.5 cm).*

BEHAVIOR: *This prawn goby occurs in shallow sandy areas and can be encountered close to the shore in bays. Like other prawn gobies, it forms a close association with alpheid shrimps, inhabiting their burrow alongside them. It is not a powerful swimmer but its prominent eyes protruding above its head help alert it to danger, at which time it will retreat with the shrimps to their burrow.*

DIET: *Requires a diet of meaty foods, such as vitamin-enriched brine shrimps. Chop larger items finely as necessary.*

AQUARIUM: *Will settle well in a reef tank and will generally stay close to the substrate.*

COMPATIBILITY: *Males may be aggressive toward each other, but generally the yellow prawn goby is very inoffensive and is more likely to be bullied by other aquarium occupants.*

BLUE-FINNED SHRIMP GOBY

(Y-BAR SHRIMP GOBY)

Cryptocentrus fasciatus

As its name suggests, the blue-finned shrimp goby has bright blue markings on its clear fins, which tend to stand out from its darker body color. The appearance of individuals is variable, however, with some being blackish rather than dark brown in color. There are irregular white patches on the body, typically forming blotches along the base of the dorsal fin, and there is a white area above the eyes. In some cases the white areas are tinged with yellow. Bright blue streaks are evident on the sides of the face.

DISTRIBUTION: *Ranges from East Africa across the Indian Ocean southward as far as Australia's Great Barrier Reef, extending east as far as the islands of Melanesia in the Pacific.*

SIZE: *5.5 in (14 cm).*

▼ *YELLOW PRAWN GOBY (CRYPTOCENTRUS CINCTUS)*

▲ *LUTHER'S PRAWN GOBY (CRYPTOCENTRUS LUTHERI)*

BEHAVIOR: *This shrimp goby lives in sandy areas, where its patterning helps it merge into the background. When danger threatens, it retreats into a burrow excavated and inhabited by alpheid shrimps. More than one goby may occupy a burrow.*

DIET: *Feeds on small prey. Should therefore be provided with a meaty diet, and larger food items should be finely chopped before being offered.*

AQUARIUM: *Will settle well in a reef tank, particularly if it contains alpheid shrimps and a suitable sandy substrate for burrowing purposes.*

COMPATIBILITY: *Inoffensive by nature. Companions have to be chosen carefully, particularly in view of the relatively small size of this fish, which makes it vulnerable to predators.*

LUTHER'S PRAWN GOBY

Cryptocentrus lutheri

In terms of its overall color, Luther's prawn goby is dark. It can be identified by the light and dark banding on its body, running from below the front of the dorsal fin to the base of the caudal peduncle. The fins are relatively clear in color, but there are prominent bright bluish markings on the sides of the head. The area around the eyes and the front of the head is often grayish.

DISTRIBUTION: *Restricted to the western part of the Indian Ocean, from the Red Sea to the Persian Gulf. May extend to East Africa, although its range in that region is uncertain.*

SIZE: *4.5 in (11 cm).*

BEHAVIOR: *Unlike some prawn gobies, this species is found in areas strewn with rubble as well as sandy areas of the reef. It occurs in association with alpheid shrimps, which create burrows that they share with the gobies.*

DIET: *Offer small or finely chopped meaty foods. Vitamin-enriched brine shrimps are often favored.*

AQUARIUM: *Suitable for a reef aquarium. Although it will retreat into burrows dug by alpheid shrimps in these surroundings at first, it is likely to be more conspicuous once it has settled in the aquarium.*

COMPATIBILITY: *Males may prove territorial, but generally Luther's prawn goby is inoffensive by nature.*

DRACULA SHRIMP GOBY

Stonogobiops dracula

Members of this genus tend to be very striking in coloration. The Dracula shrimp goby displays four vivid bands that run at a slight angle along the length of its body in the direction of the tail. The black coloration extends up into the first part of the dorsal fin as

▲ DRACULA SHRIMP GOBY (STONOGOBIOPS DRACULA)

well. The intervening broader areas of the body are white, and the fins are whitish. There is a yellow area on the head extending up from the mouth to between the eyes.

DISTRIBUTION: *Restricted to the western part of the Indian Ocean, around Seychelles, and extending northeastward to the Maldives, which lie off the southwest coast of India.*

SIZE: *2.75 in (7 cm).*

BEHAVIOR: *This species occurs in pairs. It is encountered over sandy areas of the reef and also in parts where there is rubble. The Dracula shrimp goby tends to remain close to a burrow created by the shrimp* Alpheus randalli, *the species with which it associates. By remaining close to the entrance of the burrow, it can retreat inside at the first sign of any threat.*

DIET: *Needs a meaty diet of finely chopped foods, along with smaller items such as vitamin-enriched brine shrimps.*

AQUARIUM: *Very suitable for inclusion in a reef tank, especially if kept together with alpheid shrimps.*

COMPATIBILITY: *Males will disagree if kept together, so obtain a true pair for the aquarium at the outset. Do not mix these prawn gobies with companions that may bully them or deprive them of their food.*

BLACK-RAY PRAWN GOBY

(FILAMENT-FINNED PRAWN GOBY; HI-FIN PRAWN GOBY)

Stonogobiops nematodes

An unmistakable fish, owing to the first two rays at the front of the dorsal fin, which are significantly elongated and are black in color. There are four reddish brown bands encircling the body, running from in front of the dorsal fin to the caudal peduncle. The last of these bands is vertical rather than being angular like the three that lie farther forward on the body. The front of the head is a brilliant rich shade of yellow, and there is a yellowish hue evident along the top of the body in the white areas that separate the darker bands.

DISTRIBUTION: *Ranges from the area around Seychelles northward via Indonesia to the Philippine Islands.*

SIZE: *2 in (5 cm).*

BEHAVIOR: *This relatively small fish associates with the shrimp* Alpheus randalli *on the reef, where it tends to occur over dark sandy areas. Pairs form a close bond, rarely straying far from each other or from the protection of their burrow. If disturbed in the open, this fish may try to bury itself under the sand.*

DIET: *Provide vitamin-enriched brine shrimps, mysid shrimps, and finely chopped meaty foods for this small fish.*

AQUARIUM: *Suitable for a reef tank, where it will ignore other aquarium occupants—the exception perhaps being at feeding time, when it may become more assertive.*

COMPATIBILITY: *Keep a single mated pair together. Males will fight if placed in the same aquarium. Other peaceful fish will make suitable companions for it, but it is sensible not to add too many other bottom dwelling fish to the aquarium alongside this prawn goby.*

▼ BLACK-RAY PRAWN GOBY (STONOGOBIOPS NEMATODES)

DARTFISH FAMILY MICRODESMIDAE

The elongated shape of dartfish has led to them being given the alternative common name of wormfish. They are closely related to gobies. Their most obvious characteristic is probably the raised area at the front of the split dorsal fin. It helps keep them anchored in place under a rock, for example, out of reach of predators. They will also flick this area of the dorsal fin regularly in the open in order to communicate with other dartfish. They are generally shy fish, but can settle well in aquarium surroundings.

FIRE GOBY

(MAGNIFICENT DARTFISH)

Nemateleotris magnifica

The front part of the body of the fire goby is whitish, but it becomes lemon-yellow in the vicinity of the head. The rear half is orange-brown, becoming darker toward the tail. Its most distinctive feature is the very long slightly backward-sloping front part of the dorsal fin, which is yellowish with narrow red and blue stripes at the leading edge. There is also a white stripe extending from the front of this fin down to the center of the head between the eyes.

DISTRIBUTION: *Ranges widely from East Africa across the Indian Ocean, extending north to the Ryukyu Islands near Japan and south as far as New Caledonia and the Austral Islands. Ranges through the Pacific from Micronesia eastward as far as the Hawaiian, Marquesan, and Pitcairn Islands.*

SIZE: *3.5 in (9 cm).*

BEHAVIOR: *This fish is likely to be seen hovering in the water, off the bottom, with its head pointed in the direction of the water column so that it can filter out edible particles of zooplankton flowing past. It will seek the sanctuary of holes in the reef when not feeding and will shelter there from predators.*

DIET: *Can be induced to feed relatively easily on live brine shrimps. Also requires very finely chopped meaty foods, which can be wafted near the*

▼ *FIRE GOBY (NEMATELEOTRIS MAGNIFICA)*

fish on the current. A varied diet seems to be essential to allow this fish to maintain its attractive coloration.

AQUARIUM: *Ideally suited to a reef tank, provided that it is not mixed with larger, more dominant companions. Otherwise, it is likely to hide away and refuse to eat. It must be kept in a covered aquarium, however, because it jumps readily (particularly if disturbed) and may leap out of the tank.*

COMPATIBILITY: *In spite of its small size, the fire goby is exceedingly intolerant of the company of its own kind in aquarium surroundings, so keep individuals separate from each other to prevent serious fighting. True pairs are more likely to be compatible in an aquarium.*

ELEGANT FIREFISH

(PURPLE FIREFISH)

Nemateleotris decora

The elegant firefish is an extremely attractive species, although it is very variable in appearance. The body is often predominantly yellow with a purplish stripe running along the back. The stripe continues onto the dorsal fin, which extends down onto the caudal peduncle. There are also black and red markings evident in the fins alongside the purple coloration, and some individuals display more prominent red flashes at the top and the bottom of the caudal fin.

▲ *ELEGANT FIREFISH (NEMATELEOTRIS DECORA)*

DISTRIBUTION: *Ranges from the waters around Mauritius in the Indian Ocean northward to the Ryukyu Islands and southward down to New Caledonia. Extends eastward through the Pacific as far as the island of Samoa.*

SIZE: *3.5 in (9 cm).*

BEHAVIOR: *Hovers above the surface of the reef seeking zooplankton, which forms the basis of its diet. If alarmed, it will retreat into a suitable hiding place. For that reason, it is important to include some coral rubble on the base of the aquarium with rockwork nearby as well. This will give the fish a sense of security.*

DIET: *Live brine shrimps will help acclimatize this fish. It can be persuaded to take particles of food from the floor of the aquarium. Meaty items should be provided and should be chopped very finely so that they can be eaten easily and so that they will be wafted on the current in the tank. May ultimately take some marine flake.*

AQUARIUM: *Suitable for a reef tank but should not be mixed with larger fish that may bully it, because it will then hide away and refuse to feed.*

COMPATIBILITY: *Aggressive toward its own kind. Keep only singly or in compatible pairs.*

Surgeonfish, Tangs, and Unicornfish Family Acanthuridae

The unusual name of surgeonfish—or, less commonly, doctorfish—given to many of the members of the family Acanthuridae stems from the presence of a sharp scalpel-like projection on each side of the tail. Together with sharp spines incorporated in the dorsal and anal fins, which are venomous in some fish, they are used for defensive purposes and can cause an unpleasant injury. For this reason, be very careful when servicing the aquarium. Surgeonfish also need to be caught with care to prevent the projections and spines from getting caught in the mesh of the net. When transporting them, it is also essential to ensure they are well packed in order to avoid puncturing a travel bag inadvertently.

Members of the group known as unicornfish are easily distinguished by a hornlike projection on the head. The other common name of tang that is applied to some of them is an abbreviation of the German word *seetang*, meaning seaweed. This form of marine alga features prominently in the natural diets of many species of tangs. They will also scavenge among detritus, taking small invertebrates as well, and should therefore be given the same opportunity in aquarium surroundings.

There are approximately 72 recognized species in the family Acanthuridae. They are very lively fish and require well-oxygenated surroundings, preferably with good water movement. Spawning is most unlikely in aquarium surroundings and has not been recorded to date. On the reef, however, it is not unusual for spawning to be a mass event, although some species develop pair bonds.

Unfortunately, acanthurids can be prone to ick (white spot) and must therefore be quarantined carefully. Medication can be a problem with this group of fish, because they rely to an extent on commensal bacteria in their digestive tract, which help them break down plant material. Medicines that depress bacteria may therefore prevent the fish from being able to feed properly.

But assuming they are healthy, even acanthurids that are reluctant to feed will usually start doing so when offered live brine shrimps. It is important to provide some food of this type, partly because it will aid the growth of young fish, which tend to be more omnivorous than adults. Generally acanthurids are natural browsers and spend much of the time feeding. As a result, they need to be fed three or four times a day in an aquarium.

◄ *Japan Surgeonfish (Acanthurus japonicus).* *See page 203.*

▲ *Achilles Tang (Acanthurus achilles)*

ACHILLES TANG

Acanthurus achilles

The Achilles tang can be identified by its dark brown body color with a dark bluish overlay. Its most striking feature, however, is the large orange spot tapering toward the tail on each side of the body. The caudal fin itself is also brightly colored, and there is a narrow orange stripe that reaches around the base of the dorsal fin at the rear of the body. A light blue area is evident around the lower edge of the gills. Juveniles are duller and lack the bright orange spot seen in adults.

Synonym: *A. aterrimus.*

Distribution: *Ranges from the Torres Straits between New Guinea and northern Australia to the Caroline Islands farther north and eastward across the Pacific to the Hawaiian, Marquesan, and Ducie Islands, reaching the southern tip of Baja California, Mexico.*

Size: *9.5 in (24 cm).*

Behavior: *Often seen in groups on areas of the reef where there is a strong current.*

Diet: *Feeds mainly on filamentous and fleshy macroalgae and will acclimatize more easily if these are growing in the tank. Offer a wide range of herbivorous foods in any event.*

Aquarium: *Requires spacious surroundings, with a good water movement. Usually safe for a reef tank, but may occasionally nip at coral.*

Compatibility: *Not especially social with its own kind in aquaria and should be kept on its own unless the aquarium is very big.*

BLACK-SPOT SURGEONFISH

Acanthurus bariene

The brown body color of this fish is broken by a yellow band on the top of the head above the eyes, extending back along the dorsal fin. Similar coloration is also present along the lower and upper edges of the caudal fin, which becomes lyre shaped with age. There is a black spot on each side of the body in the vicinity of the caudal peduncle. Once mature, the male also develops a more convex forehead that bulges out beyond its lips.

Distribution: *East Africa, extending down to Mozambique through the Indian Ocean to the Maldives off the southwest coast of India and into the western Pacific.*

SIZE: *19.75 in (50 cm).*

BEHAVIOR: *Adults are seen either on their own or in pairs, usually in relatively deep water below 100 ft (30 m) in open areas of the reef. By contrast, youngsters are found in much more shallow and sheltered areas, often among soft corals.*

DIET: *Feeds on algae that it grazes off rocks. Offer a herbivorous diet, including fresh items such as a little organic broccoli.*

AQUARIUM: *Can be kept safely in a reef tank, although the tank needs to be relatively large to reflect the size of this fish.*

COMPATIBILITY: *Unlikely to agree well together, so keep separate.*

BLUE TANG

(ATLANTIC BLUE TANG)

Acanthurus coeruleus

Blue coloration predominates in adults of this species, with wavy horizontal patterning present on the sides of the body. The spine along the side of the caudal peduncle is highlighted in yellow. In contrast, juveniles are mainly yellow overall, with the caudal fin being the last part of the body to change to blue.

SYNONYMS: *A. brevis; A. broussonnetii; A. caeruleus.*

DISTRIBUTION: *Ranges from the vicinity of New York on the East Coast eastward to Bermuda and through the Gulf of Mexico as far south as Brazil. Also found around Ascension Island farther out in the Atlantic Ocean.*

SIZE: *15.5 in (39 cm).*

BEHAVIOR: *Young of this species have been known to form an unusually close relationship with green turtles* (Chelonia mydas). *They will groom the reptiles (removing parasites from their bodies, often concentrating on their flippers) at particular sites on the reef, known as cleaning stations. They may be joined by other fish, such as the Pacific sergeant majors,* (Abudefduf saxatilis).

DIET: *Feeds essentially on algae of various types, which should be present in its diet, along with other sources of green food. This can even include a little dry seaweed, sold under the name sushi noir.*

AQUARIUM: *This fish is suitable for a reef tank, although it may occasionally nibble at coral.*

COMPATIBILITY: *Needs spacious surroundings and should be housed individually, because fights involving the caudal spines can cause serious injuries.*

WHITE-SPOTTED SURGEONFISH

Acanthurus guttatus

As its name suggests, much of the rear part of the body of this fish is covered in white spots. The background is a dark brownish color. The caudal fin is white, and there is a broad white stripe encircling the body behind the pectoral fins and another one passing down behind the eye to the level of the lower jaw. The pelvic fins are a contrasting bright yellow.

DISTRIBUTION: *Ranges from the western part of the Indian Ocean eastward as far as the Hawaiian, Marquesan, and Tuamotu Islands. Extends northward*

▼ BLUE TANG (ACANTHURUS COERULEUS)

▲ White-spotted Surgeonfish (Acanthurus guttatus)

to the Ryukyu Islands near Japan, and its southerly limits are marked by New Caledonia and Rapa.

Size: *10.25 in (26 cm).*

Behavior: *Seen in shoals on the seaward side of the reef in shallow water, where there are often breaking waves. It has been suggested that its spots resemble the air bubbles seen in water crashing over the reef. They help provide protection from predators by disguising the outline of the fish.*

Diet: *Feeds on algae of various types and will benefit from the presence of existing algal growth when moved to a new aquarium. More likely to take a thawed herbivorous food at first than a corresponding dry food.*

Aquarium: *Safe for a reef tank, but needs good water movement in the aquarium.*

Compatibility: *Although these fish are found in shoals in the wild, it is safer to keep individuals apart in an aquarium setting.*

JAPAN SURGEONFISH (POWDER-BROWN SURGEONFISH)

Acanthurus japonicus

A prominent white area extends from the eyes to the jaws. Overall, the body is brownish in color. There is an orange band at the rear of the dorsal fin, while the caudal peduncle is yellow. The yellow color often extends to the caudal fin, which is otherwise powdery blue. The pectoral fins have extremely conspicuous yellow bases.

Distribution: *Extends from the Ryukyu Islands close to Japan southward via the Philippines as far as the island of Sulawesi, Indonesia.*

Size: *8.5 in (21 cm).*

Behavior: *This surgeonfish occurs in shallow waters over open areas of the reef. Its body coloration may vary a little, depending partly on its mood. Some individuals are significantly brighter than others.*

Diet: *Feeds on various types of algae. It is not a particularly easy surgeonfish to acclimatize, so be sure to have a tank in which there is plenty of natural food available before acquiring this species. This should make the transition period much easier.*

Aquarium: *Suitable for a reef tank and will benefit from the presence of retreats. It may sometimes nibble at stony coral polyps.*

Compatibility: *Shy, especially at first, so should be kept individually and not mixed with other related species.*

POWDER-BLUE SURGEONFISH

Acanthurus leucosternon

The vivid coloration of this surgeonfish makes it unmistakable. The body coloration is powder blue, as its name suggests, and the fish's face is blackish with a white band across the throat. These markings contrast with the bright yellow coloration that predominates in the dorsal fin and extends as far as the caudal peduncle. The base of the pectoral fins is also yellow. Females grow to a significantly larger size than males.

Synonyms: *A. delisiani; A. leucocheilus.*

Distribution: *Ranges across the Indian Ocean from east Africa via the Andaman Islands as far as southwest Indonesia and Christmas Island, which lies south of Sumatra.*

Size: *10 in (25 cm).*

▼ Powder-blue Surgeonfish (Acanthurus leucosternon)

BEHAVIOR: *Tends to inhabit relatively shallow waters and is often seen over open areas of the reef. It can be encountered in large groups, but in aquaria the powder-blue surgeonfish proves to be one of the most aggressive surgeonfish.*

DIET: *Feeds mainly on algae. The acclimatization of new arrivals is helped if algae are already growing in the aquarium. Supplement with suitable prepared diets containing* Spirulina *and other species. Offer other types of greenstuff too, including pieces of organic broccoli.*

AQUARIUM: *Usually suitable for a reef tank, since it will not generally harm typical occupants found there, although occasionally it may nibble at stony coral polyps.*

COMPATIBILITY: *Must be kept apart from its own kind because it is highly aggressive, especially in the confines of an aquarium. If stressed, this species is very likely to develop ick (white spot), to which surgeonfish as a group are very susceptible.*

LINED SURGEONFISH
(CLOWN SURGEONFISH)

Acanthurus lineatus

Horizontal yellow and blue stripes running along the sides of the body from the eyes help identify this species. The patterning on the face is similar, with faint dark edging present on the blue stripes, although they are random in appearance, allowing individuals to be distinguished easily. There is a powder-bluish area extending along the underside of the body, while the dorsal and anal fins have narrow black and pale blue striping running along their length.

DISTRIBUTION: *East Africa via the Mascarene Islands through the Indian Ocean to the Philippines, reaching northern Japan. Ranges southward as far as the Great Barrier Reef off Australia's east coast, and to New Caledonia. Extends eastward to the Hawaiian and Marquesan Islands and the Tuamotu Archipelago.*

SIZE: *15 in (38 cm).*

BEHAVIOR: *This surgeonfish is very lively by nature and it occurs in the outer surge zone on reefs, where the sea is very fast flowing. Males are highly territorial and they watch over groups of females. They have a particularly large venomous spine on each side of the caudal peduncle.*

DIET: *Feeds primarily on algae but may also occasionally eat crustaceans. Having algae already established in the aquarium will simplify the acclimatization process. Then introduce prepared herbivorous foods to the diet, as well as fresh food such as organic broccoli.*

AQUARIUM: *Not to be trusted with crustaceans and may occasionally damage stony coral in a reef tank. Must have plenty of open space .*

COMPATIBILITY: *Definitely should not be mixed with its own kind or with other members of the family; can even be belligerent toward unrelated fish that display similar patterning.*

EPAULETTE SURGEONFISH

Acanthurus nigricauda

This surgeonfish is essentially dark brown in color, although there may be some variation through its wide range, with a purplish gray variant also having been documented. The black band that runs horizontally behind the gills is always absent in juveniles. Pale markings are apparent on the pectoral fins and on the edge of the caudal fin.

DISTRIBUTION: *East Africa via the Mascarene Islands, extending northward as far as the Ryukyu Islands near Japan and south to the Great Barrier Reef off Australia's east coast. Ranges as far east as the Tuamotu Archipelago in the Pacific.*

SIZE: *15.75 in (40 cm).*

BEHAVIOR: *Unlike most members of the family, this surgeonfish tends to be found in sandy areas of the reef, where coral does not predominate.*

DIET: *Likely to feed partly on crustaceans and other invertebrates. Algae are less significant in its diet, reflecting the anatomy of its stomach—the stomach walls are well muscled for grinding up hard-shelled food.*

AQUARIUM: *Can be problematic in a reef aquarium alongside crustaceans and hard coral.*

COMPATIBILITY: *Should be kept apart from other surgeonfish unless the aquarium is very large. Its dull coloration means that it is one of the less popular surgeonfish, although its care presents no particular problems.*

ORANGE-SPOT SURGEONFISH

Acanthurus olivaceus

The coloration of this surgeonfish changes dramatically with age. Whereas juveniles are bright yellow in color, sometimes with a hint of a black bar behind the eyes on each side of the head, adults are markedly different in appearance. The front part of their body is light gray with a darker gray area behind. A horizontal broad bright orange stripe, edged with a purplish black, extends back from behind the eyes. The caudal fin becomes lyre

▼ *EPAULETTE SURGEONFISH (ACANTHURUS NIGRICAUDA)*

▲ *ORANGE-SPOT SURGEONFISH (ACANTHURUS OLIVACEUS)—JUVENILE*

shaped. It has a light bluish gray border, although the lyres themselves are dark. There may be blue markings present on the lower lips as well.

SYNONYMS: *A. chrysosoma; A. eparei; A. humeralis.*

DISTRIBUTION: *Ranges from the Cocos (Keeling) and Christmas Islands in the eastern Indian Ocean northward to southern Japan and south to Lord Howe Island, which lies off the east coast of Australia. Ranges across the Pacific as far as the Hawaiian Islands and the Tuamotu Archipelago.*

SIZE: *15.75 in (40 cm).*

BEHAVIOR: *The young occur, often in groups, in more sheltered areas in relatively shallow waters—less than 10 ft (3 m) in depth. Adults are encountered on bare areas of the reef, frequently among rocks and on their own, but sometimes associating in groups. It is often possible to distinguish the sexes, since adult males have more convex mouths than the females.*

DIET: *Feeds naturally on algae growing on the rocks and small edible particles gathered on the sand. Herbivorous foods of various types are suitable for this species.*

AQUARIUM: *Does not normally prove disruptive in a reef aquarium and can help keep algal growth in check, while not attacking coral. Benefits from an aquarium with relatively little decor and a broad sandy base on which it can browse for edible items.*

COMPATIBILITY: *One of the more tolerant members of the group, but do not mix with its own kind. A juvenile and adult may be suitable to house together, however, if you have a very large aquarium. Normally unlikely to attack other related species, but choose companions with care.*

CHOCOLATE SURGEONFISH

Acanthurus pyroferus

This is another acanthurid that alters dramatically in appearance as it matures. Juveniles are a rich shade of yellow overall, often with relatively faint sky-blue stripes highlighting the edges of the gills and the region around the mouth, while the area surrounding the eyes is also blue. These markings enable them to mimic the lemonpeel angelfish (*Centropyge flavissimus*) in waters around Guam. In areas where *C. flavissimus* is absent, such as Palau, juvenile chocolate surgeonfish have pearl-like

▼ *CHOCOLATE SURGEONFISH (ACANTHURUS PYROFERUS)—JUVENILE*

markings and are also significantly darker over the rear of the body, so they look like the pearl-scale angelfish (*C. vrolikii*). This form of mimicry gives the juvenile chocolate surgeonfish increased protection against predators at this early stage, because angelfish are less susceptible to predation. Chocolate surgeonfish then start to attain the basic brownish yellow adult coloration when they grow larger. Their caudal fin also ceases to be rounded and develops the characteristic lyre-tailed appearance of adults. Their bodies darken around the head and take on a chocolate color, with the rear of the body and caudal fin becoming similarly colored. A white area extends from the eyes to the snout, and the rear edge of the caudal fin is yellowish.

SYNONYMS: *A. armiger; A. celebicus; A. fuscus.*

DISTRIBUTION: *From Seychelles through the Pacific region, reaching the Marquesan Islands and the Tuamotu Archipelago. Ranges north as far as southern Japan and south to the Great Barrier Reef and New Caledonia.*

SIZE: *10 in (25 cm).*

BEHAVIOR: *Can sometimes be found in silty water as well as in areas of the reef where there may be a range of sand, rubble, and coral present. Usually encountered individually and not in groups.*

DIET: *Feeds primarily on algae and other plant matter. Will take herbivorous diets in aquarium surroundings and will also browse on fresh greenstuff such as organic zucchini.*

AQUARIUM: *Relatively safe to include in a reef setup, where it helps control algal growth, although it may occasionally nibble at stony coral.*

COMPATIBILITY: *It may be possible to keep a juvenile and an adult together in a spacious aquarium, but adults should generally be housed on their own and not with other acanthurids.*

SOHAL SURGEONFISH

Acanthurus sohal

The predominant pattern on the body of this surgeonfish consists of horizontal stripes—usually pale blue and brown. There are scattered yellowish areas on the body and the pectoral fins. The patterning in the facial area is paler than elsewhere, and its underparts are a pale grayish shade. The dorsal and anal fins and and the lyre-tailed caudal fin are narrowly edged in bright blue, while the very prominent spine present on each side of the caudal peduncle is highlighted in orange.

SYNONYM: *A. carinatus.*

DISTRIBUTION: *Restricted to the Red Sea and the Arabian Gulf.*

SIZE: *15.75 in (40 cm).*

BEHAVIOR: *Tends to be found on the seaward fringes of reefs where there are strong currents. It is highly territorial by nature, and in the wild it has been observed attacking unrelated fish, such as triggerfish, that stray into its feeding grounds. It can use its caudal spines to inflict severe injuries on its opponents.*

▼ SOHAL SURGEONFISH (ACANTHURUS SOHAL)

▲ *YELLOW-FIN SURGEONFISH (ACANTHURUS XANTHOPTERUS)*

DIET: *Feeds on algae, in spite of its aggressive nature.* Spirulina *algae should feature in its diet, along with other similar foods. Both thawed and dried herbivorous foods are acceptable, although the former are likely to be more palatable.*

AQUARIUM: *Can be included in a reef tank, since it is unlikely to cause harm to the occupants, although it may nibble at spongy coral polyps.*

COMPATIBILITY: *Must be kept separate from both other acanthurids and fish with similar feeding habits in aquarium surroundings, because it is likely to become extremely aggressive in a confined space.*

YELLOW-FIN SURGEONFISH

Acanthurus xanthopterus

One of the less brightly colored members of the family, the yellow-fin surgeonfish has a purplish gray body and a pale yellow area on the face. The pectoral fins are also yellowish. There is narrow yellow edging around the dorsal, anal, and caudal fins, which explains this fish's common name. The caudal fin is otherwise mainly purplish, however, with a white area at its base. The caudal spines are relatively small.

SYNONYM: *A. crestonis.*

DISTRIBUTION: *Ranges widely from East Africa across the Pacific to the Hawaiian Islands, south along the Great Barrier Reef and New Caledonia, and east to the Marquesan Islands. Also from the lower Gulf of California and Clipperton Island to the coast of Panama in Central America, extending to the Galápagos Islands. Its northernmost distribution is southern Japan.*

SIZE: *27.5 in (70 cm).*

BEHAVIOR: *This large surgeonfish is a little less aggressive than some of its relatives and often occurs in shoals. Juveniles congregate in sheltered inshore areas in relatively shallow waters, while adults are likely to be found in deeper yet fairly tranquil areas of the reef, such as lagoons.*

DIET: *Feeds on algae and other vegetable matter, but is likely to prove omnivorous in its feeding habits. In the wild it can be caught on a baited hook. In addition to herbivorous foods, therefore, in aquarium surroundings it can be offered small pieces of fish and crustaceans such as prawns.*

AQUARIUM: *May prey on some of the inhabitants of a reef tank as well as browsing on algae.*

COMPATIBILITY: *Its size means that this species should be kept in an aquarium on its own. It should not be mixed with related species because of the risk of aggression.*

CHEVRON TANG

(BLACK SURGEONFISH)

Ctenochaetus hawaiiensis

Juveniles of this species are especially colorful, having a very vivid orange-red body color crossed by an irregular pattern of purplish lines that often form chevrons. Solid purple markings are also present on the dorsal, caudal, and anal fins. The brilliant coloration disappears as the fish matures, when its appearance is transformed and becomes much duller. Although adults may appear black, their bodies are, in fact, covered by a series of fine greenish lines on a dark background.

▲ CHEVRON TANG (CTENOCHAETUS HAWAIIENSIS)

DISTRIBUTION: *Restricted to the central part of the Pacific Ocean from the islands of Micronesia eastward via Marcus and Wake Islands to the Hawaiian Islands and south to Pitcairn Island.*
SIZE: *10 cm (25 cm).*

BEHAVIOR: *Juveniles occur in relatively deep water, typically hiding among corals, while adults may occur in more rocky areas. This species has not been well studied in the wild and appears to be relatively scarce.*

DIET: *Algae and other vegetable matter should form the basis of its diet. In common with related species, this fish needs to be fed small quantities several times during the day, because it naturally appears to be a browser that eats throughout much of the day.*

AQUARIUM: *Not usually a problem in a reef tank.*

COMPATIBILITY: *Although it is not an especially aggressive species, do not mix it with other surgeonfish, including its own kind. Can form part of a community aquarium.*

▼ STRIATED SURGEONFISH (CTENOCHAETUS STRIATUS)

STRIATED SURGEONFISH

Ctenochaetus striatus

Juveniles of this species are recognizable by a series of between eight and 12 pale stripes angled vertically backward across the body. They disappear in adult fish, which have dark olive-brown coloring offset against some orange spotting on the head and the yellow of the pectoral fins. A well-concealed sharp spine lies in a groove on each side of the caudal peduncle, with its tip directed toward the head.

SYNONYM: *Acanthurus striatus.*

DISTRIBUTION: *Has a wide range from East Africa through most of the Pacific, but is not present around the Hawaiian Islands nor some others lying farther south, including Jarvis, Malden, Marquesan, and Easter Islands.*

SIZE: *10.25 in (26 cm).*

BEHAVIOR: *Sometimes observed in large groups, even in the company of other species, although often seen individually. It is not tied to coral-rich areas of the reef and may venture down to below depths of 100 ft (30 m).*

DIET: *Feeds on blue-green algae and diatoms as well as invertebrates, aided by its movable teeth with their sharp, backward-curving tips. A diet based primarily on herbivorous foods suits this surgeonfish well, along with occasional offerings of a little chopped meaty food.*

AQUARIUM: *Likely to prey on some of the occupants of a typical reef tank.*

COMPATIBILITY: *Should be kept apart from its own kind in an aquarium, but may be housed successfully with unrelated nonagressive fish.*

SPOTTED SURGEONFISH

(KOLE'S TANG; YELLOW-EYE SURGEONFISH)

Ctenochaetus strigosus

The distinctive golden ring that surrounds the eyes of this surgeonfish is one of its distinguishing features. A series of blue spots on the head is replaced by stripes on the body that are essentially horizontal. There is a more haphazard patterning on the fins in the case of individuals that occur in the Pacific Ocean. Those originating from the Indian Ocean, on the other hand, display an entirely spotted pattern. Juveniles are bright yellow, but their appearance alters significantly as they mature—their background color turning much browner as they develop.

SYNONYM: *Acanthurus strigosus.*

DISTRIBUTION: *Ranges widely through the Indo-Pacific region and is especially well known around Johnston Island and the Hawaiian Islands.*

▲ *Spotted Surgeonfish (Ctenochaetus strigosus)*

Size: *7 in (18 cm).*

Behavior: *A solitary species that finds food by sweeping through accumulations of debris for edible particles. It is equipped with comblike teeth, which it can use as a filter when seeking food.*

Diet: *Browses readily on algae in the aquarium. This needs to be supplemented with other prepared sources of food, including frozen herbivorous items and fresh food such as organic zucchini, which should be sliced into thin rings.*

Aquarium: *An ideal choice for the reef aquarium, because it should not molest other occupants.*

Compatibility: *Needs to be kept individually to avoid conflict with members of the same genus; also liable to be bullied by other surgeonfish.*

SPOTTED UNICORNFISH

(PALE-TAIL UNICORNFISH)

Naso brevirostris

This unmistakable species has a hornlike projection on its head, resembling the horn of the mythical unicorn. However, it is absent in younger individuals, whose dark spots on the head and body also disappear on maturity. The color of adults can be variable. It is paler on the head than on the body and ranges from olive-brown to bluish gray. There may be some lines evident on the body, but the lips are bluish in all cases.

Distribution: *Ranges from the Red Sea and East Africa to the Hawaiian and Marquesan Islands, reaching as far east as Ducie Islands in the Pacific. Extends to southern Japan in the north and to Lord Howe Island off Australia's east coast in the south.*

Size: *23.75 in (60 cm).*

Behavior: *A very active fish, most likely to be observed along the steep sides of a reef, sometimes venturing alongside a rocky coastline. It needs a large covered aquarium with plenty of open space for swimming.*

Diet: *Juveniles feed mainly on algae in the wild and require a corresponding diet in the aquarium, preferably incorporating some natural food that they*

▼ *Spotted Unicornfish (Naso brevirostris)*

can graze in the aquarium. Adults, on the other hand, feed on zooplankton and should therefore receive a primarily meaty diet, incorporating items such as brine shrimps.

AQUARIUM: *Should not prove to be disruptive in a reef aquarium.*

COMPATIBILITY: *Avoid the company of its own kind and similar species, although unicornfish will not harass unrelated fish.*

ORANGE-SPINE UNICORNFISH

Naso lituratus

This is a particularly beautiful fish. Males can be distinguished by the filaments that trail from the tips of the caudal fin. The basic body color is mauve with a yellow area surrounding the eyes. Stripes extend down to the lips on each side of the face and are divided by a dark area that extends down to the jaws. The lips themselves are highlighted in orange, and there are two prominent orange spots next to the white area on the caudal peduncle. The anal fin also has an orange hue. The caudal fin has a a yellow bar close to the end, while the blue dorsal fin has a black line separating its tip and base. Similar striped patterning is apparent across the cheeks. The population from the Indian Ocean is now regarded as a distinct species, known as *N. elegans*.

DISTRIBUTION: *Ranges from New Caledonia and the Great Barrier Reef northward to Honshu, Japan, and eastward across the Pacific to the Hawaiian and Marquesan Islands, extending south to Pitcairn Island.*

▲ *ORANGE-SPINE UNICORNFISH (NASO LITURATUS)*

SIZE: *18 in (46 cm).*

BEHAVIOR: *Sometimes seen in groups ranging over rubble just as often as coral, favoring both seaward reefs and lagoons. The bright orange coloration on the caudal peduncle serves to highlight the spines there, which are held permanently erect. This fish therefore must be caught and handled with care. It requires a spacious aquarium with suitable retreats.*

DIET: *An algae feeder, favoring brown varieties such as sargassum (available from some healthfood stores), which can be used to supplement standard herbivorous foods. May also eat some brine shrimps, mirroring the plankton that also features in its natural diet.*

AQUARIUM: *Should not prove destructive in a reef aquarium, apart from grazing on algae. May occasionally nip at coral.*

COMPATIBILITY: *Likely to be aggressive toward its own kind, but can be mixed with other unrelated fish.*

HUMP-NOSE UNICORNFISH

Naso tuberosus

A highly distinctive and not unattractive species, in spite of its rather bizarre appearance and plain coloration. This unicornfish is predominantly whitish in color, with black spotting on the upper part of the body behind the eyes. The pectoral fins are dark, as is the tip of the caudal fin. There is a distinctive hump on the back some distance from the head, and a broad swollen area extends down below the eyes in the direction of the upper jaw. This jaw is much larger than that of other unicornfish.

Distribution: *Restricted to the vicinity of the western Indian Ocean, occurring from the coast of Mozambique eastward to the islands of Réunion and Mauritius and northward as far as the Seychelles.*

Size: *23.75 in (60 cm).*

Behavior: *Almost nothing has been recorded about this particular unicornfish, although the size of the hump is thought to increase with age and to be especially evident in males. It is believed that juveniles at least may sometimes wander into coastal areas and even into the mouths of rivers where they encounter brackish water conditions.*

Diet: *A typical mixed diet is recommended, with the emphasis on plant matter as a substitute for the algae on which this fish feeds naturally, along with brine shrimps on occasion.*

Aquarium: *Suitable for a reef aquarium, where it browses on algae. Unlikely to cause any serious damage to corals.*

Compatibility: *Keep these unicornfish separate from each other and related acanthurids. They can, however, be mixed with other species in a large aquarium.*

BLUE-SPINE UNICORNFISH

Naso unicornis

The overall coloration of this unicornfish is bluish green, and there are distinctive bluish spots over the caudal peduncle, where its spines are located. The spines are held permanently erect. There may also be some blackish spots, particularly toward the

▼ *Hump-nose Unicornfish (Naso tuberosus)*

▲ *Blue-spine Unicornfish (Naso unicornis)*

front of the body. Its most characteristic feature, however, which is especially apparent in older individuals, is the bulbous projection present on the head just below the eyes. It tends to be largest in males, as are the caudal spines and the caudal fin filaments.

Synonym: *Acanthurus unicornis.*

Distribution: *Ranges from the Red Sea and East Africa north as far as southern Japan and south to Lord Howe Island off Australia's east coast. Its easterly range through the Pacific extends to the Hawaiian Islands, the Marquesan Islands, and the Tuamotu Archipelago.*

Size: *27.5 in (70 cm).*

Behavior: *Found in areas where there is a strong current. It is an active fish that needs plenty of swimming space accompanied by water movement. Where food is available, it is often seen in small groups.*

Diet: *Favors brown algae. It may be possible to acquire dried seaweed to supplement herbivorous foods in aquarium surroundings.*

Aquarium: *Will destroy algae in a reef aquarium, but rarely attacks coral. Unlikely to disturb other sessile reef occupants.*

Compatibility: *Do not mix with other acanthurids. Can be kept with unrelated species in a suitably large setup.*

BIG-NOSE UNICORNFISH

Naso vlamingii

The nasal projection of this particular unicornfish is highlighted by a horizontal blue band that also connects the eyes. The remainder of the head—apart from the blue lips—is green, broken by an irregular pattern of pale spots. The body is blue, becoming darker toward the rear. There is a series of blue spots on the upper parts that forms variable stripes lower down the sides of the body before breaking into spots again. The caudal fin is light blue at its base, and this coloration extends around the sides of the fin and runs into the filaments. The rest of the tail is black.

DISTRIBUTION: *Extends from East Africa northward as far as southern Japan. Its southerly range extends down to Australia's Great Barrier Reef and New Caledonia, while Its easterly range reaches the Line and Marquesan Islands as well as the Tuamotu Archipelago.*

SIZE: *23.75 in (60 cm).*

BEHAVIOR: *Has the ability to disguise its blue markings to merge into the background. Usually encountered as pairs or individually, frequenting the slopes around the edge of the reef to feed on plankton during the day.*

DIET: *Requires a combination of foods such as mysid shrimps and brine shrimps—enriched with vitamins if possible—as well as herbivorous foods.*

AQUARIUM: *Reasonably safe in a reef aquarium, since it is unlikely to cause any significant damage to coral, for example.*

COMPATIBILITY: *Do not try to mix this unicornfish with related species. A single individual may be housed successfully with other fish, however, but not with other acanthurids.*

PALETTE SURGEONFISH

(PACIFIC BLUE TANG)

Paracanthurus hepatus

The appearance of this surgeonfish varies across its range, although blue coloration predominates. A black stripe runs back from the eyes below the level of the dorsal fin. Near the caudal peduncle it joins with another streak that runs forward, curving at its tip close to the pectoral fin on each side of the body. The caudal fin itself is yellow with black markings along its upper and lower edges. Once these surgeonfish reach about 8 inches (20 cm) long, however, their coloration begins to change dramatically—the belly becomes yellowish, and the black stripes disappear. The upperparts of the body turn brownish, and green appears on the face. This range of colors, resembling an artist's palette, explains the common name for the fish.

DISTRIBUTION: *From East Africa and the Mascarene Islands north to southern Japan. In a southerly direction it can be found on the southern part of Australia's Great Barrier Reef. Extends eastward through the Pacific Ocean to Samoa.*

SIZE: *12.25 in (31 cm).*

BEHAVIOR: *Occurs on the seaward side of the reef where there are strong currents. The young seek the protection of coral if danger threatens, and hide among the polyps.*

DIET: *Naturally feeds predominantly on zooplankton, so a corresponding diet in aquarium surroundings should include vitamin-enhanced brine shrimps as well as some vegetarian food.*

AQUARIUM: *Can usually be included in a reef tank without causing any problems. Will benefit from retreats, since it is often rather shy by nature, especially at first.*

COMPATIBILITY: *It is possible to keep juveniles together, but aim to keep three as a minimum—bullying is then less likely than if just two are housed together. They are ultimately likely to need separating, however, unless the aquarium is very large, because they become more territorial with age. They can be mixed with placid companions but not with other members of their family.*

▼ PALETTE SURGEONFISH (PARACANTHURUS HEPATUS)

▲ Indian Ocean Sail-fin Tang (Zebrasoma desjardinii)

INDIAN OCEAN SAIL-FIN TANG
(RED SEA SAIL-FIN TANG)

Zebrasoma desjardinii

The relatively high dorsal fin of this tang resembles a sail, which explains the common name for the fish. The body markings are bold, with a series of alternating blue and yellow-edged dark vertical stripes. The underside is covered with yellow spots. Brownish stripes broken by whitish spots run through the eye and the pectoral fins, and the intervening areas are bluish white.

Distribution: *Ranges from the Red Sea down the eastern coast of Africa as far as the South African state of Natal. Extends across the Indian Ocean via India to the Cocos (Keeling) Islands and eastward as far as Java, although it is absent from the waters around Christmas Island.*

Size: *15.75 in (40 cm).*

Behavior: *Juveniles are likely to be seen in sheltered inner areas of the reef, but adults range more widely and can be found in rougher water on the seaward side. They may venture to depths below 100 ft (30 m).*

Diet: *Predominantly herbivorous, feeding on algae, but in aquarium surroundings will also take some invertebrates, such as brine shrimps. Prepared foods containing algae are to be recommended, along with other items such as dried seaweed, which is sold as sushi nori.*

Aquarium: *Can be beneficial in a reef tank, helping curb unwanted algal growth. Unlikely to be a problem, although it may occasionally nip at the polyps of some stony corals.*

Compatibility: *Keep these tangs individually, although a singleton should agree with unrelated species in such surroundings without any problems.*

YELLOW TANG

Zebrasoma flavescens

One of the most stunning and instantly recognizable members of the genus, the yellow tang is entirely yellow in color apart from the whitish blue spine that is evident on each side of the caudal peduncle. However, its bright coloration will only be maintained by an appropriate diet that includes natural coloring agents. Without such a diet, its body will become much paler. When buying these fish, check for any signs of obvious pitting on the head in the region of the eyes. It is a sign of head erosion, a serious condition that is usually combined with damage to the lateral line.

Distribution: *Occurs in the vicinity of various Pacific islands, including the Ryukyus near to Japan, as well as the Marianas, Marshall, Marcus, and Wake Islands. Common in the Hawaiian Islands, where most exports originate.*

Size: *8 in (20 cm).*

Behavior: *Closely linked with areas of coral on the reef and seen in groups. The appearance of this fish can alter to reflect its mood. For example, the the appearance of a temporary white band on the body indicates fear.*

Diet: *Feeds mainly on filamentous algae, probing as necessary with its snout for this purpose. Offer a herbivorous diet based on both thawed and dry foods. The provision of living algae in the aquarium is beneficial.*

Aquarium: *Suitable for a reef tank, since it will help keep algal growth in check in the aquarium. Unlikely to create problems, although it may nip at corals if hungry.*

Compatibility: *Unfortunately, these tangs need to be housed on their own in the average setup, since they do not generally agree in groups in these surroundings. If you have a very large tank, introduce all the fish at the same time to minimize any risk of subsequent territorial disputes. They are less likely to disagree as part of a larger group than if just two or three individuals are housed together.*

▼ Yellow Tang (Zebrasoma flavescens)

SPOTTED TANG

Zebrasoma gemmatum

As its name suggests, this tang is blackish brown in color with fine white spots, roughly arranged in horizontal rows, extending over its body onto the fins.

DISTRIBUTION: *Found in the western part of the Indian Ocean. Occurs off the coast of Mozambique and South Africa eastward to Madagascar, Mauritius, and Réunion Island.*

SIZE: *8.75 in (22 cm).*

BEHAVIOR: *Roams over both rocky and coral areas. Juveniles tend to be found in shallower areas of water than adults.*

DIET: *Primarily herbivorous, so offer suitable food such as thawed and dried saltwater diets containing algae. May also take seaweed and pieces of organic broccoli.*

AQUARIUM: *Suitable for a reef aquarium, since it will feed on algae but is unlikely to cause any damage.*

COMPATIBILITY: *One of the most territorial members of the genus, the spotted tang needs to be kept apart from its own and related species.*

TWO-TONE TANG
(BROWN TANG)

Zebrasoma scopas

Young of this species can be recognized by yellow bars on the body. They also display more evident yellow speckling than the adults, and as a result are brighter in color. This fish is called the two-tone tang because its yellowish upper parts contrast with the fine pattern of green horizontal streaking on the underside. The area between the eyes and the snout in particular is finely spotted. This species acquired the alternative common name of brown tang simply because this is the color of its body after death rather than while living, although there may be a slightly brownish hue evident on the underparts of a healthy individual.

DISTRIBUTION: *Extends over a wide area from the coast of East Africa to the Mascarene Islands, reaching as far north as southern Japan and as far south as Lord Howe and Rapa Islands off Australia's east coast. It ranges across the Pacific to the Tuamotu Archipelago.*

SIZE: *8 in (20 cm).*

BEHAVIOR: *Closely allied to areas of the reef where coral predominates, this tang has developed special teeth in the pharyngeal area at the back of the throat to assist it in eating strands of filamentous algae. May be seen feeding in groups of as many as 20 individuals, and often spawns communally as well.*

DIET: *Filamentous algae preferred, so ensure that you provide plenty of* Spirulina *as part of this tang's regular diet. It will browse on other greenstuff as well. When introducing an individual to the aquarium, a lush growth of suitable food is beneficial. It will help maintain the fish's coloration as well as giving some protection against head and lateral-line erosion disease, to which tangs can be very susceptible.*

AQUARIUM: *Suitable for a reef aquarium, since it helps curb unwanted algal growth and is unlikely to disturb the other occupants.*

COMPATIBILITY: *Unfortunately, just as with other tangs, although this species associates in groups on the reef, introducing more than a single individual to the aquarium (unless the tank is very large) is not usually possible This is partly because tangs can be territorial about their feeding grounds, seeking to defend areas where algal growth is abundant. In addition, interactions between members of a group are transient on the reef, but there is much greater potential for bullying to occur in aquarium surroundings because the fish are confined in closer proximity to each other.*

SAIL-FIN TANG

Zebrasoma veliferum

The species is very similar in profile to the Indian Ocean sail-fin tang, *Z. desjardinii*, which it replaces in the Pacific region. It has a very high dorsal fin, resembling a sail edged with a fine band of white on the rear. Juveniles tend to be more vividly marked than adults, displaying striking dark brown and yellow vertical banding in the vicinity of the head, with the contrast reducing farther down the body. The base of the caudal fin is orange-yellow. The

▼ *TWO-TONE TANG (ZEBRASOMA SCOPAS)*

▲ SAIL-FIN TANG (ZEBRASOMA VELIFERUM)

yellow markings tend to disappear in older fish, however, and the body patterning becomes brown and white—although there can be marked individual variations among different populations. The face itself may be white, for example, or a pale shade of grayish brown, broken by fine dots.

DISTRIBUTION: *Extends from Indonesia northward to southern Japan and south to Australia's Great Barrier Reef and nearby New Caledonia. Extends across the Pacific to the Hawaiian Islands and the Tuamotu Archipelago.*

SIZE: *15.75 in (40 cm).*

BEHAVIOR: *Juveniles tend to seek out more protected areas on the reef, hiding among rocks and coral if danger threatens. Adults favor lagoon and seaward reefs, extending down to depths of 100 ft (30 m). Studies have shown that the pharyngeal teeth in the vicinity of the throat are larger and less numerous in the sail-fin tang than in other* Zebrasoma *species.*

DIET: *The pattern of its pharyngeal teeth suggests that this tang tends to eat coarser vegetable food than its relatives—such as sargassum—although it will browse on other types of algae. Fresh food is important for maintaining the fish's coloration, and dry seaweed should be offered regularly in small amounts.*

AQUARIUM: *Suitable for inclusion in a reef tank and will keep algal growth in check. May occasionally nibble at coral polyps, especially if hungry.*

COMPATIBILITY: *Although often considered to be the most tolerant member of the genus, it is still not possible to keep two of these tangs together in the same aquarium unless it is very large. Otherwise, bullying will be inevitable.*

YELLOW-TAIL TANG

Zebrasoma xanthurum

In this species the bright yellow coloration of the caudal fin stands out against the vibrant blue of the body. The pectoral fins are also tipped with yellow. The yellow-tail is an unmistakable tang and it has become more readily available over recent years. However, as is the case with some other species, the natural vibrancy of its color will be lost unless it receives a well-balanced diet that helps maintain its appearance, as it does in the wild.

DISTRIBUTION: *Ranges from the Red Sea to the Persian Gulf, although it has also been reported farther west from the Maldives in the Indian Ocean.*

SIZE: *8.75 in (22 cm).*

BEHAVIOR: *Occurs in groups in both rocky areas of the reef and among areas of coral.*

DIET: *Fine pharyngeal teeth suggest a diet based naturally on filamentous algae. Herbivorous foods containing* Spirulina *are recommended. Both thawed and dry foods are acceptable. In common with related species, this tang needs feeding several times each day and ideally should also have the opportunity to browse on algae in the aquarium.*

AQUARIUM: *Suitable for a reef tank. This tang will keep algal growth in check without attacking sessile invertebrates, apart from perhaps occasionally nibbling at coral polyps.*

COMPATIBILITY: *Belligerent toward its own kind. It should not be mixed with related species either. If it is to be housed with other fish, it is a good idea to introduce this species last, because it has a somewhat dominant nature.*

▼ YELLOW-TAIL TANG (ZEBRASOMA XANTHURUM)

Moorish Idol Family Zanclidae

The Moorish idol is the only member of its family and is a stunningly beautiful fish. Unfortunately, it is not a species suitable for the novice saltwater enthusiast, because it will not adapt readily to substitute diets.

MOORISH IDOL

Zanclus cornutus

The adult fish is unmistakable—the dorsal fin spines form a long, narrow, filament that extends back beyond the caudal fin. Its body has a series of vertical stripes, and around the mouth is mainly white apart from a yellow marking with a narrow dark border across the snout. A broad black band encompasses the eyes and is separated from another at the rear of the body by a white area, part of which is suffused with yellow. A very narrow white stripe runs through this area. The caudal peduncle is yellow, but the caudal fin itself is black with white edges.

Synonym: *Z. canescens.*

Distribution: *Ranges throughout much of the Indo-Pacific region from East Africa north to southern Japan, south as far as Australia's southeast coast, and east via the Hawaiian Islands to the coast of California and south to Peru.*

Size: *9 in (23 cm).*

Behavior: *Most commonly observed in groups of two or three swimming widely over the reef, pausing in search of food. Adults start to attain their highly distinctive appearance once they reach about 3 in (7.5 cm) long.*

Diet: *Feeds naturally on invertebrates growing on the reef, so offer live rock that contains suitable creatures. Provide supplements, for example, vitamin-enriched brine shrimps and finely chopped meaty items such as pieces of squid. Be prepared to experiment in order to find something that will be acceptable. Some* Spirulina *and other algae may also be eaten.*

Aquarium: *Can be kept in a reef tank, but may eat some of the creatures.*

Compatibility: *It is possible to keep several together in a large aquarium. Incorporate retreats and large open areas for swimming.*

▼ *Moorish Idol (Zanclus cornutus)*

RABBITFISH FAMILY SIGANIDAE

The name of this family is a reflection of the shape of the upper lip, which is similar to a rabbit's. The group is closely allied to surgeonfish and tangs. Rabbitfish are protected by venomous spines on their fins, giving rise to their alternative name of "spinefoot." Their care is relatively simple. They adapt well to aquarium life and are often suitable for inclusion in a reef aquarium.

GOLD-SPOTTED RABBITFISH
(ORANGE-SPOTTED SPINEFOOT)

Siganus guttatus

The orange-spotted patterning of this fish is distinctive. The markings extend over the entire body and into the fins, while the background color is turquoise blue. Take care when servicing an aquarium housing these fish, however, because their dorsal spines are venomous. A distinctive gold spot is present on each side of the body, close to the rear of the dorsal fin.

SYNONYM: *S. concatenatus.*

DISTRIBUTION: *Confined to the eastern part of the Indian Ocean, ranging into the western Pacific, from the Andaman Islands to Thailand, Malaysia, and Indonesia, to Irian Jaya on New Guinea. Ranges northward via the Philippines and Palau to Taiwan and China and as far as the Ryukyu Archipelago south of Japan.*

SIZE: *16.5 in (42 cm).*

BEHAVIOR: *Tends to be found on reefs close to the coastline, often in mangrove areas. It is not unusual for adults to venture into the mouths of rivers at high tide. On the reef they remain relatively close to the surface and are unlikely to be encountered below 20 ft (6 m). The young seek cover in the relative safety of seagrass. These rabbitfish are highly social by nature, occurring in shoals of a dozen or more individuals.*

DIET: *Provide algae-rich herbivorous foods (preferably alongside algae growing in the aquarium). A little dried seaweed can also be offered.*

AQUARIUM: *Can be safely housed in a reef aquarium as a general rule, but this fish may occasionally attack coral.*

COMPATIBILITY: *Not especially social with its own kind in the confines of an aquarium, particularly once adult. The young need spacious accommodation in any case, since they grow to a potentially large size. Will not interfere with other unrelated fish, so suitable for a community tank.*

▲ GOLD-SPOTTED RABBITFISH (SIGANUS GUTTATUS)

MAGNIFICENT RABBITFISH

Siganus magnificus

Orange edging to the caudal, dorsal, and anal fins, together with pectorals that are almost entirely orange, helps distinguish this species. There is also a narrow orange band running down between the eyes and a broad blackish area beneath, extending through the eyes. The body is a whitish silvery shade, with a prominent chocolate-brown area on each side of the body below the dorsal fin.

▼ MAGNIFICENT RABBITFISH (SIGANUS MAGNIFICUS)

▲ *MASKED RABBITFISH (SIGANUS PUELLUS)*

SYNONYM: *Lo magnificus.*

DISTRIBUTION: *Occurs in the eastern Indian Ocean, ranging from Thailand and the Similan Islands east to Indonesia, extending as far south as Java.*

SIZE: *9 in (23 cm).*

BEHAVIOR: *Little has been documented about the behavior of this species in the wild. It is most likely to be encountered in true pairs.*

DIET: *Algae and small invertebrates. Usually feeds readily on substitute diets, for example,* Spirulina *augmented with vitamin-enriched brine shrimps and finely chopped meaty foods such as clams or shrimps.*

AQUARIUM: *Will not usually harm other occupants in a reef tank, although may occasionally nibble at coral polyps.*

COMPATIBILITY: *Pairs are compatible, but visual sexing is not possible, so it may be safer to keep them individually. Not aggressive toward tankmates.*

MASKED RABBITFISH

(MASKED SPINEFOOT)

Siganus puellus

This rabbitfish displays the typically angular dark band passing through the eyes that is a feature of many of its relatives. What sets it apart, however, is the horizontal pale blue streaking running along the sides of its body, set against a bright yellow background and becoming paler and more silvery on the flanks. Similar blue markings are also apparent on the forehead, with the upper part of the mask often displaying darker black spots.

SYNONYMS: *S. hexacanthus; S. sevenlineatus; S. zoniceps.*

DISTRIBUTION: *Ranges from the vicinity of the Cocos (Keeling Islands), which lie to the south of Sumatra, through the South China Sea and north as far as the Ryukyu Islands off southern Japan. Extends across the Pacific via the Gilbert Islands to Tonga and southward as far as the southern end of Australia's Great Barrier Reef.*

SIZE: *15 in (38 cm).*

BEHAVIOR: *Most likely to be encountered in shallow water where coral is abundant. The young form large shoals, especially where* Acropora *species coral is prevalent. As they mature, however, they tend to become more solitary and are seen in individual pairs. This rabbitfish's protective spines are especially venomous, so take particular care when placing your hand in the aquarium that you do not inadvertently come into contact with them.*

DIET: *Changes with age, with filamentous algae such as* Spirulina *predominating in the diet of juveniles. The food intake of adults is more varied and includes both sponges and tunicates. Juveniles should therefore be given mainly herbivorous foods, but adults should be offered larger amounts of chopped meaty foods.*

AQUARIUM: *Juveniles are suitable for a reef tank, but adults are likely to cause damage in these surroundings.*

COMPATIBILITY: *Juveniles will agree well together but they become more intolerant of the company of their own kind as they mature. However, it should be possible to keep pairs together in a suitably spacious aquarium. They are unlikely to harm other fish, although they can defend themselves with their spines if attacked.*

BICOLORED FOX-FACE

Siganus uspi

Rabbitfish as a group display considerable variation in their coloring. The bicolored fox-face is one of the less brightly colored members of its genus. Its appearance is unmistakable, however, with the front area of its body being a dark purplish brown. The rear part, including the ends of both the dorsal and the ventral fins, is bright yellow, as is the entire caudal fin.

SYNONYMS: *Lo uspi; Siganus uspae.*

DISTRIBUTION: *Has a very restricted distribution, centered on the waters around Fiji. Thie species may also be recorded occasionally from nearby New Caledonia.*

SIZE: *9.5 in (24 cm).*

BEHAVIOR: *Occurs among hard coral, typically on the edges of the reef where there are adjacent areas of deep water. Like other members of the group, this fish is well protected against potential predators by its dorsal spines, which contain venom. Juveniles associate in shoals, whereas adults are more solitary and are encountered in pairs.*

▼ BICOLORED FOX-FACE (SIGANUS USPI)

▲ VIRGATE RABBITFISH (SIGANUS VIRGATUS)

DIET: *Feeds mainly on seaweeds in the wild. A herbivorous diet will suit it well, along with regular offerings of a type of dry seaweed called sushi noir, which is usually obtainable from outlets that stock Japanese foods.*

AQUARIUM: *This fish is generally considered safe for inclusion in a reef aquarium.*

COMPATIBILITY: *Starting out with juveniles gives these fish the opportunity to pair off before they mature and need to be separated. They are likely to become aggressive at this stage if housed in groups.*

VIRGATE RABBITFISH
(BAR-HEAD SPINEFOOT)

Siganus virgatus

There are two chocolate stripes on each side of the head of this rabbitfish. One runs obliquely through the eye, while the other extends down from the back to the base of the pectoral fin. The head and the top of the body are both yellow, and there is a variable pattern of blue markings forming stripes, especially across the forehead, as well as spots. The rest of the body is a pale bluish shade overall.

SYNONYM: *S. notostictus.*

DISTRIBUTION: *Ranges from the eastern Indian ocean, from the southern tip of India, eastward via Sri Lanka and the Andamans to Thailand and northward via the Philippines to China and ultimately to the Ryukyu Islands. Southward, its range extends down through Indonesia to the coast of Australia's Northern Territory.*

SIZE: *12 in (30 cm).*

BEHAVIOR: *Tends to be found relatively close to shore in shallow waters and may be encountered occasionally in estuarine waters. The spines on* the dorsal and anal fins help protect this fish. Adults associate together in pairs, but their breeding habits appear to be undocumented.

DIET: *Feeds largely on seaweed. A herbivorous diet that substitutes for this form of alga is therefore needed. As with other herbivorous species, this rabbitfish will benefit from being fed several times a day, since its natural feeding habits are those of an instinctive browser.*

AQUARIUM: *Should be safe to include in a reef aquarium.*

COMPATIBILITY: *Keep adults only in true pairs. However, even a single individual will not interfere with other unrelated aquarium occupants.*

FOX-FACE RABBITFISH

Siganus vulpinus

From the front of the dorsal spines to the lower jaw a dark band extends right down the face at an angle. A white area on the throat is separated from another white area behind by a black band that broadens out on the underside of the body. Yellow predominates right across the rest of the body, although there is some darker suffusion running along the upper sides below the dorsal fin. The species is, in fact, almost identical to the one-spot fox-face (*S. unimaculatus*), which has a discrete black spot on each side of the body. Some taxonomists even regard them as being variants of the same species.

SYNONYM: *Lo vulpinus.*

DISTRIBUTION: *Occurs in the western Pacific from Indonesia and New Guinea north to the western Philippines and south to Australia's Great Barrier Reef. Recorded from New Caledonia and the Caroline and Marshall Islands eastward to Tonga. There are also unverified reports from other islands in the region, including Kiribati and Vanuatu.*

SIZE: *9.5 in (24 cm).*

BEHAVIOR: *Found in areas of the reef where coral predominates, with juveniles sometimes seen in groups. Adult fish are much more territorial by nature. It is known as the fox-face because of the shape of its jaws and also because of the dark coloration on this part of the body. It uses its mouthparts to reach algae growing around the bases of the coral.*

DIET: *Relatively easy to acclimatize to aquarium foodstuffs, as with other members of the genus. Offer a herbivorous diet, which can be supplemented occasionally with fresh sources of organic greenfood such as broccoli. Algae growing in the tank are also likely to be eaten.*

AQUARIUM: *Relatively safe in a reef tank, but may feed on coral polyps if hungry. Must be fed small amounts several times daily.*

COMPATIBILITY: *Has gained a reputation as one of the least tolerant members of the genus (especially once adult) toward others of its own kind. True pairs will agree, but apart from watching their behavior, there is no visual way of distinguishing between the sexes, although females generally tend to attain a slightly larger size overall.*

▼ *FOX-FACE RABBITFISH (SIGANUS VULPINUS)*

BATFISH AND SPADEFISH FAMILY EPHIPPIDAE

The size of these fish can mean there are difficulties in housing them successfully in the home aquarium. In any event, certain species are not always easy to acclimatize to such surroundings. Batfish are the most commonly kept members of the family Ephippidae and require a deep aquarium to take account of their height even when young. Some species, such as the long-fin batfish, are true personality fish and can become very tame—they will even recognize their owners and can be trained to take food from the hand. The appearance of juveniles—which tend to be most commonly offered in aquatic stores—differs markedly from that of older individuals of the same species.

ATLANTIC SPADEFISH

Chaetodipterus faber

Spadefish are so called because of their body shape, which resembles the outline of the spade symbol on playing cards. This particular species is silvery in color. It has a series of dark vertical stripes running across its body when young, but they become paler with age. The markings generally tend to be more apparent toward the rear of the body, and their exact shape and positioning vary slightly from individual to individual.

DISTRIBUTION: *Occurs in the western part of the Atlantic, extending from the coast of New England and Massachusetts southward through the Gulf of Mexico down as far as Brazil.*

SIZE: *36 in (91 cm).*

BEHAVIOR: *Like batfish, juvenile spadefish favor brackish water. They often swim at an angle to disguise their presence, appearing like a floating leaf being swept along by the current. Adults associate in shoals in open water, sometimes forming very large aggregations of as many as 500 individuals, particularly off the coast of Florida.*

DIET: *Feeds on both plankton and invertebrates such as mollusks and crustaceans. Offer meaty foods such as shrimps, cut into pieces of appropriate size. May also take some herbivorous foods.*

▲ *Atlantic Spadefish (Chaetodipterus faber)*

AQUARIUM: *Not suitable for a reef tank, since it preys on a variety of the occupants typically found there.*

COMPATIBILITY: *Small Atlantic spadefish are often nervous, whereas adults are much bolder. Unfortunately, the size of the adults precludes all but the most dedicated hobbyist from keeping a shoal of these fish together.*

DUSKY BATFISH

(RED-FIN BATFISH; PINNATE BATFISH)

Platax pinnatus

Juveniles are especially striking in appearance, with a bright orange or even crimson border around the entire edge of their bodies, contrasting with the brownish black color of the body itself. This patterning acts as camouflage, so they look rather like leaves drifting through the water at this stage, or perhaps flatworms (platyhelminths) rather than fish. In contrast, adults are far less colorful. They are silvery and have shorter fins.

▲ *DUSKY BATFISH (PLATAX PINNATUS)*

DISTRIBUTION: *Its exact distribution is unclear, but it occurs in the western part of the Pacific, ranging from Australia northward as far as the Ryukyu Islands close to Japan. Although it reportedly occurs in the Indian Ocean, this has not been confirmed.*

SIZE: *17.75 in (45 cm).*

BEHAVIOR: *Young dusky batfish are frequently found close inshore, often in mangrove areas. They often appear to drift through the water at an angle, which helps conceal their presence from predators. This behavior can also be seen in those housed in aquaria and is not a cause for concern, unlike the situation with most aquarium fish. Also occurs in sheltered parts of the reef, often found in relatively shallow water, where the young batfish are more inclined to hide from potential danger. Adults range into much deeper water and tend to be encountered on their own.*

DIET: *Feeds on zooplankton and algae in the wild, but it is very difficult to acclimatize this fish to substitute aquarium foods. A tank with good algal growth may help establish it, along with offerings of herbivorous foods, brine shrimps, and other meaty foods chopped into small pieces. Be prepared to try a wide range of foods in small amounts to see what individual fish will eat.*

AQUARIUM: *Feeds naturally on various invertebrates, so it is not recommended for a reef tank.*

COMPATIBILITY: *Provide plenty of seclusion with retreats around the aquarium. Keeping several individuals together may encourage feeding. Live rock as decor can be provided in the hope of persuading this batfish to continue feeding during the critical initial stages of acclimatization.*

LONG-FIN BATFISH

(TIERA BATFISH)

Platax teira

This is a particularly majestic fish with a very tall body. Juveniles display alternating brown and pale yellow vertical stripes on their bodies, with a broader brown area at the rear of the body. Their markings vary somewhat, depending on the individual. Adults are less brightly colored. Their appearance is much grayer rather than brown, and their body shape has a more rounded profile. The very pronounced dorsal and ventral fins become shorter as these fish mature. Large examples may have a hump on their heads.

SYNONYM: *P. terra.*

DISTRIBUTION: *Ranges from East Africa and the Red Sea eastward through the Indian Ocean as far as New Guinea. Its northward range extends to the Ryukyu Islands close to Japan, while in the south they are found around the east coast of Australia.*

SIZE: *27.5 in (70 cm).*

BEHAVIOR: *It is thought that the brown appearance of juveniles helps conceal their presence among the roots of the mangroves that they frequent at this age, although they also occur on sheltered parts of the reef. As they grow older, these batfish wander farther afield, moving to seaward reefs where they can be observed down to depths of at least 60 ft (20 m).*

DIET: *Omnivorous, feeding quite readily both on algae and meaty foods. Requires a suitable herbivorous diet, with the opportunity to browse on algae in the aquarium, too. May take a wide range of chopped meaty foods, including squid, as well as brine shrimps.*

AQUARIUM: *Not safe for a reef tank, since it is likely to feed on various invertebrates that are usually present in such surroundings.*

COMPATIBILITY: *Can be kept in relatively small groups when young or included in a community tank provided there are no fin-nipping species present.*

TRIGGERFISH FAMILY BALISTIDAE

If you are seeking fish that have big personalities to include in your aquarium, then triggerfish are a good choice. Unfortunately, their size at maturity and their aggressive nature mean they must not be mixed with other species. They have acquired their common name because of a defensive adaptation in the form of spines on the front of the dorsal fin. These spines are raised if danger threatens, making the fish very difficult to dislodge from a cave or similar retreat. This can sometimes make a triggerfish difficult to catch in the aquarium if hiding places have not been removed first.

Triggerfish are generally robust by nature and easy to cater for, since they are relatively unfussy in their feeding habits. From this point of view, they represent a good introduction to the hobby. Efficient filtration in the tank is necessary, however, because of the waste they produce. If not filtered, it can potentially lead to serious deterioration in water quality in the aquarium.

ORANGE-LINED TRIGGERFISH

(UNDULATE TRIGGERFISH)

Balistapus undulatus

The wavy orange stripes on the body of this triggerfish give the species its common name. They are separated by intervening blue bands of a similar width. The patterning also extends to the caudal fin, while the soft parts of the pectoral, dorsal, and anal fins are all orange. The distinctive pattern varies in each individual fish.

▼ *Orange-lined Triggerfish (Balistapus undulatus)*

DISTRIBUTION: *Extends from the Red Sea south down the eastern coast of Africa as far as Natal. Ranges eastward across the Pacific as far as the Line, Marquesan, and Tuamotu Islands. Its northerly range extends to southern Japan, and to the south it can be seen around Australia's Great Barrier Reef.*

SIZE: *12 in (30 cm).*

BEHAVIOR: *Found in areas of the reef where coral is plentiful. This species is territorial by nature. Spawning takes place in a small pit dug in an area of sand, into which the eggs are then deposited and guarded.*

DIET: *Feeds naturally on a wide range of foods from echinoderms to other fish. Offer meaty foods several times each day. Usually feeds readily.*

AQUARIUM: *Not suitable for a reef tank, because it will destroy many of the inhabitants.*

COMPATIBILITY: *Highly aggressive. Needs to be kept on its own in a spacious aquarium.*

GRAY TRIGGERFISH

Balistes capriscus

One of the less colorful triggerfish, this species is grayish, as its name suggests, with three faint but broad vertical bands running down the side of its body. Pale blue spotting is evident on the upper side of the body, and a small pale stripe can be seen immediately below the lower jaw.

SYNONYMS: *B. carolinensis; B. moribundus; B. powellii; B. spilotopterygius; B. taeniopterus.*

DISTRIBUTION: *Extends over a wide area of the western Atlantic from Nova Scotia off Canada's eastern coast via Bermuda and the northern part of the Gulf of Mexico down the South American coast to Argentina. In the eastern Atlantic it can be found from the Mediterranean Sea southward down the west coast of Africa as far as Angola.*

▼ *GRAY TRIGGERFISH (BALISTES CAPRISCUS)*

▲ *QUEEN TRIGGERFISH (BALISTES VETULA)*

SIZE: *24 in (60 cm).*

BEHAVIOR: *Tends to occur quite close to the shore, venturing into harbors as well as bays, and is also encountered over reefs. Often seen singly, but sometimes may associate in small groups, particularly when spawning.*

DIET: *Naturally feeds on crustaceans and mollusks. Offer similar meaty foods. May be tamed sufficiently to take such foods from the hand, but take care to avoid being bitten.*

AQUARIUM: *Likely to be disruptive in a reef tank.*

COMPATIBILITY: *Should be kept on its own, because it can prove to be aggressive, especially when adult.*

QUEEN TRIGGERFISH

Balistes vetula

One of the largest triggerfish, the queen triggerfish can weigh up to 12 pounds (5.5 kg) and is caught for food in various parts of its range. Two distinctive blue bands extend across the lower face, the bottom one forming a ring around the jaws. A prominent blue band is also evident on the caudal peduncle. The body tends to be greenish with variable dark markings on the flanks.

DISTRIBUTION: *Occurs in the Atlantic, ranging from the coast of Massachusetts through the Gulf of Mexico and the Caribbean south as far as southeastern Mexico. An eastern population occurs around Ascension Island, Cape Verde, and the Azores, extending to the African coast off southern Angola.*

SIZE: *24 in (60 cm).*

BEHAVIOR: *Can sometimes be seen in spectacular shoals, although in aquarium surroundings this triggerfish becomes highly aggressive as it matures. May be encountered in a variety of reef habitats, but tends to be more solitary over sandy areas, perhaps because food is scarcer there.*

DIET: *Hunts invertebrates, including sea urchins, which it flips over in order to attack the vulnerable undersides of their bodies. Will feed readily on a range of meaty foods and will benefit from being given algae.*

AQUARIUM: *A large well-designed tank is required. The heating and power cables should be positioned out of reach to avoid them being attacked by this triggerfish with its powerful teeth. Not suitable for a reef tank.*

COMPATIBILITY: *Juveniles may be housed temporarily with fish of similar size but will ultimately need to be separated and kept individually.*

CLOWN TRIGGERFISH

Balistoides conspicillum

The clown triggerfish is one of the most distinctive members of the family, thanks in part to the bright orange area surrounding the mouthparts and the very unusual white patches on the underside of the body. There is also a bright yellow horizontal stripe just below the eyes, while a series of sharp raised projections is evident on the caudal peduncle.

DISTRIBUTION: *The coast of East Africa, extending down to the vicinity of Durban in South Africa. Ranges eastward via Indonesia as far as the Samoan Islands in the Pacific. Its northerly range extends to southern Japan, and it is found as far south as New Caledonia.*

SIZE: *19.75 in (50 cm).*

BEHAVIOR: *Occurs on the seaward side of reefs, often close to deep water. It is invariably solitary and is not commonly seen, in spite of its wide range. A young individual will grow rapidly and can be tamed, although it is capable of inflicting a painful nip on the hand when the aquarium is being serviced.*

DIET: *Feeds on crustaceans, tunicates, and mollusks. Offer both chopped meaty items and prepared foods as well as some marine algae.*

AQUARIUM: *Not suitable for a reef tank, since it is likely to prey on the other occupants.*

COMPATIBILITY: *Must be housed individually because it is aggressive and unpredictable. Although the young may appear more tolerant at first, they are capable of killing tankmates.*

TITAN TRIGGERFISH

Balistoides viridescens

The lower jaw and throat are white, while the upper jaw is pale yellow. The main area of the body is blackish with white markings behind. The titan triggerfish displays the characteristic

▼ CLOWN TRIGGERFISH (BALISTOIDES CONSPICILLUM)

shape of these fish—it has a broad sloping profile from the eyes down to the mouth. It is the largest member of the family.

Distribution: *Extends from the Red Sea down the African coast to Mozambique and eastward across the Pacific as far as the Line and Tuamotu Islands. Its northerly range reaches to southern Japan, and it can be found as far south as New Caledonia.*

Size: *29.5 in (75 cm).*

Behavior: *Most likely to be encountered individually or sometimes in pairs on the seaward side of the reef. Juveniles seek out more sheltered areas, sometimes hiding among coral. Females may resort to attacking divers who approach too close to their nests.*

Diet: *Eats a range of crustaceans, corals, mollusks, and similar invertebrates. A typical meaty diet suits this species well.*

Aquarium: *Will prey on many of the inhabitants.*

Compatibility: *Keep singly. A very large aquarium is essential, even for a single individual of this species.*

PINK-TAIL TRIGGERFISH

Melichthys vidua

Only the terminal part of the caudal fin in this fish is pink, and its base is white. The dorsal and anal fins are a slightly darker pinkish shade with black edging, while the body is bluish green,

▲ *Pink-tail Triggerfish (Melichthys vidua)*

becoming yellowish on the face around the jaws. Juveniles have dark streaking extending from the eyes.

Distribution: *East Africa, south to Durban, South Africa, and east across the Pacific as far the Hawaiian, Marquesan, and Tuamotu Islands. Range north to southern Japan, and south to the Great Barrier Reef, Australia.*

Size: *15.75 in (40 cm).*

Behavior: *Most likely to be found in deep water on the seaward side of the reef where there are strong currents and in areas where there is coral growth. Proves less destructive than related species in an aquarium.*

Diet: *Feeds largely on algae, but also takes some meaty foods.*

Aquarium: *Unusually for triggerfish, the pink-tail can be housed in a spacious reef tank, although it may hunt small crustaceans that are present.*

Compatibility: *Relatively docile, so can be kept as part of a community tank with nonaggressive fish of a similar size.*

RED-TOOTHED TRIGGERFISH

(NIGER TRIGGERFISH)

Odonus niger

Although less colorful than many of its relatives, the red-toothed triggerfish is a spectacular species when seen in profile. It has a high dorsal fin and a lyre-tailed caudal fin. Its overall color is bluish—lighter in some cases than in others—with a green area below the eyes that extends down to the throat. The area around the lips is blue, setting off its red teeth, and blue lines run from the mouth to the eyes and extend back in the direction of the pectoral fins.

DISTRIBUTION: *Red Sea southward down the African coast to Durban. Also extends eastward via Australia's Great Barrier Reef and Micronesia to the Marquesan and Society Islands in the Pacific. Range reaches north as far as southern Japan.*

SIZE: *19.75 in (50 cm).*

BEHAVIOR: *Often associates in groups once adult and favors the outer part of the reef where there are strong currents. Juveniles are shyer than adults by nature and tend to hide away in suitable retreats.*

DIET: *A typical mix of chopped meaty foods and some algae. Plankton features in its natural diet.*

AQUARIUM: *Not to be trusted entirely in a reef tank, particularly because sponges form part of its natural diet.*

COMPATIBILITY: *Introduce juveniles to the aquarium at the same time. They will hopefully agree subsequently. Can be kept also with nonaggressive companions of a similar size.*

YELLOW-MARGIN TRIGGERFISH

Pseudobalistes flavimarginatus

Yellow coloration is often most evident around the jaws of this triggerfish. The head is generally pale in color compared with the body. The large scales of this species display dark blotches.

▼ *YELLOW-MARGIN TRIGGERFISH* (PSEUDOBALISTES FLAVIMARGINATUS)

▲ *BLUE-LINED TRIGGERFISH* (PSEUDOBALISTES FUSCUS)

DISTRIBUTION: *Red Sea southward down the African coast to Natal in South Africa. Ranges eastward to Indonesia, northward as far as southern Japan, and farther east to the Tuamotu Archipelago.*
SIZE: *24 in (60 cm).*

BEHAVIOR: *Usually seen singly, sometimes in pairs. The yellow-margin triggerfish is an agile swimmer in spite of its size and is even capable of moving backward if danger threatens. Spawning occurs in a pit in the ground, and females defend their nests ferociously.*

DIET: *A diet based on chopped meaty foods is required.*

AQUARIUM: *Unsuitable for a reef aquarium, since it feeds on invertebrates of all types, including corals, crustaceans, and even sea urchins.*

COMPATIBILITY: *Aggressive by nature, especially as it matures, so this fish should be kept on its own.*

BLUE-LINED TRIGGERFISH

(YELLOW-SPOTTED TRIGGERFISH)

Pseudobalistes fuscus

Yellow stripes and some spots on a blue background identify these triggerfish. Their colorful patterning is highly individual and tends to fade with age, with the adults becoming duller in color. Young fish measuring less than 2 inches (5 cm) long have distinctive dark saddles on their bodies, which disappear as they approach maturity.

DISTRIBUTION: *Red Sea southward down the African coast to Durban. Most common in the Indo-Pacific region, although continuing through the Pacific to the Society Islands. Ranges north to southern Japan and southward to Australia's Great Barrier Reef.*

SIZE: *21.75 in (55 cm).*

BEHAVIOR: *Prefers relatively open areas of the reef and often combs the sandy bottom for food. Will use its mouth to blow jets of water to displace the sand while seeking food at the bottom. Often proves very destructive in aquarium surroundings.*

DIET: *Offer a variety of chopped meaty foods. Normally eats a range of invertebrates and sometimes even other fish.*

AQUARIUM: *Not suitable for a reef tank, since it will prey on the occupants.*

COMPATIBILITY: *Aggressive by nature. Needs a large aquarium, typically about 200 gallons (760 l) in volume. Risky to include with other fish, even those of a similar size.*

PICASSO TRIGGERFISH

(BLACK-BAR TRIGGERFISH; HUMUHUMU TRIGGERFISH)

Rhinecanthus aculeatus

Strikingly abstract patterning, consisting of blue and brown stripes connecting the eyes and extending down the cheeks, is a feature of this triggerfish. The lips are yellow with a blue border, while yellow stripes extend back from this region to the pectoral fins. The rest of the face is white.

DISTRIBUTION: *Found in the Red Sea southward as far as the coast of South Africa and eastward through the Pacific to the Hawaiian, Marquesan, and Tuamotu Islands. Ranges north to southern Japan and south to Lord Howe Island off Australia's east coast. Also occurs on the eastern side of Africa, southward from Senegal.*

SIZE: *12 in (30 cm).*

BEHAVIOR: *The Picasso triggerfish has the habit of sometimes sleeping on its side. Juveniles can be easily tamed but tend to become aggressive as they grow older. This triggerfish is often seen in open areas of the reef.*

DIET: *Offer a range of meaty foods. Also needs algae, which form part of its natural diet.*

AQUARIUM: *Not suitable for a reef tank, because it will prey on a number of the invertebrates present.*

COMPATIBILITY: *Young may be tolerant with each other if placed in a large aquarium at the same time, but they will become more aggressive and territorial with age.*

ARABIAN PICASSO TRIGGERFISH

Rhinecanthus assasi

There can be confusion between this species and the Picasso triggerfish, but the three yellow and brown pairs of horizontal lines extending back from the caudal fin help identify the Arabian Picasso triggerfish at a glance. It also has a more greenish face and a brown moustachelike marking extending back on each side of the face from the mouth.

▼ *PICASSO TRIGGERFISH (RHINECANTHUS ACULEATUS)*

▲ ARABIAN PICASSO TRIGGERFISH (RHINECANTHUS ASSASI)

DISTRIBUTION: *Restricted to the western part of the Indian Ocean, extending from the Red Sea to the Gulf of Oman and the Persian Gulf.*

SIZE: *12 in (30 cm).*

BEHAVIOR: *Similar in habits to the Picasso triggerfish, occurring in relatively shallow and open stretches of water on the reef.*

DIET: *Provide a meaty diet, since this species hunts invertebrates of various types in its natural habitat.*

AQUARIUM: *Cannot be included safely in a reef setup, because it will hunt many of the occupants.*

COMPATIBILITY: *Becomes increasingly aggressive with age, so it is not usually possible to keep more than one individual safely in an aquarium. May also disturb the decor and attack accessible equipment.*

BLACK-BELLY TRIGGERFISH

Rhinecanthus verrucosus

Recognizable by the very distinctive black blotch on the lower area of the body, this species also displays a series of blue and brownish yellow stripes above the eyes. The rest of the body tends to be yellowish, with a narrow reddish stripe running back from the mouth to the base of the pectoral fins.

DISTRIBUTION: *Extends from the Chagos Archipelago south of India eastward via Indonesia as far as the Solomon Islands. Its northerly range extends to southern Japan, while its southerly distribution reaches Vanuatu.*

SIZE: *9 in (23 cm).*

BEHAVIOR: *Ranges over open areas of the reef. Displays the typical split dorsal fin configuration that is a feature of triggerfish. The front part of the fin can be raised to hold the fish in place in a rocky retreat. The fish*

▼ BLACK-BELLY TRIGGERFISH (RHINECANTHUS VERRUCOSUS)

◀ SCIMITAR TRIGGERFISH (SUFFLAMEN BURSA)

may inflate its body slightly, too, so that it can remain anchored in its retreat, safe from most predators. At other times this part of the fin lies out of sight in a groove running down the center of the body.

DIET: *Feed on meaty foods such as shrimps or cockles.*

AQUARIUM: *Not safe for a reef aquarium, because it will prey on many of the occupants.*

COMPATIBILITY: *Aggressive and territorial, although juveniles may agree for a while if they are introduced to the aquarium at the same time.*

SCIMITAR TRIGGERFISH
(BOOMERANG TRIGGERFISH; PALLID TRIGGERFISH)

Sufflamen bursa

An elegant-looking fish, predominantly white in color, which explains the alternative name of pallid triggerfish. It has a suffusion of yellow on its back just behind the eyes. A slightly curved yellowish tan stripe runs through the eyes and another one extends to the pectoral fin, creating the impression of a scythe or scimitar. The caudal fin is bluish in color.

DISTRIBUTION: *Ranges from East Africa across the Pacific to the Hawaiian, Marquesan, and Ducie Islands. Extends northward as far as southern Japan and ranges south to Australia's Great Barrier Reef.*

SIZE: *10 in (25 cm).*

BEHAVIOR: *Seen on the seaward aspect of reefs, occurring from close to the surface down to a depth of about 295 ft (90 m).*

DIET: *Eats a variety of meaty foods readily.*

AQUARIUM: *Not safe for inclusion in a reef aquarium.*

COMPATIBILITY: *Juveniles in particular tend to be less aggressive than many other species of triggerfish, but adults often still retain strong territorial instincts. They can be tamed to feed from the hand quite easily, but take care not to be bitten by their sharp teeth.*

RED-TAIL TRIGGERFISH

Xanthichthys mento

This is an unusually patterned triggerfish in which individual scales are highlighted by dark edging. Only males have the characteristic red tail, and they are also more colorful overall than females, with a yellow rather than grayish body color.

DISTRIBUTION: *Occurs in the Pacific region, ranging as far north as the Ryukyu Islands near Japan, eastward via Marcus and Wake Islands to the Hawaiian Islands. Also found from southern California southward via the Revilla Gigedo Islands and Clipperton to the Galápagos, Pitcairn, and Easter Islands.*

SIZE: *11.5 in (29 cm).*

BEHAVIOR: *Active by nature, this triggerfish is often encountered in shoals on the outer edges of reefs where there is a strong current. For this reason, water movement is important in the aquarium. May prove shy in aquarium surroundings at first but will soon adapt to this environment.*

DIET: *Offer a variety of meaty foods and provide algae as well.*

AQUARIUM: *Relatively safe to include in a reef tank, although it may prey on shrimps and similar crustaceans.*

COMPATIBILITY: *Far more adaptable and less aggressive than most triggerfish. It is possible to keep groups of a single male with two females in a large aquarium, which must have a volume of more than 100 gallons (380 l). Can also be housed with other fish of a similar size, provided the newcomers are introduced to the aquarium after the red-tail triggerfish is established—it is then less likely to be bullied.*

▼ RED-TAIL TRIGGERFISH (XANTHICHTHYS MENTO)

FILEFISH FAMILY MONACANTHIDAE

The members of the family Monacanthidae, the filefish, vary significantly in size. Some species become too large for an average aquarium. Filefish are so called because the surface of their bodies is rough. They vary in appearance, although in common with triggerfish in the family Balistidae, the front part of their dorsal fin is strengthened to form a vertical spine. They can use it to anchor themselves in a rocky retreat if threatened by a predator. The spine may be relatively inconspicuous, however, lying flat along the top of its body, especially once the fish is settled in its quarters. It will be raised only if the fish feels under threat, but it is important always to take care not to get caught by the spine when you are servicing the aquarium.

SCRAWLED FILEFISH

Aluterus scriptus

Scrawled filefish are long bodied, and their eyes are set well back from their jaws. Juveniles tend to display a yellowish brown background color, whereas the adults are more of a tan shade. There are dark spots on the body as well as a series of irregular blue lines and spots that give each fish an individual patterning.

▼ *Scrawled Filefish (Aluterus scriptus)*

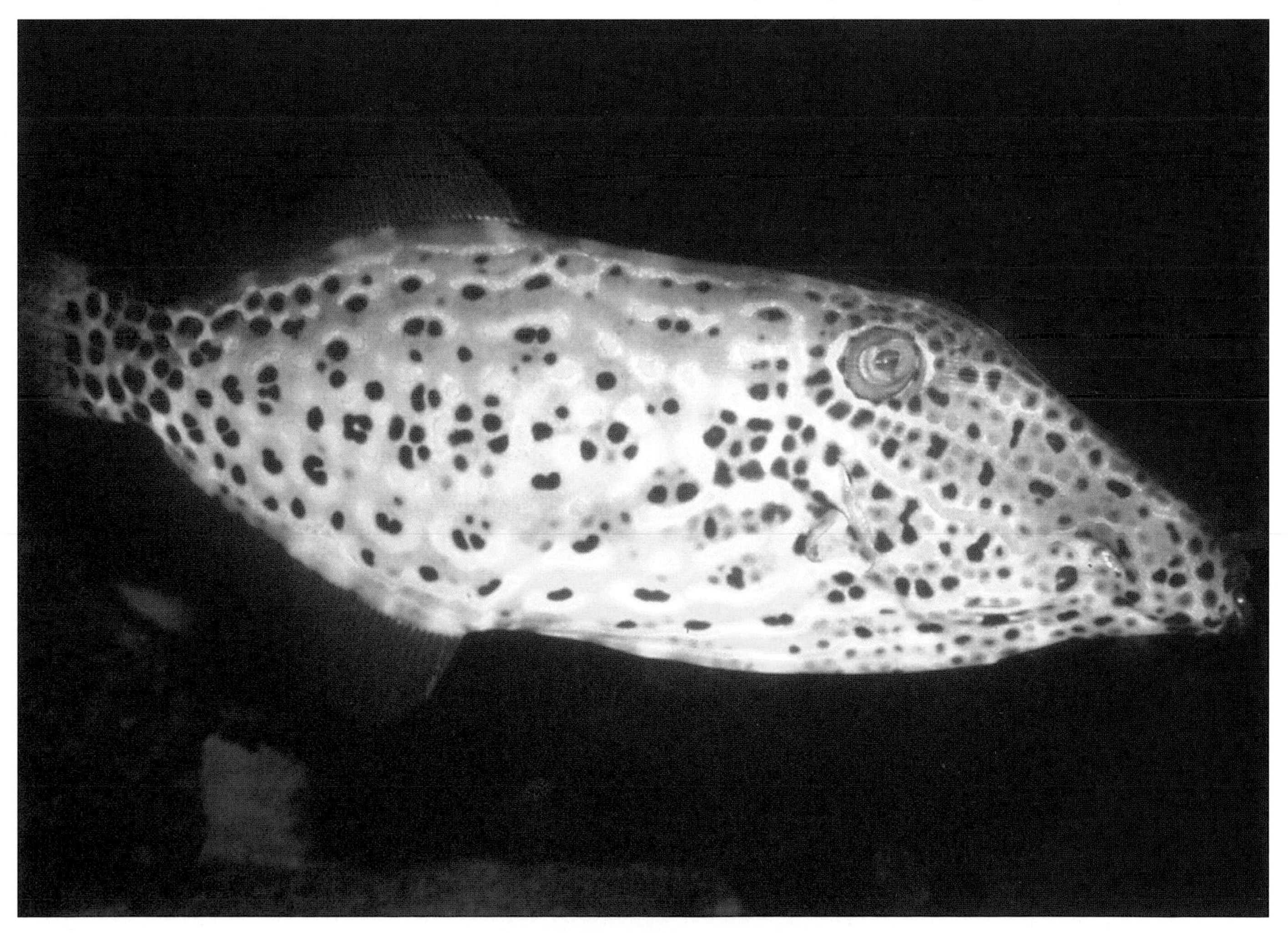

▲ HONEYCOMB FILEFISH (CANTHERHINES PARDALIS)

These markings also help disguise the fish's presence on the reef. It is not an especially strong swimmer and has a rounded caudal fin.

SYNONYMS: *A. renosus; A. venosus.*

DISTRIBUTION: *Has a very wide circumtropical range. Occurs in the warmer waters of the Pacific and Atlantic Oceans. Extends from the vicinity of Nova Scotia, Canada, south through the Gulf of Mexico as far as Brazil. Also present around various islands in the eastern Atlantic and occurs as far south as South Africa. Present in the Pacific from the southern part of Australia's Great Barrier Reef, extending northward to southern Japan. Ranges eastward to Easter Island. Also present in waters from the Gulf of California south as far as Colombia in northern South America.*

SIZE: *43 in (110 cm).*

BEHAVIOR: *Most likely to be encountered on the seaward side of the reef and also on lagoons. May be seen near the surface, often just floating but sometimes concealing itself under flotsam. In this species a clamped caudal fin is not necessarily a sign of ill health, and the color of this fish can also lighten and darken dramatically without being a cause for concern.*

DIET: *Omnivorous, eating a range of foods from algae to invertebrates. Can therefore be maintained easily on a range of both meaty and herbivorous foods in aquarium surroundings.*

AQUARIUM: *Cannot be trusted in an aquarium that houses invertebrates.*

COMPATIBILITY: *Not usually aggressive, but the potentially large size of the scrawled filefish means that considerable thought needs to be given to its long-term accommodation before acquiring one.*

HONEYCOMB FILEFISH

Cantherhines pardalis

The underlying body color of this filefish is basically brownish, often broken up by an elaborate bluish honeycombed pattern that extends over the entire body. On the other hand, this filefish can vary its coloration markedly, appearing either as a dark shade of brown or as a combination of gray and brown. There are some pale yellowish stripes just below the eye and a series of irregular blue and brown stripes present over the rest of the face. The area around the lips is bluer in color. When raised, the spine created by the first part of the dorsal fin extends back from above the eyes, while a white area is always present at the rear of the second dorsal fin.

DISTRIBUTION: *Ranges from the Red Sea southward along the east African coast to South Africa and through the Indian Ocean. Extends north to southern Japan and reaches southeastern Oceania in the Pacific. Also present from Annobón Island in the Gulf of Guinea south to South Africa in the eastern Atlantic.*

SIZE: *10 in (25 cm).*

BEHAVIOR: *A naturally shy species about which little has been documented. Tends to be seen on its own, sometimes relatively close to the surface.*

DIET: *Can be offered both herbivorous and meaty foods, but the latter must be chopped into small pieces because, in common with all filefish, this species has a small mouth.*

AQUARIUM: *Unsuitable for a reef aquarium because it may prey on other occupants.*

COMPATIBILITY: *Not generally aggressive by nature so can be kept as part of a community aquarium.*

TASSELED FILEFISH

(PRICKLY FILEFISH)

Chaetodermis penicilligerus

A yellowish green background color is a feature of this species, broken by an irregular pattern of narrow horizontal brown stripes. These markings contrast with the clear coloration of the dorsal, anal, and caudal fins, which are highlighted by small dark spots. Its most distinctive features, however, are the tassels that are present around the body. There are black blotches evident behind the eyes in juveniles of this species, although they fade with age.

DISTRIBUTION: *Ranges from the Bay of Bengal to Malaysia, extending northward to southern Japan and south to Australia's Great Barrier Reef.*

SIZE: *12.25 in (31 cm).*

BEHAVIOR: *Its unusual appearance provides this fish with excellent camouflage among the seaweed on the shallow coastal reefs where it occurs. Shy by nature, it will hide away, relying on subterfuge rather than speed to escape from potential predators.*

DIET: *Feeds on various invertebrates, so offer finely chopped meaty foods in aquarium surroundings. Will usually feed readily, provided it is not disturbed by more active companions. Must be fed several times daily, since it tends not to eat voraciously.*

▼ *TASSELED FILEFISH* (CHAETODERMIS PENICILLIGERUS)

▲ *RED SEA LONG-NOSE FILEFISH* (OXYMONACANTHUS HALLI)

AQUARIUM: *Not suitable for a reef aquarium because of its feeding habits, although it can be housed successfully with large anemones such as* Stichodactyla *species.*

COMPATIBILITY: *Tankmates should be chosen with care to prevent them from nibbling at the tassels of this filefish. This means that pufferfish, to which the filefish are related, are not suitable.*

RED SEA LONG-NOSE FILEFISH

Oxymonacanthus halli

The Red Sea long-nose filefish is very similar in appearance to the long-nose filefish (*O. longirostris*). It has a relatively narrow elongated shape. Its body color is light blue, broken by a series of yellow spots that extend over the entire body and tend to become smaller toward the rear. It has a much larger black spot on the caudal fin than *O. longirostris* and attains a smaller size.

DISTRIBUTION: *As its name suggests, this fish is restricted to the Red Sea .*

SIZE: *2.75 in (7 cm).*

BEHAVIOR: *Occurs in areas of the reef where coral predominates. Its small size helps conceal its presence. Like other filefish, it is not a fast swimmer.*

DIET: *Feeds naturally on the polyps of* Acropora *species coral, and it can be difficult to persuade this fish to take alternative foods. Try providing finely chopped meaty foods and offer them several times during the day. Offer herbivorous foods as well, since algae also feature in its diet.*

AQUARIUM: *Not suitable for a reef tank, since it will prey on the coral polyps.*

COMPATIBILITY: *Suitable for a community aquarium alongside other nonaggressive species.*

LONG-NOSE FILEFISH

(HARLEQUIN FILEFISH; BEAKED LEATHER-JACKET)

Oxymonacanthus longirostris

This species is very similar in appearance to its Red Sea relative, *O. halli*, but it has a smaller black spot on the caudal fin. Its body patterning extends to the area of the iris surrounding the pupil, where there are alternating orange and bluish bands separated by fine black lines. It is possible to sex long-nose filefish by the very evident orange patch edged with black that is present on the underside of the male's body, adjacent to the tiny ventral fins. This area of the body is gray in females.

DISTRIBUTION: *From East Africa southward down the coast as far as Mozambique, extending through the Indian Ocean and northward to the Ryukyu Islands. Ranges in a southerly direction as far as Australia's Great Barrier Reef. Its range through the Pacific extends to Samoa and Tonga.*

SIZE: *4.75 in (12 cm).*

BEHAVIOR: *The long-nose filefish is found over areas of the reef where* Acropora *coral predominates. The coral provides not only a source of food but also an area in which the fish can nest. This species makes strong pair bonds. The female lays her greenish eggs underneath the coral, hiding them among algae. It is normal for this filefish to adopt a vertical posture when resting or sometimes when feeding, which it tends to do only during the daytime.*

DIET: Acropora *coral polyps feature prominently in the diet of the long-nose filefish, but it may browse on algae as well. Can be weaned onto substitute diets consisting of both finely chopped meaty foods and herbivorous rations.*

AQUARIUM: *Unsafe for a reef tank alongside coral.*

COMPATIBILITY: *Not aggressive, even occurring in small groups in the wild, but a little shy by nature. Choose quiet, inoffensive companions, preferably fish of similar size. This should then help encourage the filefish to feed more readily.*

BLACK-SADDLE FILEFISH

(MIMIC FILEFISH)

Paraluteres prionurus

This is a strikingly marked species, with vermiculated (wavy) brownish markings extending from the jaws to the eyes. Three relatively broad blackish brown bands extend along the back down to the caudal fin, broken by narrower yellow bands. The lower sides of the body are mainly whitish, but appearing pale green in some cases. There are variable brown streaks and stripes in this region that are replaced by spots toward the caudal fin, and there are blue stripes present in the throat area.

▼ LONG-NOSE FILEFISH (OXYMONACANTHUS LONGIROSTRIS)

▲ BLACK-SADDLE FILEFISH (PARALUTERES PRIONURUS)

DISTRIBUTION: *East Africa, ranging as far south as the Aliwal Shoal off South Africa. Extends through the Indian Ocean, reaching Australia's Great Barrier Reef to the south, while southern Japan marks the northerly extent of its range. It is found through the Pacific as far east as the islands of Fiji and Tonga.*

SIZE: *4.5 in (11 cm).*

BEHAVIOR: *This fish has as a very similar appearance to, and therefore mimics, the saddled toby* (Canthigaster valentini)*, a pufferfish whose flesh contains a deadly toxin. As a result, it is largely avoided by predators. May be encountered close to the surface on the reef, occasionally over lagoon areas.*

DIET: *Should receive a diet of both finely chopped meaty foods and some herbivorous items. Often seeks food on the substrate.*

AQUARIUM: *Will feed on a range of creatures occurring in a reef aquarium, including gastropods, so cannot be included safely in these surroundings.*

COMPATIBILITY: *Requires inoffensive companions. Will be persecuted by the fish it mimics if they are placed together in the same aquarium. Male black-saddle filefish may disagree as well, so it may be better to keep these fish on their own as part of a community tank unless you are offered a compatible adult pair.*

RED-TAIL FILEFISH

Pervagor melanocephalus

This is another very colorful filefish, although individuals show variations in their appearance. Some—notably those originating from the waters around the Hawaiian Islands—display more vivid red coloration than others. In other individuals the body is more of an orange shade, with the head being bluish black or even brownish black.

DISTRIBUTION: *Ranges through the Indian Ocean down to Australia's Great Barrier Reef, extending eastward across the Pacific to Hawaii. Its northern range probably extends close to southern Japan.*

SIZE: *6.25 in (16 cm).*

BEHAVIOR: *This species is often found in deeper water than many other filefish, down to a depth of at least 130 ft (40 m). Keeping the background decor bright in the aquarium will help emphasize the natural coloration of this fish, because it will adjust its body color to match its surroundings, as it does in the wild. Since it is a slow swimmer, this ability provides an important means of escaping the attention of would-be predators.*

DIET: *Unfortunately, it can be difficult to wean this species onto artificial diets, but brine shrimps are often useful for this purpose. May also browse on algae growing in the aquarium, and can be offered meaty foods chopped into small pieces as well as herbivorous items.*

AQUARIUM: *Unsuitable for a reef tank, since it will prey on coral polyps.*

COMPATIBILITY: *Males tend to disagree but will thrive in true pairs. They can also be kept singly in a collection alongside other fish that are placid by nature. Under such conditions it will be inoffensive.*

▼ RED-TAIL FILEFISH (PERVAGOR MELANOCEPHALUS)

BOXFISH AND THEIR RELATIVES FAMILY OSTRACIIDAE

These distinctive fish are known as boxfish because of their body shape, which is far from the typical sleek profile of many fish. Their body shape resembles a small box, which limits their ability to avoid predators by speed. However, their bodies usually contain a deadly toxin that gives them protection against their enemies. But this can be a potential problem in the aquarium, because if the fish become stressed in these confined surroundings, they are likely to release the toxin and poison all the occupants of the tank, including themselves. This is most likely to happen with a newly acquired individual or when bullying is occurring. It is therefore very important to choose companions carefully for all members of this group.

SCRAWLED COWFISH

Acanthostracion quadricornis

In general, cowfish can be identified not only by their boxlike shape but also by the hornlike projections that are present just above the eyes. A second smaller sharp pair is present on each side of the body, pointing in the direction of the tail in front of the anal fin. This feature is the reason for the scrawled cowfish's scientific name *quadricornis*, which means "four-horned." This species also has a long caudal peduncle. The basic body coloration is a dull shade of yellow, with variable sky-blue spotting. There is also an underlying hexagonal patterning apparent over certain parts of the body, depending on the light.

▼ *SCRAWLED COWFISH (ACANTHOSTRACION QUADRICORNIS)*

SYNONYM: *A. tricornis.*

DISTRIBUTION: *Occurs in the Atlantic from the coast of Massachusetts southward through the Caribbean and the Gulf of Mexico as far as southeastern Brazil. Also found farther east around the island of Bermuda and has been recorded from South Africa.*

SIZE: *21.75 in (55 cm).*

BEHAVIOR: *Occurs in the shallows, often in areas where seagrass predominates. The seagrass provides this rather nervous, slow-swimming fish with cover. If threatened, it tends to freeze in the hope of escaping detection. Scrawled cowfish reproduce by scattering eggs in open water.*

DIET: *Feeds naturally on a range of sessile invertebrates, including gorgonians, and hunts slow-moving crustaceans and similar creatures. This fish needs to be offered chopped meaty food such as clams and shrimps, but it may nibble at marine algae as well, so some herbivorous food is also recommended.*

AQUARIUM: *Its feeding habits mean that this species is not to be trusted in a reef tank.*

COMPATIBILITY: *Suitable for a tank with inoffensive companions. Usually agrees well with its own kind, although occasionally individuals may chase each other. This is probably territorial behavior by the males, although it is not possible to distinguish the sexes visually.*

LONG-HORN COWFISH

Lactoria cornuta

One of the most appealing members of the group, this cowfish is instantly identifiable thanks to the pair of long projections that resemble horns on its head. Young individuals are especially attractive. They are bright yellow in color and have proportionately longer horns than the adults. Their coloration can become more variable with age, helping disguise their presence against their surroundings. They may even appear olive in some cases, and spots of various colors can also be apparent on their bodies.

▲ LONG-HORN COWFISH (LACTORIA CORNUTA)

SYNONYM: *Ostracion cornutus.*

DISTRIBUTION: *Ranges from the Red Sea and East Africa through the Indo-Pacific region south to the vicinity of Lord Howe Island off Australia's east coast and north as far as southern Japan. Extends through the Pacific to the Marquesan Islands and the Tuamotu Archipelago.*

SIZE: *18 in (46 cm).*

BEHAVIOR: *Juveniles tend to be found closer inshore than adults, even venturing into the brackish waters of river estuaries. Adults occur in weedy areas of the reef where they can conceal themselves easily. They tend to be solitary by nature, whereas juveniles may associate in groups.*

DIET: *Feeds primarily by using its mouthparts to blow away the sandy substrate in order to locate edible items. Requires a mixed diet of both finely chopped meaty foods and a herbivorous ration. Food should be provided several times a day.*

AQUARIUM: *Not really to be trusted in a reef aquarium, since it may inflict damage on the other occupants.*

COMPATIBILITY: *A nervous species, especially when first transferred to new surroundings. It may even attempt to swim through the side of the tank, injuring its horns, but the biggest danger is that it may release its deadly toxin if stressed. This will occur also if it is being harassed by other tank occupants. Should one of these fish appear ill, transfer it to a separate aquarium because, as with other members of the group, its toxin will be released if it dies.*

YELLOW BOXFISH

(POLKADOT BOXFISH)

Ostracion cubicus

The appearance of these boxfish varies quite significantly, depending on their age, with juveniles being significantly more colorful than adults. The basic body coloration tends to be yellowish, with a distinctive greenish tinge developing as they grow older. The spots in the case of juveniles are black, whereas older fish have white spots with dark borders. The spotted patterning differs from fish to fish, so that recognizing individuals on this basis is straightforward. The largest specimens are often bluish in color, but they retain yellow markings between their protective body plates. Various regional variations in appearance have also been documented.

SYNONYMS: *O. cubicum; O. tuberculatus.*

DISTRIBUTION: *Ranges widely from the Red Sea and East Africa throughout the Indo-Pacific region, extending to the Ryukyu Archipelago in the north and southward to Lord Howe Island off Australia's southeast coast. Its easterly distribution in the Pacific reaches as far as the Hawaiian Islands and the Tuamotu Archipelago. There is also an Atlantic population confined to the southern coast of South Africa.*

SIZE: *17.75 in (45 cm).*

BEHAVIOR: *Occurs in sheltered areas of the reef. The young are often encountered among* Acropora *species coral, which may give them some protection as well as allowing them to feed on algae often found growing at the base of this coral. Observed individually rather than in groups.*

▲ YELLOW BOXFISH (OSTRACION CUBICUS)

DIET: *Algae are the main items in its diet, but this boxfish may also feed on various invertebrates, ranging from sponges to mollusks. Herbivorous foods should therefore be the predominant items in its diet, augmented with some finely chopped meaty foods. Will also browse on algae growing in the aquarium.*

AQUARIUM: *Its feeding habits make this species unsuitable for a reef tank.*

COMPATIBILITY: *Males are likely to disagree, and since these fish cannot be sexed easily, it is safest to keep them individually. They can, however, be housed with other essentially nonaggressive companions.*

WHITE-SPOTTED BOXFISH

Ostracion meleagris

In the white-spotted boxfish the two sexes can be distinguished easily by their different coloration. Adult females, as well as juveniles of both sexes, are predominantly brown in color (or occasionally green) with a distinctive pattern of white spots covering the body. Males alter dramatically in color as they mature, however, developing a rich blue area on each side of the body, broken up by an irregular pattern of orange spots and bands. There are also bluish markings around the jaws. The body shape of these fish also gives an indication of their state of health, as with other members of the group. Avoid individuals that have concave underparts, because they may be ill or not feeding well. Healthy boxfish have a straight base to their bodies.

SYNONYMS: *O. camurum; O. clippertonense; O. lentiginosus; O. punctatus; O. sebae.*

DISTRIBUTION: *Ranges from East Africa through the Indo-Pacific region, extending southward to New Caledonia and northward to southern Japan. Extends as far east as the Hawaiian Islands and the Tuamotu Archipelago in the Pacific.*

SIZE: *10 in (25 cm).*

BEHAVIOR: *Although sometimes encountered close to the surface, this boxfish has been observed on the reef down to depths of 100 ft (30 m).*

▼ WHITE-SPOTTED BOXFISH (OSTRACION MELEAGRIS)

◀ *THORN-BACK BOXFISH (TETROSOMUS GIBBOSUS)*

SIZE: *4.5 in (11 cm).*

BEHAVOIR: *This species occurs in areas of the reef where coral is dominant, although it never appears to be especially common in such surroundings. Individuals are solitary by nature.*

DIET: *A mixed diet of chopped meaty and herbivorous foods will suit this species in aquarium surroundings. As with other boxfish, however, avoid feeding flake or similar foods that float. This fish is not equipped to feed at the surface. It can swallow air, which will affect its buoyancy, causing it to swim at an abnormal angle.*

AQUARIUM: *Not safe to include in a reef tank because of its feeding habits.*

COMPATIBILITY: *Keep either in true pairs or individually alongside inoffensive companions to minimize any risk of its deadly toxin being released into the water. If you suspect this has happened, transfer all the fish to new surroundings without delay, keeping the boxfish isolated in a separate tank.*

It is encountered individually, rather than in pairs or groups. Unfortunately, it tends to release its deadly ostracitoxin very readily, which obviously protects it on the reef but will be fatal both to itself and all the other occupants of an aquarium.

DIET: *Can be difficult to wean onto aquarium foods, although brine shrimps may help, as well as algae growing in the aquarium. Requires both meaty items such as chopped shrimps and herbivorous foods.*

AQUARIUM: *Because it normally feeds on sponges, polychaetes, and similar invertebrates, it is not a suitable species for a reef tank.*

COMPATIBILITY: *Do not house two males together. The result is likely to be fatal. Can be dangerous mixing this fish with other species because of its toxin, which tends to be released more readily than that of other boxfish.*

RETICULATE BOXFISH

Ostracion solorensis

The common name of this species comes from the striped markings on its body. The reticulated patterning is especially apparent on the front area of the head and on the lower part of the body. The upper sides of its body are marked mainly with horizontal stripes that merge into spotted patterning over the top of the body. The basic background color is dark brown, lightening markedly to yellow on the cheeks, mouthparts, and underside of the body, where brown spotted patterning predominates. Young fish are more brightly colored than adults.

DISTRIBUTION: *Ranges from the vicinity of Christmas Island in the Indian Ocean via Indonesia, northward to the Philippines, and south to the northern part of the Great Barrier Reef. Its easterly range reaches as far as Fiji and Tonga.*

THORN-BACK BOXFISH
(HUMPBACK TURRETFISH; HOVERCRAFT BOXFISH)

Tetrosomus gibbosus

Yellowish coloration predominates on the body of this species, which may occasionally display a mustard-colored hue. There are likely to be pale blue spots and lines forming an irregular pattern alongside some chocolate-brown markings. Perhaps its most obvious feature is the raised triangular area on its back, which gives it its common name. It has a sharp tip and is located in front of the small transparent dorsal fin. Older specimens tend to darken significantly in color, although this species tends not to grow very large in aquarium surroundings.

DISTRIBUTION: *Ranges from the Red Sea and East Africa through the Indo-Pacific north as far as Japan and south to the northern coast of Australia. Has also spread up from the Red Sea through the Suez Canal into the Mediterranean.*

SIZE: *12 in (30 cm).*

BEHAVIOR: *Often found in coastal areas where seagrass is present, which provides a retreat for this relatively slow-swimming fish. The raised area on its back serves as protection, making it harder for would-be predators to swallow it easily. The manner in which it swims suggests that it is floating on air, hence its alternative common name of hovercraft boxfish.*

DIET: *Feeds on invertebrates, so requires a meaty diet in aquarium surroundings. Chop up pieces of shrimp and similar items.*

AQUARIUM: *Unsuitable for a reef aquarium because of its feeding habits.*

COMPATIBILITY: *May be kept with other fish that will not persecute it, but take particular care to avoid any that may nip at its vulnerable fins.*

PUFFERFISH FAMILY TETRAODONTIDAE

Although related to boxfish, pufferfish have a very soft and pliable body covering rather than a hard casing. Pufferfish are so called because of their ability to inflate themselves as a means of avoiding predators. They draw either water or air into their bodies for this purpose and, as a result, they undergo a dramatic increase in size. This makes it much harder for a predator to swallow them. They also have another means of defense—rather than exuding deadly toxins like members of the boxfish family, they often have poisonous flesh. This means that predators will rarely bother them.

While true puffers can attain a relatively large size, smaller members of the family—often known as tobies—are much more manageable and can be ideal for inclusion in a typical community aquarium, although they may sometimes nip at the long fins of companions. As a group, these fish generally prove to be robust and easily managed, adapting without difficulty in many cases to alternative diets.

▼ *MASKED PUFFER (AROTHRON DIADEMATUS)*

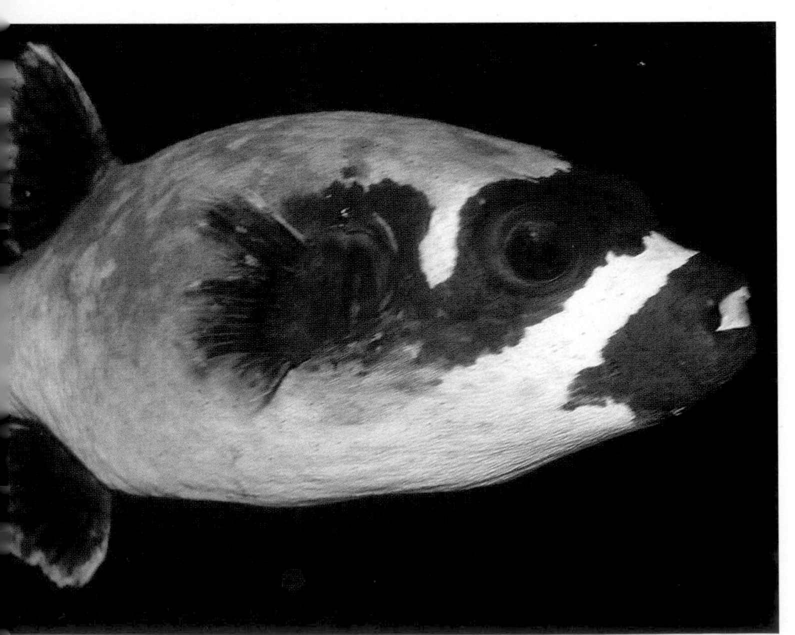

MASKED PUFFER

(PANDA PUFFER)

Arothron diadematus

This species is very similar to its relative, the black-spotted puffer (*A. nigropunctatus*). It has a black mask over the eyes and a separate black area on the lips, divided by an intervening white band. There is another prominent black area on each side of the body, surrounding the pectoral fins, and the fins are also black—apart from the caudal fin, which is mottled. The body is grayish white with speckled blackish markings, which vary from individual to individual. In some cases they form large blotches.

DISTRIBUTION: *Restricted to the area of the Red Sea and is replaced farther east by the black-spotted puffer* (A. nigropunctatus).

SIZE: *12 in (30 cm).*

BEHAVIOR: *This species ranges widely over the areas of the reef where live coral predominates. It feeds naturally on the polyps of* Acropora *species and other corals. Also hunts mollusks and crustaceans, crushing their shells with its powerful teeth, and browses on algae.*

DIET: *Provide a range of meaty foods as well as some herbivorous items. It is important to try to offer some hard-shelled items, such as small shrimps, so that the puffer can wear down its teeth. Otherwise its teeth can become overgrown and will have to be trimmed back to allow the fish to continue eating.*

AQUARIUM: *Not suitable for a reef aquarium, since it is likely to prey on many of the occupants.*

COMPATIBILITY: *Quite docile by nature toward its own kind and other fish in spite of its fierce teeth.*

WHITE-SPOTTED PUFFER

Arothron hispidus

A pale greenish brown background color typifies this pufferfish, although some tend to have a darker appearance than others, and there may be regional variations through its wide range. The brownish patches are distributed over the body, which is also broken by white spots. These spots also form a random pattern. The area at the base of the pectoral fins is typically greenish with a narrow yellow border, and there is similar coloration around the eyes.

▲ WHITE-SPOTTED PUFFER (AROTHRON HISPIDUS)

DISTRIBUTION: *Ranges from the Red Sea and East Africa through the Indo-Pacific north to southern Japan and south to Lord Howe Island off the east coast of Australia. Extends across the Pacific beyond the Hawaiian Islands to Baja California and from there through the Gulf of California to Panama.*

SIZE: *19.75 in (50 cm).*

BEHAVIOR: *While young of this species often venture into estuarine waters where they frequent areas of weed that provide them with protection from predators, adults tend to be found on open areas of the reef down to depths of 165 ft (50 m). They are usually observed on their own and are territorial by nature.*

DIET: *Feeds naturally on a wide variety of items ranging from algae to echinoderms. A varied diet suits these puffers well in aquarium surroundings. It should be based largely on meaty foods but should also include some plant matter.*

AQUARIUM: *Definitely not recommended for a reef aquarium because it will prove to be very destructive.*

COMPATIBILITY: *Its territorial nature means that this species is not suitable to be kept with others of its own kind. Its potential large size may also influence the choice of companions. Since these puffers can become very tame, it may be better to keep an individual on its own.*

MAP PUFFER

Arothron mappa

Very variable black lines on a grayish or white background help distinguish this species, with the area around the pectoral fins tending to be a more solid shade of black. There may be fine yellow markings there as well, bordering the black areas. As in the case of many puffers, the patterning is highly individual. Not only can this species inflate itself if threatened, but it can also defend itself with a series of spines projecting from its body.

▼ MAP PUFFER (AROTHRON MAPPA)

DISTRIBUTION: *From East Africa down as far as Natal in South Africa and through the Indo-Pacific, north to the Ryukyu Islands near Japan, and south to New Caledonia. Ranges through the Pacific to Samoa.*
SIZE: *25.75 in (65 cm).*

BEHAVIOR: *A relatively shy species, the map puffer rarely strays far from rocky areas on the reef, into which it can retreat if danger threatens. Tends to prefer calm water and hunts a variety of invertebrates.*

DIET: *Also feeds on algae, so provide a mixed diet based predominantly on meaty items, such as chopped shrimps, and a herbivorous ration as well.*

AQUARIUM: *As a predator on invertebrates, this species is unsuitable for inclusion in a reef tank.*

COMPATIBILITY: *Potentially grows to a large size, so its accommodation must be planned accordingly. Being solitary by nature, this large puffer is best housed alone in the aquarium.*

GUINEA-FOWL PUFFER

Arothron meleagris

The unusual name of this pufferfish stems from its spotted appearance, which is said to resemble the pattern seen on the flanks of a guinea fowl. Its body is dark—usually grayish black in color— and is covered with a series of white spots that may extend at least partially across some of the fins. It has a prickly texture, which presumably offers some protection against predators. The pupil is bordered by a narrow orange ring, providing some color to the eyes.

DISTRIBUTION: *Extends over a wide area from East Africa to the vicinity of Durban in South Africa. Also ranges throughout the Indo-Pacific region, reaching as far north as the Ryukyu Islands and south to Lord Howe Island off Australia's east coast. Occurs across the Pacific and is found around Easter Island and off the western coast of Mexico, from Guaymas southward as far as Ecuador in South America.*

SIZE: *19.75 in (50 cm).*

BEHAVIOR: *Closely allied to areas of the reef where living coral is abundant, this puffer uses its powerful teeth to break off pieces that it can eat. Coral forms the major component of its diet but other invertebrates, including mollusks and sponges, are also consumed as well as some algae. Aquarium equipment, especially electrical cabling, needs to be kept out of reach of this and other puffers as they become bigger, because they may try to bite through the flex and receive a deadly shock. An external panel heater will be a safer option.*

DIET: *A mix based primarily on meaty foods with some herbivorous foods will suit this fish well.*

AQUARIUM: *Will cause serious damage if introduced to a reef tank.*

COMPATIBILITY: *Not especially aggressive, but its size really dictates that this fish should be kept on its own.*

BLACK-SPOTTED PUFFER

Arothron nigropunctatus

Although similar to the masked puffer (*A. diadematus*), this species can display considerable variation in coloration, like many pufferfish. This can create taxonomic difficulties and means that identification of individuals can be problematic.

▼ *BLACK-SPOTTED PUFFER (AROTHRON NIGROPUNCTATUS)*

There are regional differences in color as well as individual variations within populations. The background coloration can vary from bluish gray to brown. The spots may also range from gold to black and rare individuals may be encountered that are entirely gold or black in color without any spots at all.

SYNONYMS: *A. citrinellus; A. ophryas.*

DISTRIBUTION: *Extends from East Africa across the Indian Ocean and the Pacific as far as Micronesia and Samoa. Its northerly range extends to southern Japan, while its southernmost distribution reaches as far as Australia's southeastern coast.*

SIZE: *13 in (33 cm).*

BEHAVIOR: *Found on coral reefs, where it hunts a wide range of invertebrate prey in addition to feeding on the coral itself.*

DIET: *Easily catered for with a range of chopped meaty foods. Try to include invertebrates such as mollusks in their shells, to help wear down the fish's teeth. May be reluctant to feed at first but it should soon regain its appetite and is ultimately likely to become quite bold when seeking food.*

AQUARIUM: *Will wreak havoc on the occupants of a reef tank.*

COMPATIBILITY: *Usually associates well with its own kind and is not aggressive by nature.*

BENNETT'S SHARP-NOSE PUFFER

(BENNETT'S TOBY)

Canthigaster bennetti

A very colorful species, Bennett's sharp-nose puffer, or toby, has green predominating on the upperparts with a paler whitish area below. Vivid orange and striking sky-blue spots and streaks are present over the green area of the body, fading in intensity on the underparts. There is also a small black area along the back, broken by sky-blue lines, while the caudal fin is reddish.

SYNONYMS: *C. benetti; C. constellatus.*

DISTRIBUTION: *Ranges from East Africa down the coast to the vicinity of Port Alfred in South Africa. Extends across the Indo-Pacific region northward via southern Taiwan, reaching Tanabe Bay in Japan. Its southernmost distribution is off the southeast coast of Australia.*

SIZE: *4 in (10 cm).*

BEHAVIOR: *This species inhabits sheltered areas of the reef and is usually encountered in pairs. Algae make up the bulk of its diet and may also be used as a nesting receptacle for the eggs of this fish.*

DIET: *This and other tobies will be established more easily if there is already a good growth of filamentous algae in the aquarium. It requires a diet based primarily on herbivorous foods, which can be supplemented with finely chopped meaty items such as squid and krill.*

▲ *BENNETT'S SHARP-NOSE PUFFER (CANTHIGASTER BENNETTI)*

AQUARIUM: *Likely to cause some damage within a reef aquarium.*

COMPATIBILITY: *True pairs will generally agree well, but males are likely to be aggressive toward each other. Unfortunately, it is not easy to sex these fish visually, although it may be possible to identify males by their slightly larger size. At a store, however, this can only be a guide because the ages of the tobies will be unknown.*

CROWNED TOBY

(THREE-BARRED TOBY)

Canthigaster coronata

The terms "toby" and "puffer" are often used interchangeably for members of the genus *Canthigaster*, with the result that this species is also sometimes known as the crowned puffer. This can be confusing, since tobies are generally much smaller fish than puffers. The crowned toby is very attractively patterned, with alternating yellow and blue lines radiating out from around the eyes and a pale green area extending down to the upper jaw.

▼ *CROWNED TOBY (CANTHIGASTER CORONATA)*

There are three white bars around the back that are separated by broader dark brown bands, edged primarily with orange. The underside of the body is whitish, while the caudal fin has a combination of light blue and yellow markings.

SYNONYMS: *C. axiologus; C. cinctus.*

DISTRIBUTION: *Ranges from the Red Sea down the East African coast to Sodwana Bay, South Africa, and through the Indo-Pacific region, extending northward to southern Japan and southward to southeastern Australia. Its easterly range extends as far as the Hawaiian Islands.*

SIZE: *5.5 in (14 cm).*

BEHAVIOR: *Often seen in deeper water than some of its relatives, typically at depths below 20 ft (6 m), and has been recorded down to a depth of 330 ft (100 m). Tends to occur in relatively open areas of the reef, sometimes among rubble, which provides some protection from predators.*

DIET: *Feeds on a very wide range of invertebrates and browses on algae. Easily catered for in aquarium surroundings.*

AQUARIUM: *Definitely not suitable for inclusion in a reef aquarium.*

COMPATIBILITY: *True pairs will agree together, although occasionally a male may chase his would-be mate. Two males will be much more aggressive toward each other. It may therefore be safer to include just a single individual with nonaggressive species in a community aquarium.*

PEARL TOBY

Canthigaster margaritata

Spotted patterning predominates in this particular toby, which has relatively subdued coloration overall. Its upperparts are dark with whitish pearl-like spots that tend to form lines above the eyes and along the back. The underparts are white.

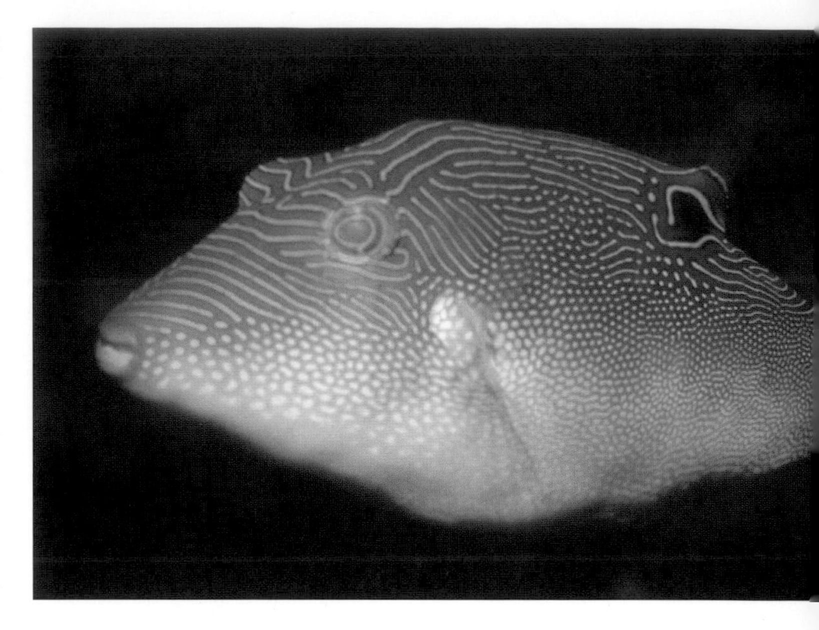

▲ PAPUAN TOBY (CANTHIGASTER PAPUA)

DISTRIBUTION: *The exact range of this species is unclear, but it occurs in the Red Sea, extending southward along the African coast as far as Inhaca Island off the coast of Mozambique. It occurs through the Indian Ocean, is present around the shores of Sri Lanka, and reaches the Ogasawara Islands near Japan.*

SIZE: *4.75 in (12 cm).*

BEHAVIOR: *The pearl toby is encountered in shallow waters on the reef. It is also not uncommonly seen in rock pools that form when the tide goes out.*

DIET: *Omnivorous, although it prefers invertebrate prey. Chopped meaty items, together with some herbivorous foods, should therefore be provided in aquarium surroundings.*

AQUARIUM: *Not suitable for a reef tank in view of its feeding habits.*

COMPATIBILITY: *It is often better to keep this fish singly in the aquarium alongside other nonaggressive companions rather than risk housing two individuals together when their genders are unknown. A true pair should, however, prove compatible.*

PAPUAN TOBY

Canthigaster papua

One of the less brightly colored members of the genus, the Papuan toby is a dark shade of blackish brown. It has variable light blue spots over the whole body. In some areas the spots are elongated into small stripes, particularly around the mouth and the eyes. The eyes themselves are encircled with orange, and the lips are whitish.

◀ PEARL TOBY (CANTHIGASTER MARGARITATA)

DISTRIBUTION: *Ranges from the Maldives off the southwestern tip of India, northward to the Philippines and Palau, and from New Guinea southward to the Great Barrier Reef and east to New Caledonia.*

SIZE: *6 in (15 cm).*

BEHAVIOR: *Relatively little has been recorded about the habits of this species, but it frequents clear water stretches on the reef where coral is abundant.*

DIET: *A mixed diet, consisting of both meaty foods and herbivorous items, will suit this toby. It may also browse on algae growing in the aquarium.*

AQUARIUM: *Avoid mixing this fish with the occupants of a reef tank, because it is likely to prey on any invertebrates that are present.*

COMPATIBILITY: *Pairs will probably agree well, but it may be better to keep this toby singly with other nonaggressive fish, rather than risk conflict.*

SADDLED TOBY

(VALENTINI'S SHARP-NOSE PUFFER; VALENTINI TOBY; BLACK-SADDLED TOBY)

Canthigaster valentini

This is another banded species, with four prominent dark bands extending along the back from the eyes, broken by narrower, paler markings. There may also be a series of fine light blue horizontal lines extending down the back, with horizontal stripes evident across the face. The sides of the body are spotted, and the spots on the head are smaller than those toward the rear of the body. Sexing is possible in this species, since the males are more colorful than the females. They have blue-green iridescent lines extending back from the eyes with a prominent bluish gray area in front of the vent. Lines of this color are present on the lower jaw, too. Males also grow to a slightly larger size overall than females. The appearance of this toby is mimicked for defensive purposes by the black-saddle filefish (*Paraluteres prionurus*).

DISTRIBUTION: *Ranges from the Red Sea southward to the coast of Durban in South Africa. Found north as far as southern Japan and south as far as Lord Howe Island off Australia's east coast. Its easterly distribution extends as far as the Tuamotu Islands.*

SIZE: *4.5 in (11 cm).*

BEHAVIOR: *These tobies may form shoals of as many as 100 individuals. They are sometimes joined by black-saddle filefish* (Paraluteres prionurus), *which seek cover among the tobies. Mature females establish territories, however, and a single male typically mates with several females.*

DIET: *Algae feature prominently in the diet of this species, along with a smaller proportion of invertebrates such as mollusks, echinoderms, and polychaetes.*

AQUARIUM: *Completely unsuitable for a reef tank because of its predatory feeding habits.*

COMPATIBILITY: *Should be kept only in true pairs or singly. Provide retreats so that a female will not be harassed by her mate. This species has spawned successfully in the home aquarium. The eggs may be concealed among filamentous algae.*

▼ *SADDLED TOBY (CANTHIGASTER VALENTINI)*

Porcupinefish Family Diodontidae

The spines present on fish in the family Diodontidae account for their common name of porcupinefish, although in reality they are much shorter than those of a porcupine. Because of their short spines, some of the 17 members of the group are known as burrfish. In all cases, the spines help protect them from predators. Like their relatives the pufferfish, or puffers, members of this group are slow swimmers.

Porcupinefish share the ability of puffers to inflate their bodies if frightened, which means that they must be caught carefully. Otherwise, they may try to inflate themselves while out of water and draw air into their bodies, which can be fatal. Move them into a container, therefore, rather than lifting them out of the aquarium in a net. Porcupinefish can become very tame—even feeding from the hand—but beware of their sharp teeth, which can inflict a painful bite. Electrical cabling in the aquarium must also be adequately protected.

BIRD-BEAK BURRFISH

Cyclichthys orbicularis

The small bird-beak burrfish has a very compact body shape, with a rounded head and slightly protuberant lips. Juveniles have a pelagic lifestyle (meaning that they live in the open sea) and can be identified by their spotted appearance. In contrast, adults have prominent blotches varying from brown to gray on their bodies, but they retain some black spots. Their appearance is highly individual. Their underparts are whitish, and their short spines are kept permanently raised. This species also has relatively large eyes, probably because of its nocturnal lifestyle.

Synonyms: *Diodon caeruleus; D. orbicularis.*

Distribution: *Extends from the Red Sea and East Africa through the Indian Ocean and via the Philippines northward as far as southern Japan. Extends southward to Australia and New Caledonia in the Pacific. Another population can be found in the southeastern area of the Atlantic Ocean, extending to the coast of South Africa.*

Size: *12 in (30 cm).*

Behavior: *The bird-beak burrfish prefers to swim over open areas, emerging from cover to hunt during darkness and swooping down on*

▼ *Bird-beak Burrfish (Cyclichthys orbicularis)*

invertebrates that it catches in the open. It can be encountered at depths down to 560 ft (170 m).

DIET: *Feed on meaty foods such as chopped krill or clams.*

AQUARIUM: *Unsuitable for a reef tank because of its predatory nature.*

COMPATIBILITY: *Generally not to be trusted with other fish, especially since this burrfish may become aggressive only after dark, so you will not see signs of this behavior, with potentially fatal consequences.*

LONG-SPINE PORCUPINEFISH
(BALLOON PORCUPINEFISH)

Diodon holocanthus

There is a marked difference in appearance between juvenile and adult long-spine porcupinefish. The young display a spotted patterning overall, which is especially pronounced on the underparts, whereas adults have a blotched pattern and their brown markings are interspersed with spots. There is a brown bar extending across the head, with similar markings running through the eyes. Its spines are normally kept flat, but the fish does not have to inflate its body in order to raise them—they can be moved independently.

SYNONYMS: *D. maculifer; D. multimaculatus; D. novemaculatus; D. paraholocanthus; D. pilosus; D. quadrimaculatus; D. sexmaculatus.*

DISTRIBUTION: *Ranges from the coast of Florida and the Bahamas south through the Gulf of Mexico to Brazil. Also occurs farther east in the Atlantic down to South Africa. Present in the Indian Ocean from the southern part of the Red Sea southward to Madagascar, Mauritius, and Réunion Island. Its northerly distribution extends to southern Japan, while its southernmost range is Lord Howe Island off Australia's east coast. Found through the Pacific to the Hawaiian Islands, Easter Island, and the Galápagos Islands. Also present in the eastern Pacific from the coast of southern California south as far as Colombia.*

▲ LONG-SPINE PORCUPINEFISH (DIODON HOLOCANTHUS)

SIZE: *19.75 in (50 cm).*

BEHAVIOR: *Juveniles may remain pelagic (living in open sea) until they grow up to 3.5 in (9 cm) long. This may help explain the wide distribution of the species, because this fish is not a strong swimmer. The long-spine porcupinefish occurs in a range of habitats, sometimes in shallow waters, where it often conceals itself among seagrass. It also ventures into estuaries, although it may range down to depths of 330 ft (100 m). It prefers rocky areas where it can hide away during the daytime. This means it will become more active in the evening, which is the best time to offer food. Take care not to encourage the puffer to inflate its body by frightening it when you are servicing the aquarium, especially at first.*

DIET: *Hunts a range of mollusks and crabs as well as feeding on sea urchins and other fish in the wild. A meaty diet featuring items such as chopped krill is therefore recommended. This porcupinefish may also eat some herbivorous food.*

AQUARIUM: *Definitely not a candidate for a reef tank because of its feeding habits.*

COMPATIBILITY: *Can be predatory toward smaller fish, ambushing them at night, and may attack bigger occupants, too. It can be kept safely with its own kind, however.*

MISCELLANEOUS FISH

As well as the species already described in this book, there are many other representatives of saltwater families that may be available from time to time. It is always vital to check out the habits and requirements of any individual fish, however, before choosing it for an aquarium. Otherwise you could find that it is highly disruptive—it may even prey on its companions. It may also rapidly outgrow its surroundings. In addition, you need to be sure whether or not it is social with its own kind, because a single individual of a species that normally displays strong shoaling instincts can prove to be uncharacteristically nervous in an aquarium on its own, when deprived of the sensory information and protection that comes from being part of a group.

RED-BACK SAND TILEFISH

(SKUNK TILEFISH; RED-STRIPE TILEFISH)

Hoplolatilus marcosi Family Malacanthidae

Tilefish form the family Malacanthidae. They are frequently colorful and quite small in size. In the case of this species, a reddish stripe extends down each side of the body from the jaws and passes through the eyes. The stripe varies in width along its length according to the individual specimen. The stripe curves up toward the dorsal fin on the fish's back. It broadens on reaching the caudal fin to create a darker triangle edged with white top and bottom. In some fish from the deepwater areas of the reef, however, this entire striped area may appear blackish, while the remainder of the body is white. The long low dorsal fin along the body has an orange tint along its upper edge.

DISTRIBUTION: *Found in the western Pacific from Indonesia and Papua New Guinea northward to the Philippines and Palau. Extends eastward via the Solomon Islands to Tonga.*

▼ *RED-BACK SAND TILEFISH (HOPLOLATILUS MARCOSI)*

SIZE: *4.75 in (12 cm).*

BEHAVIOR: *The red-back sand tilefish occurs in deepwater, in the vicinity of the drop-off areas of reefs down to depths of at least 265 ft (80 m). May also be found occasionally in shallower waters, where it constructs mounds of rubble to use as retreats.*

DIET: *Needs a meaty diet. This can include frozen foods, which will need to be thoroughly defrosted prior to use.*

AQUARIUM: *Active by nature, this tilefish needs plenty of space for swimming. The aquarium must also be kept covered, since this fish can jump well. Subdued lighting conditions, corresponding to those where it occurs on the reef, are recommended.*

COMPATIBILITY: *May prey on very small crustaceans, but otherwise it is suitable for a reef aquarium. Ideally, should be kept in small groups with others of its own kind.*

PURPLE SAND TILEFISH

Hoplolatilus purpureus Family Malacanthidae

As its name suggests, this tilefish is predominantly purple in terms of its overall coloration. This is usually of a darker shade on the top and sides of the head and for a variable distance along the base of the long dorsal fin. As in related species, the dorsal fin is low and extends along much of the back, almost reaching the caudal fin. The sides and rear of the body tend to be paler. There are two prominent red areas with black edging extending across each lobe of the caudal fin, with the central area here being pale.

▲ *PURPLE SAND TILEFISH (HOPLOLATILUS PURPUREUS)*

DISTRIBUTION: *Occurs in western and central parts of the Pacific Ocean, and is known from around the Philippines and the Solomon Islands.*

SIZE: *5 in (13 cm).*

BEHAVIOR: *This tilefish tends to frequent the edge of the reef where it gives way to deepwater. It is often encountered in pairs in these surroundings. Believed to feed on zooplankton.*

DIET: *Offer a range of suitable meaty substitutes, such as mysid shrimps. Tilefish generally have healthy appetites, to match their level of activity, and they should therefore be fed relatively small amounts of food several times each day.*

AQUARIUM: *Can be housed in a reef tank provided that there are no small crustaceans present.*

COMPATIBILITY: *This species can prove more territorial than some other tilefish, so aim to keep them in pairs, since squabbling is likely to break out if they are housed in larger groups. Can be trusted in the company of nonaggressive fish.*

STARCK'S TILEFISH

Hoplolatilus starcki Family Malacanthidae

Small juveniles of this species are bright blue in color but, as they mature, much of the blue coloration on their bodies turns yellowish white—apart from a patch of color on the head extending back behind the gills. This yellow coloration tends to be more vivid along the upper part of the body and is also very evident in the forked caudal fin.

DISTRIBUTION: *Ranges from the Moluccas northward to the Philippines and the Marianas and southward to Rowley Shoals off Australia's northwest coast. Extends to New Caledonia, east of Australia, and eastward as far as Pitcairn Island.*

SIZE: *6 in (15 cm).*

BEHAVIOR: *Often encountered in areas of rubble close to the steep fall-off on reefs and will retreat into burrows there if threatened. Juveniles may form shoals with young purple queen anthias* (Pseudanthias pascalus)*, relying on their similarity in coloration at this stage.*

DIET: *Feeds naturally on a variety of zooplankton but will eat a variety of meaty foods such as mysid shrimps in aquarium surroundings.*

AQUARIUM: *Can be included with nonaggressive species as part of a reef tank and is unlikely to prove disruptive in this environment.*

COMPATIBILITY: *Tends not to be as social with its own kind as are some other tilefish. Often settles better in aquarium surroundings if obtained while still in immature coloring.*

HUMP-BACK RED SNAPPER

Lutjanus gibbus Family Lutjanidae

Snappers belong to the family Lutjanidae. They tend to be active predators, which explain their name. Young are a shade of pale silvery blue, with a prominent black spot on each side of the caudal peduncle and a prominent silvery iris around the pupil. As they become older, however, their color and appearance change. They develop a reddish suffusion around the lips and the fins, with this color also developing on the iris. The profile alters with age as well, giving the fish more of a hump-backed appearance.

DISTRIBUTION: *Ranges from the Red Sea down the coast of East Africa, extending eastward through the Indian Ocean as far as the Line and Society Islands in the Pacific. Ranges northward as far as southern Japan and is found as far south as Australia.*

SIZE: *19.75 in (50 cm).*

BEHAVIOR: *Juveniles hide in beds of seagrass and occur in relatively shallow areas of water on the reef. Social by nature, they associate in large groups, especially as they grow older. Rather than actively swim, they form shoals that often drift during the day before becoming more active at dusk.*

▼ *STARCK'S TILEFISH (HOPLOLATILUS STARCKI)*

▲ *HUMP-BACK RED SNAPPER (LUTJANUS GIBBUS)*

DISTRIBUTION: *From the Red Sea down the coast of East Africa through the Indian Ocean, reaching as far north as southern Japan and south to Australia. Extends eastward across the Pacific to the Line and Marquesan Islands. There is also a separate population in the southeastern part of the Atlantic Ocean, off the South African coast in the vicinity of East London.*

SIZE: *15.75 in (40 cm).*

BEHAVIOR: *While young of this species often frequent areas of seagrass, adults are far more conspicuous and are likely to be encountered in large shoals, although they often remain close to cover. They are often seen in the vicinity of wrecks, for example, where they can dart inside if danger threatens. They may sometimes be found in deepwater, down to depths of 870 ft (265 m).*

DIET: *This snapper is a highly active predator and requires a meaty diet. It is unfussy about food. There are reports that it will also eat algae.*

AQUARIUM: *It is not safe to include this species in a reef tank, as it will probably consume many of the occupants likely to be living there.*

COMPATIBILITY: *It is dangerous to keep this fish with either smaller or more placid companions. If you want to house several of these snappers in an aquarium on their own, they will need to be transferred at the same time to prevent any individuals from gaining a territorial advantage and then persecuting any fish that are introduced later.*

DIET: *Will eat a wide range of meaty foods. Feeds naturally on other smaller fish and invertebrates in the wild.*

AQUARIUM: *Not suitable for a reef tank, since many of the occupants, from crustaceans to echinoderms, are likely to be eaten by this fish. A night light may enable you to observe its behavior to best effect.*

COMPATIBILITY: *Their social nature means that these snappers should be kept in a group in a spacious aquarium. Avoid mixing them with smaller companions. They can live for up to 18 years.*

COMMON BLUE-STRIPE SNAPPER

(BLUE-LINED SNAPPER)

Lutjanus kasmira Family Lutjanidae

The common blue-stripe snapper is one of the most colorful members of the genus *Lutjanus*, although the exact markings vary from individual to individual. As their name suggests, blue-stripe snappers typically have four narrow sky-blue horizontal stripes running down the sides of their bodies, highlighted with narrow black edging. The first of these starts just below the dorsal fin, and the fourth stripe is present beneath the eyes, running down to the vicinity of the caudal peduncle. The remainder of the body is bright yellow, aside from the underparts, which may show traces of paler, narrower, blue stripes as well as being silvery white. The fins themselves are yellow.

▼ *COMMON BLUE-STRIPE SNAPPER (LUTJANUS KASMIRA)*

EMPEROR RED SNAPPER

Lutjanus sebae Family Lutjanidae

The markings of juveniles of this species are especially bright and vibrant, but they fade significantly as they grow older. These snappers can grow to a large size, even when compared with other members of the genus. A broad, oblique, reddish black line extends from the lips up to near the base of the dorsal fin, with a similarly colored band encircling the body and extending along the underside here. A broad curled marking extends around from the dorsal fin across the caudal peduncle to the lower lobe of the caudal fin. There is also a reddish black band apparent in the upper part of this fin. As they grow older, the bodies of emperor red snappers take on a more silvery hue, and the markings ultimately disappear as the fish becomes reddish in color.

DISTRIBUTION: *Ranges from the southern part of the Red Sea down the coast of East Africa and across the Indian Ocean, reaching as far north as southern Japan. Extends south to the coast of Australia. Also found eastward from the Great Barrier Reef to New Caledonia.*

SIZE: *43 in (110 cm).*

BEHAVIOR: *The young of this species are found in shallow waters near the shore, often venturing into mangroves. On the reef they will often seek out the protection provided by sea urchins. They may then start to form shoals as they grow older, and move farther out into deeper water.*

▲ *EMPEROR RED SNAPPER (LUTJANUS SEBAE)*

DIET: *Should display a healthy appetite for meaty foods of all types. Good filtration in an aquarium is essential to cope with the waste output of these fish, particularly as they grow larger, to prevent any significant deterioration in water quality.*

AQUARIUM: *Too large and aggressive to be housed in a reef tank, where many of the occupants are this species' natural prey.*

COMPATIBILITY: *The young are quite shy, but they become much more assertive as they grow larger, and may then present a danger to other fish living alongside them.*

MIDNIGHT SNAPPER

Macolor macularis Family Lutjanidae

The young of this species are different in appearance from the adults and are more likely to be encountered in the aquarium trade. They are black and white, with a black bar extending through the eye and a horizontal black band extending back from the eye. The back is black with white areas, while on the underside of the body another black stripe extends to the caudal fin. The patterning is lost as these fish mature—adults are dark overall. There is a pale yellow area on the throat, however, which merges into dark speckling. The eyes also change in color—the pupils turn from dark in juveniles to yellow in adults.

DISTRIBUTION: *Found in the western Pacific, but the exact localities are unclear, since this species is very similar to the black snapper* (M. niger) *and there is confusion about their respective ranges. However, its distribution definitely extends northward from Australia to the Ryukyu Islands close to Japan, and eastward through the islands that make up Melanesia.*

SIZE: *24 in (60 cm).*

BEHAVIOR: *Juveniles tend to be found in more sheltered parts of the reef than adults, often occurring in association with staghorn corals, which may provide them with a measure of protection from would-be predators. Adults are often encountered on seaward slopes, ranging down to depths of at least 295 ft (90m). They appear to be more social than juveniles, and are often seen in groups. They may also associate with their close relative, the black snapper, at this stage.*

DIET: *Predatory and therefore needs a meaty diet. Young juveniles may require livefoods at first but can be weaned onto inanimate foods.*

AQUARIUM: *Not suitable for a reef tank because they will prey on a variety of the occupants, particularly as they grow larger.*

COMPATIBILITY: *Vulnerable to fin-nipping, especially while young. May then start to prey on tankmates as they themselves grow larger. Companions need to be chosen carefully, and a spacious aquarium will be essential.*

▼ *MIDNIGHT SNAPPER (MACOLOR MACULARIS)*

▲ *SAIL-FIN SNAPPER (SYMPHORICHTHYS SPILURUS)*

SAIL-FIN SNAPPER

(THREAD-FIN SNAPPER; BLUE-LINED SEA BREAM)

Symphorichthys spilurus Family Lutjanidae

As with other snappers, there is a distinctive difference in appearance between juveniles and adults of this species. In this case, however, it is the adult fish that are more brightly colored than the young. The most characteristic feature of the young is the broad black band extending back from the eye into the caudal fin, which has a narrow white edging above and below. The remainder of the body is silvery with a yellowish hue. The dorsal fin, set well back, has a long threadlike filament that trails back over the caudal fin. A similar but shorter filament is also apparent on the anal fin. Adults display a series of wavy blue lines on their bodies, with alternating yellowish stripes, and a blue-edged black spot on the top of the caudal peduncle. An orange stripe runs through the eyes, and another one, which is often edged with black, extends down each side of the body as far as the back of the gills.

DISTRIBUTION: *Found from western Australia in the eastern Indian Ocean northward via Indonesia, the Philippines, and Palau to the Ryukyu Islands off the coast of Japan. Extends around Australia to the Great Barrier Reef, ranging to New Caledonia and as far east as Tonga in the Pacific.*

SIZE: *23.75 in (60 cm).*

BEHAVIOR: *Favors sandy areas of the reef and is generally solitary, although this species is known to spawn communally. Lively by nature and alert to danger, this snapper will retreat and hide in caves if threatened.*

DIET: *Its predatory feeding habits mean that this species needs meaty foods. It is not usually difficult to feed, although items such as mysid shrimps may be more suitable for young individuals.*

Aquarium: *Not suitable for a reef tank, because it will hunt a number of the typical occupants, such as crustaceans, and it will also grow too large for the average reef setup.*

Compatibility: *Will prey on smaller companions and needs to be kept apart from its own kind. A suitable retreat needs to be provided in the aquarium for an individual. Although shy at first, juveniles can become quite tame.*

TWO-LINED MONOCLE BREAM

(TWO-LINE SPINE-CHEEK)

Scolopsis bilineata Family Nemipteridae

This bream belongs to the family Nemipteridae. There is a wide variation in the appearance of individuals throughout its range, especially among the populations present in the Indian and Pacific Oceans. Those originating from the Indian Ocean display no yellow markings on their bodies. In the aquarium trade young juveniles tend to be encountered most frequently. They have an alternating pattern of yellow and dark brown stripes on the upperpart of their bodies, with a white stripe in the midline. The underside of the body is grayish. As they mature, these stripes are reduced and the white area expands.

Distribution: *Extends from the Laccadive Islands and the Maldives off the west coast of India northward as far as southern Japan and southward as far as Lord Howe Island off the southeast coast of Australia. Ranges eastward through the Pacific as far as Fiji and Tonga.*

▲ *Two-lined Monocle Bream (Scolopsis bilineata)—juvenile*

▼ *Two-lined Monocle Bream (Scolopsis bilineata)*

Size: *9 in (23 cm).*

Behavior: *Ranges widely across the reef. The young tend to occur closer inshore than the adults, and are also often present in lagoons. Typically feeds by sucking in sand in search of anything that it can eat, and then spits out the inedible particles.*

Diet: *Requires a diet of meaty foods, which should be chopped up into small pieces. A varied diet is recommended.*

Aquarium: *May be disruptive in a reef tank because of its feeding habits. Also likely to prey on small invertebrates such as mollusks.*

Compatibility: *Can be aggressive, and therefore this species may be unsuitable for a reef tank alongside smaller, more docile fish. Its desire to dig in the substrate may sometimes be a drawback as well, but it is usually quite robust and will thrive in the company of similar species in a fish-only aquarium.*

COMMERSON'S FROGFISH
(GIANT FROGFISH)

Antennarius commerson Family Antennariidae

The colors of this fish are extremely variable and are influenced in part by its habitat. They can be creamy or brilliant yellow, even orange. Then there individuals that exhibit dark colors, varying from shades of green through brown to black. In common with other anglerfish, this species has a characteristic lure in front of its mouth, which accounts for its common name. The lure serves to attract its prey within reach.

Synonym: *A. moluccensis.*

Distribution: *Extends from the Red Sea down the east coast of Africa to South Africa. Found throughout the Indian Ocean, ranging north to southern Japan and as far south as Lord Howe and the Society Islands. Occurs right across the Pacific via the Hawaiian Islands to reach the coast of Panama in Central America.*

Size: *15 in (38 cm).*

Behavior: *This species is a sedentary predator, relying on its camouflage to blend into the background. In developed areas, it is often encountered in the vicinity of jetties and similar structures. Spawning may sometimes take place in the confines of the aquarium. Its breeding habits are unusual—the eggs are released in mucus, forming what is often described as a veil.*

Diet: *Naturally feeds on live fish and crustaceans. Weaning this fish onto alternative meaty foods can be difficult.*

Aquarium: *Not suitable for a reef tank, since it will prey on some of the occupants. It will also damage coral if it chooses to adopt a particular area of the mini-reef here as its perch.*

Compatibility: *Not to be trusted with smaller fish, including members of its own species—they may end up being eaten.*

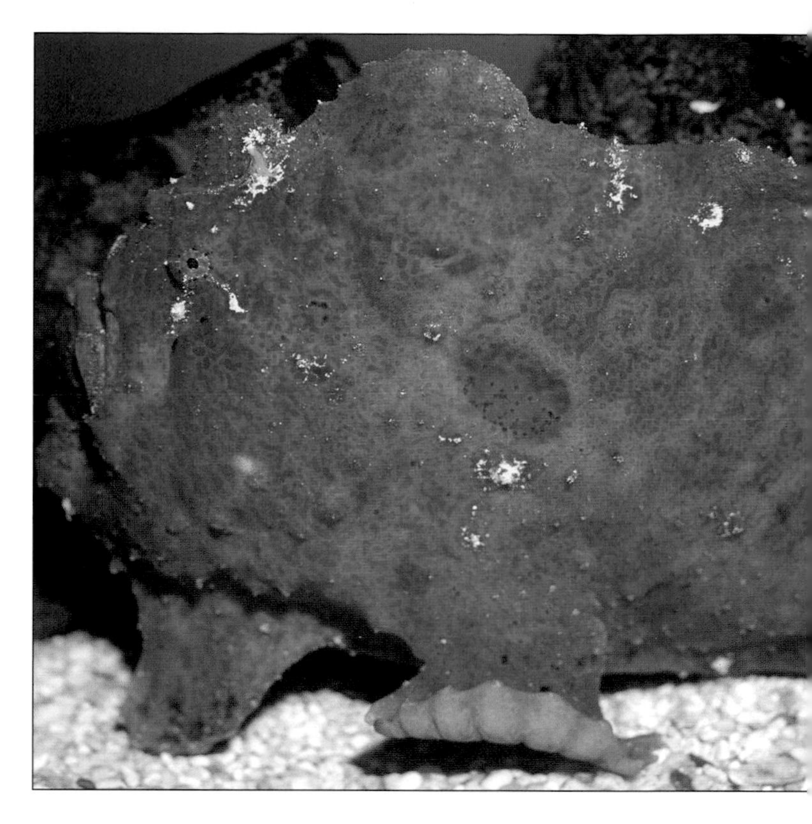

▲ *Commerson's Frogfish (Antennarius commerson)*

SHAGGY ANGLER

Antennarius hispidus Family Antennariidae

Often superbly camouflaged, like other members of this group, the shaggy angler gets its name from its overall appearance. When this fish is resting, it can be very hard to distinguish it from the general background of the reef because it blends in so well. In terms of coloration this species is orangish with tan markings

▼ *Shaggy Angler (Antennarius hispidus)*

on the body. There are blackish brown spots on the fins. The dorsal fin is tall and vertical in appearance. It possesses the lure that is characteristic of all anglerfish, and a corresponding wide mouth to scoop up any prey that is attracted within reach.

DISTRIBUTION: *Extends from East Africa across the Indian Ocean, although it seems to be found in association with continental landmasses, such as the Indian subcontinent and Malaysia, rather than islands. Occurs in the vicinity of the Moluccas, however, and ranges northward as far as Taiwan. Its southerly distribution extends to northern Australia. It may be encountered more widely across the Pacific, but is only currently recorded in this region from the vicinity of Fiji.*

SIZE: *8 in (20 cm).*

BEHAVIOR: *This anglerfish tends to favor shallow areas of the reef, but it may also be encountered sometimes in much deeper water in muddy areas. Like other anglerfish, it is relatively inactive, has a slow metabolism, and probably does not feed every day in the wild.*

DIET: *This species is an ambush predator, eating small fish and crustaceans attracted to its lure. Waggling food such as pieces of prawn with plastic forceps may help wean this and other anglerfish onto inanimate foods. Requires a meaty diet.*

AQUARIUM: *Will live in a reef aquarium provided there are no small fish or crustaceans that could be eaten.*

COMPATIBILITY: *Can be housed with other fish of similar size, but cannot be trusted with smaller companions.*

▲ *PAINTED FROGFISH (ANTENNARIUS PICTUS)*

PAINTED FROGFISH

Antennarius pictus Family Antennariidae

This frogfish, perhaps more than any other species, is highly variable in coloration. There is a good reason for this—it often associates with a variety of sponges, which may themselves attract small fish and crustaceans seeking a safe retreat on the reef. Painted frogfish often appear in shades of orange, therefore, although there are also vibrant red forms. In complete contrast, a black form with white edging along its dorsal spines is documented, which is more reminiscent of a nudibranch. The painted frogfish is one of the larger members of the family.

DISTRIBUTION: *Ranges from the Red Sea down the coast of East Africa and through the Indian Ocean. Its distribution extends across much of the Pacific too, reaching eastward as far as the Hawaiian Islands and south to the Society Islands.*

SIZE: *12 in (30 cm).*

BEHAVIOR: *Occurs in relatively shallow, sheltered areas of the reef, which is where sponges are likely to be present. It relies on its lure—a modified spine consisting of two parts, the illicium and the esca—to attract its prey.*

DIET: *Meaty foods, although time may have to be spent at first persuading this fish to take inanimate foods of this type. Chop up pieces so that the frogfish will be able to eat them easily.*

AQUARIUM: *Avoid including in a reef tank where there are prey species present. It can, however, be fascinating to observe this fish in the company of sponges.*

COMPATIBILITY: *It is not safe to house this frogfish in the company of smaller companions.*

RAZORFISH
(SHRIMPFISH)

Aeoliscus strigatus Family Centriscidae

Razorfish are related to pipefish. Like pipefish, they can be recognized by their unusually slender body shape. They are sometimes variable in color, which helps them merge into the background. In open areas of the reef they have a central blackish stripe with pale borders running down the side of the body. This makes them resemble a strand of seaweed, particularly when they adopt their characteristic vertical stance. In areas of seagrass, however, they are transformed into a greenish color, with the central stripe becoming far less prominent.

DISTRIBUTION: *Found through the Indian Ocean from the coast of Africa via Seychelles northward as far as southern Japan and southward down the eastern coast of Australia to New South Wales.*

SIZE: *6 in (15 cm).*

▼ RAZORFISH (AEOLISCUS STRIGATUS)

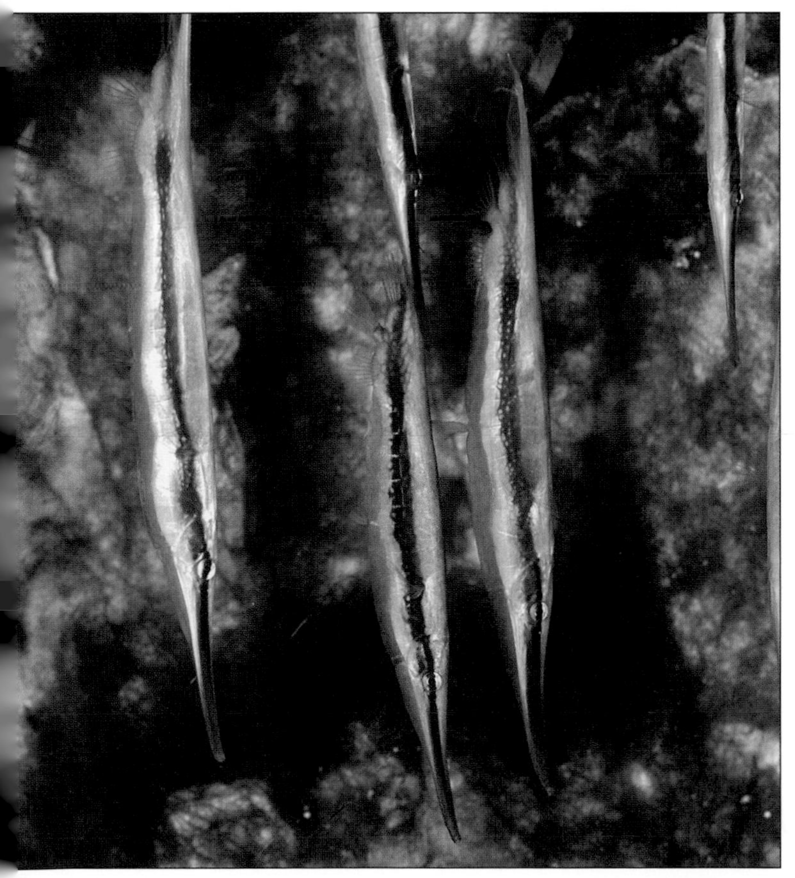

▲ CARIBBEAN TRUMPETFISH (AULOSTOMUS MACULATUS)

BEHAVIOR: *This unusual fish lives in groups and swims vertically, adopting a head-down stance, although it may take up a more horizontal position when feeding. Its body is well protected with thin plates. Razorfish breed communally, with a group of females laying their eggs in a nest that has been constructed by a single male.*

DIET: *Feeds on zooplankton and has a relatively small mouth. Providing an acceptable varied diet in the aquarium can be difficult, although brine shrimps can be useful at first. May forage for food over well-established living rock.*

AQUARIUM: *Suitable for a reef aquarium, since it is peaceful by nature and will only prey on very small invertebrates.*

COMPATIBILITY: *Can be kept with other peaceful fish, but avoid introducing boisterous companions.*

CARIBBEAN TRUMPETFISH

Aulostomus maculatus Family Aulostomidae

Trumpetfish make up the family Aulostomidae. They are related to seahorses and pipefish, a fact that is reflected in their very elongated slender body shape. Their narrow striped patterning is more evident behind the eyes—which are set some distance back behind the snout—than elsewhere on the body. The caudal peduncle is narrow and elongated and it has banded markings. There is a sensory feeler—called a barbel—present at the front of this fish's lower jaw.

DISTRIBUTION: *Ranges from southern Florida in North America down through the Caribbean region to northern South America. Extends eastward through the Atlantic as far as St. Paul's Rocks.*

SIZE: *40 in (100 cm).*

BEHAVIOR: *This trumpetfish tends to frequent relatively shallow areas on the reef, particularly where there is cover present, where it can lurk in search of prey. It swims horizontally in some cases, but also disguises itself among sea whips by adopting a vertical position.*

DIET: *Preys naturally on both fish and crustaceans and will take meaty foods. It can be difficult to persuade this fish to eat inanimate food. If necessary, try dragging food with plastic forceps or a feeding stick of some type through the water, encouraging the fish to suck it into its mouth.*

AQUARIUM: *Cannot be trusted in a reef aquarium that contains smaller fish and crustaceans.*

COMPATIBILITY: *Solitary by nature. Will not attack larger companions, however, and is ideal for a large aquarium alongside nonaggressive species.*

COMET
(MARINE BETTA)

Calloplesiops altivelis Family Plesiopidae

A member of the round-head family of fish called Plesiopidae, this attractive species has an oval body shape. It is brownish with white spots that become smaller and pale blue around the extremities of the body. There is a distinctive eyespot, or ocellus, on both sides of the body just above the caudal peduncle, which is black with a pale blue border. Aside from the arrangement of the pectoral fins and the small yellow area on the end of the caudal fin, it can be difficult to spot the actual head end of this fish. This is because the caudal fin merges in with the fins above and below on this part of the body.

DISTRIBUTION: *Extends from the Red Sea down the coast of East Africa, ranging throughout the Indo-Pacific region as far east as Tonga and the Line Islands in the Pacific.*

SIZE: *8 in (20 cm).*

BEHAVIOR: *The reason for the unusual patterning of this fish becomes clear when it is alarmed—it will retreat head-first into a crevice. It then leaves its hindquarters exposed, making it resemble a menacing moray eel waiting to lunge, thereby intimidating all but the very boldest of predators. Comets are shy by nature and tend to be nocturnal in the wild, but they will settle down well (and have even bred successfully) in aquarium surroundings.*

DIET: *Feeding in the early stages can prove to be a problem, since the comet may only eat livefoods readily, with brine shrimps being a useful stand-by. It should be possible to wean it onto inert meaty foods, but it may refuse all food at first. Every effort therefore needs to be made to encourage it to eat.*

AQUARIUM: *Its predatory habits mean that the comet is unsuited to a reef aquarium, since it may take crustaceans as well as smaller fish.*

COMPATIBILITY: *Will not disturb fish that are too large to be eaten, although individuals of this species may not always be compatible with each other.*

▼ COMET (CALLOPLESIOPS ALTIVELIS)

▲ PINEAPPLEFISH (CLEIDOPUS GLORIAMARIS)

PINEAPPLEFISH

Cleidopus gloriamaris Family Monocentridae

Unmistakable in appearance, looking just like the fruit for which it is named, the pineapplefish represents a family known as pineconefish (Monocentridae), which have changed very little in appearance for millions of years. The pineapplefish's basic coloration can vary from yellow to orange, with the body being crossed by a series of dark interconnecting lines. There is also a more solid dark area evident in the vicinity of its yellow head.

DISTRIBUTION: *Ranges through the Indian Ocean to the western part of the Pacific east of Papua New Guinea. Extends northward to southern Japan and south to eastern Australia.*

SIZE: *8.75 in (22 cm).*

BEHAVIOR: *The pineapplefish frequents dark areas on the reef and often inhabits caves. It is in the gloom that the colonies of bioluminescent bacteria present close to its eyes produce a fluorescent greenish glow. This probably helps attract invertebrates to the fish's mouth when it is hunting in these surroundings.*

DIET: *Small livefoods will be essential, particularly at first, and it may be difficult to wean this fish off them. Vitamin-enriched brine shrimps and similar items will be needed.*

AQUARIUM: *Not suitable for a brightly lit reef aquarium, particularly since it may prey on small invertebrates.*

COMPATIBILITY: *Peaceful by nature and will need a cave to use as a retreat in the aquarium.*

▲ ORIENTAL FLYING GURNARD (DACTYLOPTENA ORIENTALIS)

ORIENTAL FLYING GURNARD

Dactyloptena orientalis Family Dactylopteridae

The broad winglike pectoral fins give this fish its common name. The body itself is rather rigid, and the overall color is variable. Its coloration allows it to blend into the background and is often relatively pale with evident spotting extending from the body across its so-called wings.

DISTRIBUTION: *Ranges from the Red Sea down the coast of East Africa across the Indian Ocean north to the area around southern Japan and southward to southern Australia and New Zealand. Extends through the Pacific to Hawaii, and also to the Marquesan and Tuamotu Islands.*

SIZE: *16 in (40 cm).*

BEHAVIOR: *As its appearance suggests, this member of the family Dactylopteridae occurs in open areas of the reef. It swims slowly just above the sand, seeking food, all the while blending in well against the background.*

DIET: *Naturally hunts crustaceans and other invertebrates as well as small fish. Offer meaty foods. This fish can often eat surprisingly large pieces, thanks to its capacious mouth, although it is not always easy to wean off livefoods. Pieces of crab and prawn may be useful for this purpose.*

AQUARIUM: *This fish can be a problem in a reef tank, not just because of its feeding habits, but also because of its need for open sandy areas in which to swim.*

COMPATIBILITY: *Must not be kept with small fish, which it will prey on. Nor should it be housed with aggressive species, particularly those that are likely to nip its fins.*

◀ BARRED SOAPFISH (DIPLOPRION BIFASCIATUM)

BARRED SOAPFISH
(TWO-BANDED SOAPFISH)

Diploprion bifasciatum Family Serranidae, tribe Diploprionini

A relative of the fairy basslets and groupers, the young barred soapfish may mimic the coloration of *Meiacanthus* species blennies, which can have toxic skin secretions. It often appears yellow at this stage. As its name suggests, though, adults of the species have a black bar passing through the eyes and a much broader black marking encircling the body, from the dorsal fin down as far as the underside. The remainder of the body is a relatively pale shade of yellow, although some individuals can be almost entirely black.

DISTRIBUTION: *Ranges from the Maldives in the Indian Ocean around the coast of India to Papua New Guinea and northward as far as southern Japan. Extends southward as far as Lord Howe Island off Australia's east coast.*

SIZE: *10 in (25 cm).*

BEHAVIOR: *This soapfish is protected by a skin toxin known as grammistin. If released into the confines of the aquarium, it will be deadly to both the soapfish and other tank occupants. Net them with care, therefore, to avoid causing stress, and always avoid handling them with bare hands in case you have any minor cuts.*

DIET: *A predatory species that feeds on other fish, which it swallows whole in its large jaws. Needs a diet of meaty foods and may ultimately be persuaded to eat some flake food too.*

AQUARIUM: *Unsuitable for a reef aquarium because it will prey on smaller fish and crustaceans.*

COMPATIBILITY: *These soapfish are not aggressive toward each other providing they are of similar size, nor will they disturb larger fish of other species.*

SPOTTED SOAPFISH
(SNOWFLAKE SOAPFISH)

Pogonoperca punctata Family Serranidae, tribe Grammistini

Young of this species have spots that are significantly larger than those of the adults. Unusually, in this fish the adult patterning is effectively superimposed on the juvenile patterning, which remains evident although obscured. Many smaller, brighter, white spots will become apparent at this stage over the body, although the fins—with the exception of the dorsal and caudal fins—remain clear. There is a series of dark brown markings evident over the back but, as in the case of the spots, this barring is highly individual.

DISTRIBUTION: *Has been recorded from the vicinity of Natal, South Africa, and the Comoro Islands. Ranges across the Indian Ocean as far north as southern Japan and south via New Caledonia in the Pacific, extending to the Line, Marquesan, and Society Islands.*

SIZE: *14 in (35 cm).*

BEHAVIOR: *While the young of this species occur in relatively sheltered areas on the reef, the adults seek out mature coral and can be found at much greater depths—down to at least 665 ft (200 m). Although it occurs over a wide area, this fish seems to have very specific habitat requirements, which may account for its inconsistent distribution. Like other soapfish, it is protected by a poisonous mucus that can prove deadly to itself and other tank occupants if the fish is stressed. That is why water from the bag in which it is brought home must be thrown away and never tipped into the aquarium.*

▼ SPOTTED SOAPFISH (POGONOPERCA PUNCTATA)

DIET: *Eats meaty foods readily and usually has a large appetite.*

AQUARIUM: *Not suitable for a reef tank, since it is likely to prey on the other occupants.*

COMPATIBILITY: *Companions need to be chosen carefully to avoid the risk of them falling prey to this fish. Avoid mixing with smaller fish at any stage.*

JACK-KNIFE FISH

Equetus lanceolatus Family Sciaenidae

The coloration of these fish, which are members of the family Sciaenidae, gives them an unusual appearance. They have a black stripe running down at an angle of almost 90 degrees from the top of the dorsal fin. A second band extends right around the body to the tips of the long pelvic fins. Together, these two bands suggest two separate fish swimming through the water, with the adjacent white areas blending into the background. A third, much smaller, dark band encircles the eyes.

DISTRIBUTION: *Ranges off the coast of North Carolina in North America via the Bahamas and Bermuda, through the Caribbean and the Gulf of Mexico, as far as the central coastal region of Brazil in South America.*

▲ *JACK-KNIFE FISH (EQUETUS LANCEOLATUS)*

SIZE: *10 in (25 cm).*

BEHAVIOR: *The jack-knife fish relies on its appearance—which gives the impression that it is two separate fish—to confuse predators, rather than the ability to swim quickly away from them. It can occasionally be found close inshore, especially in the case of juveniles.*

DIET: *Unfortunately, this fish can be difficult to wean onto an artificial diet. At first, be prepared to supply small livefoods, which should include vitamin-enriched brine shrimps and black worms. If a fish is refusing all other feeding options, then young mollies may be necessary. An individual may also comb live rock for small prey. Aim to wean onto meaty foods, which will need to be chopped into tiny pieces.*

AQUARIUM: *Will prey on some of the occupants.*

COMPATIBILITY: *Tankmates need to be chosen very carefully to prevent bullying and fin damage. Vegetarian species that will not compete with it for food may be compatible. Include retreats in the aquarium. Young are agreeable with their own kind but they tend to become more territorial as they grow older.*

▲ *HIGH-HAT (PAREQUES ACUMINATUS)*

HIGH-HAT

Pareques acuminatus Family Sciaenidae

The tall first section of the dorsal fin helps explain the common name given to this drum, which belongs to the same family as the jack-knife fish. It is raised over its head, resembling a hat. The high-hat has a series of dark blackish stripes that run horizontally down the sides of the body, separated by intervening silvery areas. Young of this species have particularly long fins.

DISTRIBUTION: *Ranges from the coast of North Carolina in North America southward through the Caribbean and Gulf of Mexico to the coastal area of Rio de Janeiro in Brazil, South America.*

SIZE: *9 in (23 cm).*

BEHAVIOR: *A species that often frequents reefs around tropical islands, the high-hat is shy and must have suitable retreats in the aquarium to correspond with the caves in which it seeks safety in the wild.*

DIET: *A range of small invertebrates will be needed to encourage this fish to feed in aquarium surroundings. Live sand will help, allowing it to burrow naturally in search of worms, while vitamin-enriched brine shrimps also make a valuable addition to its diet at first. May be weaned onto meaty foods if they are chopped into tiny pieces.*

AQUARIUM: *Not trustworthy in a reef tank, since it is likely to prey on the invertebrates and perhaps even the small fish here.*

COMPATIBILITY: *Needs inoffensive companions in view of its nervous nature. Young high-hats may agree well at first, but disputes are more likely to arise as they mature, especially in a relatively small aquarium.*

TWO-STICK STINGFISH

Inimicus filamentosus Family Scorpaenidae

This relatively sedentary member of the stonefish family, like other related species, is well protected against predators by its venomous spines. It therefore needs to be maintained with great care in the aquarium, because it can inflict a very painful sting. Its coloration can be changeable, often appearing as a combination of red, white, and blackish brown, with the filaments along its back serving to break up and disguise its appearance.

▼ *TWO-STICK STINGFISH (INIMICUS FILAMENTOSUS)*

▲ YELLOW-HEAD JAWFISH (OPISTOGNATHUS AURIFRONS)

DISTRIBUTION: *Ranges from the Red Sea down the coast of East Africa and across the Indian Ocean and is present around the Maldive Islands off India's west coast. Its precise distribution is unclear, but may extend around the Philippines and possibly farther afield.*

SIZE: *10 in (25 cm).*

BEHAVIOR: *A sedentary species that remains close to the seabed, occurring on the reef both in sandy and rubble-strewn areas. If threatened, it will unfurl its broad pectoral fins.*

DIET: *Often reluctant to take inanimate foods at first. Ultimately it can be weaned onto prepared meaty foods, but never place your hand close to this fish when trying to persuade it to feed.*

AQUARIUM: *Not entirely suitable for the reef tank, because it will prey on both crustaceans and small fish that are normally resident.*

COMPATIBILITY: *Keep apart from its own kind and do not mix with other species that could become its prey.*

YELLOW-HEAD JAWFISH

Opistognathus aurifrons Family Opistognathidae

Jawfish are so called because of the way in which they depend on their jaws for digging in the substrate and for incubating their eggs. This species displays a variable amount of yellowish coloration on the head. The gill area and underparts in this region are paler than the rest of the body, which is light blue, occasionally with a slight greenish hue.

DISTRIBUTION: *Ranges from around the southern coast of Florida in North America, via the Bahamas through the Gulf of Mexico and the Caribbean to northern South America.*

SIZE: *4 in (10 cm).*

BEHAVIOR: *Will dig in the substrate with its mouthparts to excavate a burrow. Remains in the vicinity of this retreat, adopting an almost vertical position, which gives the fish better all-around visibility. If danger threatens, it will retreat back to the relative safety of its burrow. When breeding, the male jawfish will incubate the eggs in its mouth and will not feed until they hatch about a week after spawning occurred.*

DIET: *Normally eats a varied range of finely chopped meaty foods quite readily, especially those that originate from crustaceans.*

AQUARIUM: *Very suitable for a reef tank, but needs a mixed substrate—at least 3 in (7.5 cm) in depth—in order to create its burrows. Jawfish are nervous in new surroundings until they have created a burrow, so take care to ensure the aquarium is covered to keep them from jumping out.*

COMPATIBILITY: *Social by nature, so obtain several of these fish but avoid overcrowding them. Jawfish have been bred successfully in the home aquarium and they can also be mixed with other nonaggressive species.*

HARLEQUIN SWEET-LIPS
(CLOWN SWEET-LIPS)

Plectorhinchus chaetodonoides Family Haemulidae

There is a marked difference in appearance between juveniles and adults in this species. When they are very small they look rather like a flatworm, which helps them avoid being preyed on. They then display some similarity to young clownfish, thanks to their orange-brown body coloration, which is broken by white markings. As they mature, however, they change again, and develop a pattern of dark spotting on a paler background. Sweet-lips belong to the family of fish called grunts (Haemulidae) because of the sounds they are sometimes heard to make.

DISTRIBUTION: *Extends from the Maldives, which lie to the southwest of India, down to Rowley Shoals off the northwest coast of Australia and northward as far as the Ryukyu Islands near Japan. Reaches across the Pacific to New Caledonia and Fiji.*

SIZE: *28.5 in (72 cm).*

BEHAVIOR: *While the young of this species often hide among areas of coral on the reef, the significantly larger adults will roam much farther afield. They tend to be nocturnal in their feeding habits, retreating into coves and other secluded areas of the reef during hours of daylight.*

DIET: *Unfortunately, juveniles often tend to be reluctant to feed. Therefore, it is a good idea to see the fish feeding if possible before agreeing to acquire it. Otherwise, you may need to be prepared to offer live crustaceans and then slowly wean it across to inanimate meaty foods. A live sand system containing worms that the fish can dig out may also help acclimatize it.*

AQUARIUM: *The harlequin sweet-lips is not suitable for a reef tank in view of its predatory feeding habits and ultimately large size, as well as its dislike of brightly lit surroundings.*

COMPATIBILITY: *Only mix with nonaggressive species that will allow this species to feed without snapping up all the livefood first.*

▼ *HARLEQUIN SWEET-LIPS (PLECTORHINCHUS CHAETODONOIDES)*

INDIAN OCEAN ORIENTAL SWEET-LIPS

Plectorhinchus vittatus — Family Haemulidae

Young fish are more boldly marked than adults. They display a variable pattern of black blotches and stripes set against a white background, while their pectoral fins are yellow. The area beneath the dorsal fin tends to be black, as is the caudal fin, which has a white diagonal stripe running across it. The transformation to adult coloring becomes apparent once the fish have grown to about 6 inches (15cm) in length. The dark blotches will be replaced by a more even, horizontal, striped pattern running around the body. The markings will be sufficiently distinctive to allow individuals to be identified by their patterning.

DISTRIBUTION: *Ranges from East Africa through the Indian Ocean via Papua New Guinea, eastward through the western Pacific as far as New Caledonia.*

SIZE: *28 in (72 cm).*

BEHAVIOR: *Young of this species are encountered on their own in relatively sheltered areas of the reef, whereas adults tend to form up into larger groups, creating shoals.*

DIET: *Requires a meaty diet, but it can be difficult to persuade this fish to eat at the outset. Juveniles are quite nervous when feeding, and they require small worms and crustaceans.*

AQUARIUM: *Generally unsuitable for a reef tank, although this species can be used as a biological control, since it will remove unwanted fireworms from these surroundings. However, other occupants that might be eaten, including crustaceans, snails, and some starfish, should first be removed, or the rockwork should be moved into the fish's quarters.*

COMPATIBILITY: *Generally peaceful by nature if not kept with species on which it may prey. These sweet-lips can grow very large, however.*

BICOLOR GOATFISH
(DASH-AND-DOT GOATFISH)

Parupeneus barberinoides — Family Mullidae

Goatfish make up the family Mullidae, and a number of species are occasionally available to saltwater aquarists, although they are not always easy to establish. This particular species has a dark red area on the front of its body, broken by a white stripe running across the eyes. This stripe connects with a white area behind, before merging into yellow. The reddish black area on the front of the body is broken by purplish lines, with a purple spotted pattern also apparent in the yellow area running into the caudal fin. There is also an irregular black blotch on the flanks.

DISTRIBUTION: *Found in the eastern part of the Indian Ocean, extending from the Moluccan Islands of Indonesia via the Philippines to the Ryukyu Islands near Japan. Found as far south as New Caledonia. Ranges across the Pacific via Palau and the Marshall Islands to Tonga.*

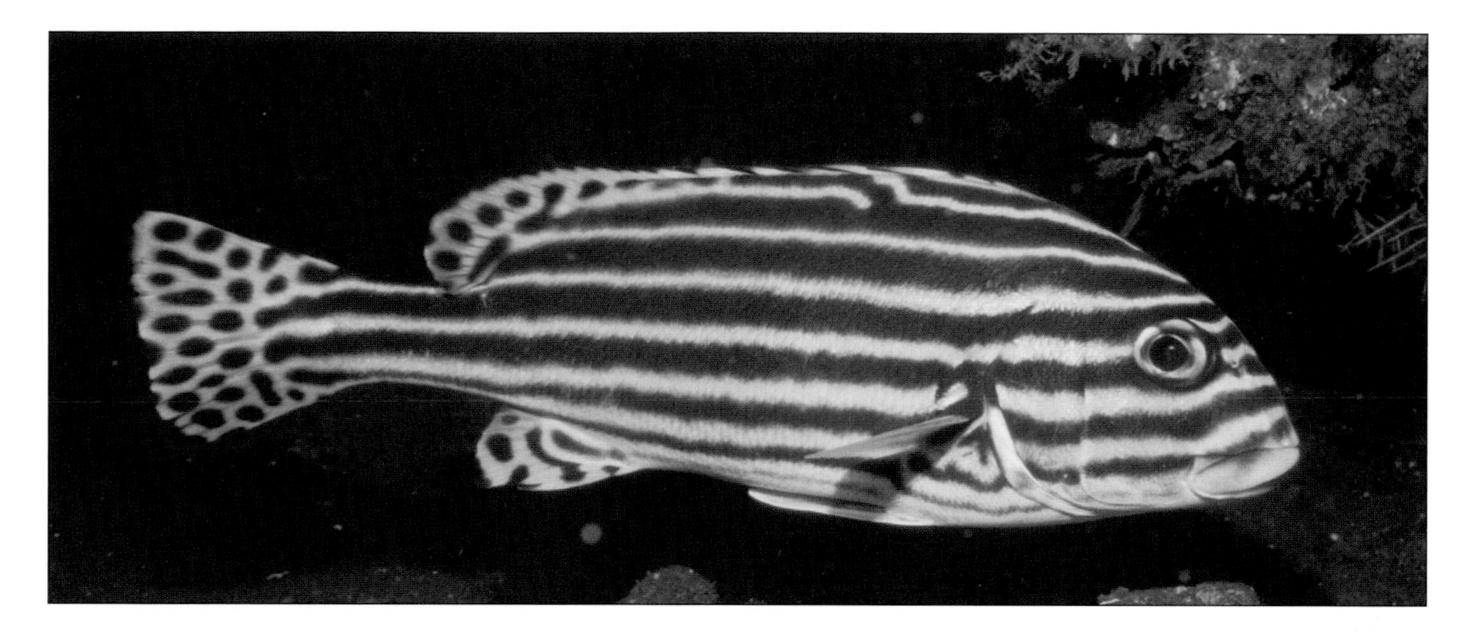

▲ Indian Ocean Oriental Sweet-lips (Plectorhinchus vittatus)

Size: *12 in (30 cm).*

Behavior: *Frequents the more sheltered areas of the reef, especially where there is a mixture of sand and rubble on the seabed. It uses the sensory barbels around its mouth to probe for edible items, especially in the vicinity of seaweed, where crustaceans may be hidden from view. Also looks for worms close to the surface in sandy areas.*

Diet: *Preys naturally on a wide range of invertebrates, which can also include small mollusks and brittlestars. Needs to be offered substitute meaty diets, and must be fed several times a day, especially while young. If weight loss becomes evident and yet the goatfish appears to be eating well, the cause may be intestinal worms, which these fish often contract. In this case, seek advice from a veterinarian about deworming.*

▼ Bicolor Goatfish (Parupeneus barberinoides)

Aquarium: *Certainly not recommended for a reef tank because of its feeding habits.*

Compatibility: *This fish is not threatening to other fish of similar size or larger. It is, however, likely to prove a persistent digger in the substrate, in which it will seek edible items.*

STRIPED EEL CATFISH
(CORAL CATFISH)

Plotosus lineatus Family Plotosidae

The majority of catfish are freshwater species, although some are known to venture into brackish water occasionally. The striped eel catfish is, however, the only species occurring in the coral reef environment. It is dark in color with yellowish white stripes on its body. There are two on each side of the body, as well as six prominent sensory barbels around the mouth. These catfish need to be caught and handled very carefully, because they are well protected by a sharp venomous spine at the beginning of the dorsal fin. This fin is fused with the caudal fin behind and the anal fin beneath the body. There are other venomous spines present on the pectoral fins.

Distribution: *Ranges from the Red Sea and East Africa, even sometimes venturing into freshwater areas of Lake Malawi and of Madagascar, across the Indian Ocean. Here it extends northward as far as southern Japan and Korea, and is found as far south as Australia. Occurs also around Palau and Yap, and eastward across the Pacific to the vicinity of Samoa.*

Size: *13 in (32 cm).*

Behavior: *The young live in close association with each other, forming large groups of perhaps as many as 100 fish. They twist and turn as they move through the water, creating the impression of a ball. Older fish often live on their own but sometimes associate in groups. They seek their food usually by digging in the seabed.*

DIET: *This catfish is very straightforward to feed and will take flake or pelleted foods as well as fresh foods.*

AQUARIUM: *Not suitable for a reef tank, because it will dig up the substrate and prove to be disruptive. It is also increasingly likely to eat the other occupants as it grows bigger.*

COMPATIBILITY: *This catfish can usually only be kept in groups when young, making them ideal for a single-species aquarium. Unfortunately, their relatively large adult size and increasing intolerance toward each other means they will need to be split up as they grow older, unless the aquarium itself is very large.*

EMBER PARROTFISH

Scarus rubroviolaceus Family Callyodontidae

This family is known as the parrotfish because of their bright coloration. Unfortunately, they are not easy to keep in aquarium surroundings because of their feeding habits and also because—as in this case—they can grow to a large size. The ember parrotfish is predominantly blue in color, but its exact coloration varies through its range. Some individuals are sky blue with a hint of green on the sides of the body, while others have a more heavily mottled body and blackish blotches. Their appearance is determined by their area of origin.

DISTRIBUTION: *Ranges down the coast of East Africa as far as Durban in South Africa. Extends northward across the Indian Ocean to the Ryukyu Islands of southern Japan, and south to the Great Barrier Reef off Australia's east coast. Widely distributed through the Pacific region, extending past the Hawaiian Islands to the Gulf of Mexico.*

SIZE: *27.5 in (70 cm).*

BEHAVIOR: *This is a shoaling species. Males can be recognized within a group by the enlarged shape of the forehead. The coloration of these parrotfish tends to lighten at night.*

▼ EMBER PARROTFISH (SCARUS RUBROVIOLACEUS)

▲ LEAF SCORPIONFISH (TAENIANOTUS TRIACANTHUS)

DIET: *Ember parrotfish feed almost entirely on algae and need to have constant access to microalgae, which they can graze in the aquarium. This can be cultured in advance, with a regular supply being grown on rocks elsewhere. They will also graze on coral skeletons as a source of calcium in their diet. Alternatively, they should be give a calcium block for this purpose. Some individuals may be persuaded to eat herbivorous diets too, along with small amounts of livefood such as vitamin-enriched brine shrimps, so be prepared to experiment as necessary.*

AQUARIUM: *This parrotfish will damage stony coral in a reef tank with its sharp teeth.*

COMPATIBILITY: *It is not generally aggressive toward other fish, although two males may not agree if they are housed together in the confines of the aquarium.*

LEAF SCORPIONFISH

Taenianotus triacanthus Family Scorpaenidae

Some variation in color is likely to be apparent among individuals in the case of the leaf scorpionfish. Specimens can range from yellow and shades of red to tan, brown, and even almost black in some cases. The leaf scorpionfish is also characterized by its particularly tall dorsal fin and correspondingly wide pectoral fins on each side of the body. These features may help disguise the presence of this relatively inactive fish on the reef, where it tends to conceal itself in weedy areas.

DISTRIBUTION: *Extends over a wide area, from East Africa across the Indian Ocean northward to the Ryukyu Islands near Japan and southward to*

Australia. Occurs throughout most of the Pacific region, and is present around both the Hawaiian Islands and the Tuamotu Islands eastward as far as the Galápagos Islands off the coast of northern South America.

SIZE: *4 in (10 cm).*

BEHAVIOR: *The name of the leaf scorpionfish comes from the way in which it sways in the current, like a leaf, and also from the venomous spines on its body. It is imperative to handle it with care for this reason. In the aquarium it may adopt a particular perch from which it will rarely stray. Its small size makes it an ideal choice for these surroundings.*

DIET: *It can be difficult to wean this fish off livefoods, which form their natural diet in the wild. If you are offering other meaty items, chop them up so that the fish will be able to eat them more easily.*

AQUARIUM: *May be a problem in terms of preying on smaller shrimps and even other fish if housed in a reef tank.*

COMPATIBILITY: *Several of these scorpionfish will agree well together in an aquarium and are likely to become quite bold and conspicuous in these surroundings. Companions need to be chosen with care, however, because leaf scorpionfish are vulnerable to fin-nipping,and they may also lose out in the competition for food with other, more agile species.*

BLUE-SPOTTED STINGRAY

(BLUE-SPOTTED RIBBONTAIL RAY)

Taeniura lymma — Family Dasyatidae

There has been a considerable growth of interest in keeping rays in aquaria, but their requirements are very specific. Furthermore, they are equipped with a venomous spine close to the base of their tail, which means that not just catching them, but even servicing their aquarium can be hazardous. The blue-spotted stingray is one of the most readily available ray species and it displays the family's typical disklike body as well as a long narrow tail. The color of its dorsal surface can vary markedly, from yellow and shades of olive green through to grayish or reddish brown—all colors that help the fish merge into the background. In all cases, there is a random pattern of blue spots evident on the upper part of the body. The underparts are white.

DISTRIBUTION: *Extends from the Red Sea and the coast of East Africa across the Indian Ocean, northward as far as southern Japan. Also found around the coast of northern Australia to the south. Ranges across the Pacific as far as the Solomon Islands.*

SIZE: *12 in (30 cm).*

BEHAVIOR: *The blue-spotted stingray is likely to be encountered on sandy areas of the reef, hunting in groups at high tide. They will then separate and retreat out to deeper water, where they will hide under ledges and other suitable retreats.*

DIET: *Predatory by nature, this fish feeds on a variety of crustaceans and mollusks and needs to be given a meaty diet. If frozen, the food needs to thawed thoroughly before being given to the fish.*

AQUARIUM: *Not safe to include in a reef tank, since many of the occupants resident there form its natural prey.*

COMPATIBILITY: *This fish can be housed with nonaggressive species, but it requires a large aquarium if it is to thrive. There should be an extensive open area in the tank, as well as plenty of suitable retreats, although—unlike some of its relatives—the blue-spotted stingray tends not to bury into the sand here.*

▼ *BLUE-SPOTTED STINGRAY (TAENIURA LYMMA)*

INVERTEBRATES AND ALGAE

It is not easy, even within the context of a saltwater aquarium, to define an invertebrate other than to say it is a creature without a backbone. The vast majority of creatures on the planet are invertebrates, and this section of the book looks at the major groups, or phyla, that are popular in the marine aquarium. Without invertebrates, specifically hard coral—the building block of the reef—this particular ecosystem with all its associated and often highly distinctive life forms, including the fish themselves, would simply not exist. These tiny invertebrates have resulted in the formation of the Great Barrier Reef, for example, which extends for a distance of about 1,260 miles (2,000 km) down the eastern coast of Australia. Algae are primitive plantlike organisms. In the context of the saltwater aquarium there are two types: those added for decoration and intrusive invaders capable of overrunning an aquarium.

▶ *A REEF AQUARIUM WITH A VARIETY OF CORALS AND SEA URCHINS, AS WELL AS FISH.*

CORALS, SEA ANEMONES, SPONGES

STONY OR HARD CORALS

Corals—with sea anemones, hydras, and jellyfish—make up the phylum Cnidaria. Stony corals often consist of a series of individual organisms rather than one large interconnecting body. They have tiny tentacles with feathery protrusions, called polyps, that trap minute food particles in the water, as well as extracting oxygen. The polyps are located in a calcium-rich exoskeleton, which gives them some protection, since they can be withdrawn from danger.

Reef aquaria must have a relatively high level of calcium in the water because the stony corals need to extract this mineral if they are to remain healthy. As a guide, about 0.14 ounces (400 mg) of special calcium hydroxide powder (often sold under the German name of Kalkwasser) should be used for every 2 pints (1 l) of water to ensure the correct concentration, although more general supplements are also now available.

The other requirement for these corals, which are typically green or pale brown in color, is clear water. This is essential to allow sunlight to reach them, because within stony corals there are microscopic single-celled plants belonging to the alga family, which are known as zooxanthellae. These algae must be exposed to bright light in order for them to carry out photosynthesis, a process by which they use sunlight to create food. The food benefits not only the algae, but is also very important for the growth and well-being of the coral. As a by-product of photosynthesis, potentially harmful carbon dioxide is used up and oxygen is released into the water.

If the water becomes murky or if pollution affects the reef, the stony coral will die, affecting many of the other occupants as well. Some butterflyfish, for example, are obligate corallivores, meaning that they feed almost exclusively on coral polyps. (The overall size of a natural reef, however, means that the fish will not damage it with their feeding habits.)

The growth of stony coral is a relatively slow process, and it typically takes thousands of years for a reef to develop. During this period new colonies of stony coral grow on top of the old, raising the level of the reef, while dead corals leave calcareous deposits behind

HARD CORALS—TANK CONDITIONS AND CARE

HABITAT An almost exclusively invertebrate tank, established for at least 4 to 6 months in which the following exceptional conditions are maintained.

Tank Size A minimum of 48 x 18 x 12 in (122 x 46 x 30 cm), or 48 U. S. gallons (40 Imp. galls/182 l)

pH	8.2–8.3
Temperature	77–79°F (25–26°C)
Ammonia	Zero
Nitrite	Zero
Nitrate	Zero
Specific Gravity	1.021–1.026
Dissolved Oxygen	7-8 ppm
Calcium	400–450 ppm
Phosphates	Zero
KH	A natural seawater level of 7°dKH
Redox Potential	350–400mv

Lighting Metal halide lighting is ideal with one 150-watt lamp at 6,500k per 2 square feet (0.18 sq. m). High-intensity fluorescent tubes in the right quantity would be acceptable—a 5-foot (1.5-m) tank would need five or six tubes plus reflectors. A continuous photoperiod of 12 hours each day.

Substrate	Avoid detritus build-up

Circulation Good water turbulence is essential using several pulse-controlled pumps.

Filtration External trickle filtration with excellent water circulation. Efficient protein skimming with additional ozone. Use the highest quality activated carbon filtration.

Water Changes Either 15 to 20% per 2 weeks without fail or a constant water change system, which would prove much more successful. Reverse osmosis or deionized water is strongly recommended (toxins in mains water usually prove fatal). Use only finest quality salt.

Extras Ultraviolet sterilizer, oxygen reactor, redox controller, dosing pumps for calcium supplements etc., platinum chiller to keep temperature stable, since lighting and possibly the weather will tend to increase it unacceptably.

FEEDING Most hard coral will benefit from feeding with small pieces of squid and lancefish once a week. Live foods such as brine shrimp and rotifers supplement the coral's symbiotic algae. Much of this food will be wasted should the fish be too numerous, because they will devour it before the corals have a chance to benefit. Should the corals reject the food, all traces need to be siphoned off promptly to prevent pollution.

HEALTH Nearly all health problems are associated with poor water quality. The flesh becomes detached from the skeleton, never to become reattached; polyps retract further back into the skeleton each day until they do not extend at all; the white skeleton becomes more exposed as the animal shrinks. It is usually impossible to stop deterioration, even by an improvement in water quality. Prevention is better than cure.

▲ *The Sun Coral (*Tubastrea aurea*) prefers to inhabit shaded areas in the reef aquarium.*

them. At the same time, tropical storms whip up waves that can break up areas of the reef. Although they can spread vegetatively, stony corals reproduce by producing gametes. In one of the most amazing natural events, the entire reef spawns simultaneously, releasing eggs and sperm. The very high concentration of sperm in the water helps ensure that the eggs are fertilized. They drift on the ocean current before being deposited elsewhere to form the basis of a new area of coral growth.

Goniopora species of stony coral with their attractive feathery tentacles, such as *G. lobata,* are among the most widely distributed groups of these invertebrates. The polyps of these corals are not individual animals, however, so if the coral is damaged for any reason an infection may engulf the entire organism. This can happen easily as the result of rough handling or even if sharp protrusions from the body of a crustacean penetrate its exoskeleton. As with most corals, bright lighting is essential for these species.

One of the most striking of all stony corals is the brilliantly colored sun coral (*Tubastrea aurea*). As its name suggests, it is a rich golden-orange color. It is unusual because it occupies shady areas on the reef and lacks zooxanthellae in its body tissues. The sun coral should thrive in an area such as beneath an overhang, although it will reveal its full glory only at night. The coral spreads by a process known as budding, in which new polyps emerge from near the base of existing polyps. As with fish, however, corals tend to grow significantly smaller in aquarium surroundings than in the wild.

The organ pipe coral (*Tubipora musica*) is reddish in color, and dead pieces of it are sometimes offered for sale as aquarium decor. It is not always easy to establish though, so choose a relatively large piece that will give the best chance of success.

One of the most distinctive of all the stony corals is the brain coral (*Leptoria* species), so called because its convoluted appearance resembles the folds, or gyri, seen on the surface of the brain. It can grow very large on the reef and varies in color from brown (the most common color) to more exotic shades of pink and green. It requires excellent lighting in order to thrive in an aquarium setting.

Rather similar in appearance are the moon corals (*Favites* species). These corals are dark green in color, although they fluoresce when exposed to ultraviolet light. They too require suitable lighting to support the algal population present in their tissues. Another group that fluoresce are members of the genus *Euphyllia*, for example, the tooth coral *E. picteti*. These species have long and slender tentacles when extended, with club-shaped or hammer-shaped tips. They require less demanding conditions than some other hard corals, but the tentacles can have a powerful sting.

When buying stony coral, try to obtain larger pieces, particularly in the case of corals in which the polyps form the entire organism rather than being individual. Avoid flooding the area close to the coral with food, because this will deter it from putting out its tentacles, and make sure that the aquarium lighting is appropriate for the species concerned. Fix the coral securely in place in the aquarium and bear in mind, especially when servicing the tank, that exposed tentacles may be capable of stinging your hand or arm. Not all corals will naturally attach to rockwork. Plate coral (*Heliofungia actiniformis*), for example, needs to be located directly on a coral sand substrate.

▲ *Pulse Coral (*Xenia *species), showing a plume of tentacles on an elongated stem.*

Soft Corals

As their name suggests, soft corals lack a calcareous exoskeleton. Instead, they anchor directly onto the substrate. They occur in a wide variety of colors and forms, which can range from white to vibrant shades of red. Identifying species of soft coral is not easy. They do not conform to any exact shape, form, or coloration, and many of the species are almost indistinguishable from each other. Good lighting is essential, as is water movement, which will waft food particles within their reach. The most important thing when purchasing these corals is to ensure that they are not damaged at their base. They can grow fast, so allow for this when putting them in place in the aquarium.

One benefit is that they are nonaggressive and will coexist in close proximity. *Sarcophyton* species are often called leather corals because of the texture and color of these animals when the polyps are withdrawn. *Sinularia* species branch into many finger-like vertical lobes, covered in shorter polyps. One exception is *S. brassica,* named the cauliflower coral for its tight rosettes of polyps that resemble cauliflower florets. Both *Sarcophyton* and *Sinularia* species are easy to keep and are ideal for the novice invertebrate keeper.

Dendronephthya species differ noticeably in that their bodies are supported by sharp calcium spicules (spiky protrusions) and they may be colored red, orange, or white. Closely related pulse corals (*Anthelia* and *Xenia* species) have a particularly delicate and beautiful form, comprising pulsing pinnate (featherlike) tentacles on elongated stems. They come in colors from dark brown to light gray. The rhythmic pulsing is believed to help the flow of water over the tentacles.

SOFT CORALS—TANK CONDITIONS AND CARE

HABITAT For tank and water conditions see Mushroom Corals, page 273.

Lighting Soft corals are very forgiving when it comes to lighting. They will do quite well under only a few fluorescent lights. However, the stronger the lighting, the better the animal will look and grow. Watch them blossom under intense metal halide lighting.

FEEDING A few drops of juice from some meaty marine fare can be introduced at night when the corals are feeding, but this is not essential, especially in a mixed aquarium where the fish are also being fed.

HEALTH Soft corals suffer from bacterial infections caused by poor water conditions. Rotting of the basal attachment point is fairly common, as is blackening of the tissues. Both these undesirable situations can be improved by restoring good water quality and increased water circulation over the affected areas. Once this is achieved, many species will heal themselves, given time.

MUSHROOM CORALS

Mushroom corals are often referred to as false corals because they occupy the middle ground between anemones and corals. Marine aquarists will know these creatures by a variety of common names: coral anemones, disk anemones, plate anemones, mushroom polyps, mushroom anemones, and so on. Whatever you choose to call them, they all belong to the family Actinodiscidae, and most species are to be found in relatively shallow tropical reef locations that are spread throughout the world. As a general rule, they live in areas of slack water current in preference to more animated areas of the reef.

Despite being found at varying depths from just below the low water mark down to 12 feet (40 m), all species possess symbiotic zooxanthellae algae within their tissues as a reliable and constant source of nourishment. Even so, mushroom corals can gain extra sustenance from more direct methods. Many species cover themselves in a mucus layer that traps waterborne nutrients and transports them to a central mouth.

Various other species have a more dramatic approach. Giant elephant ears (*Rhodactis* species) can transform their flat disks into a hollow ball in which they trap small fish and crustaceans. Between 12 and 18 hours later the hapless victim has been consumed and the disk shape is assumed once more. It is true that not all *Rhodactis* species are able to perform this feat, but the aquarist should be aware that those specimens capable of extending to 12 to 15 inches (30–38 cm) in

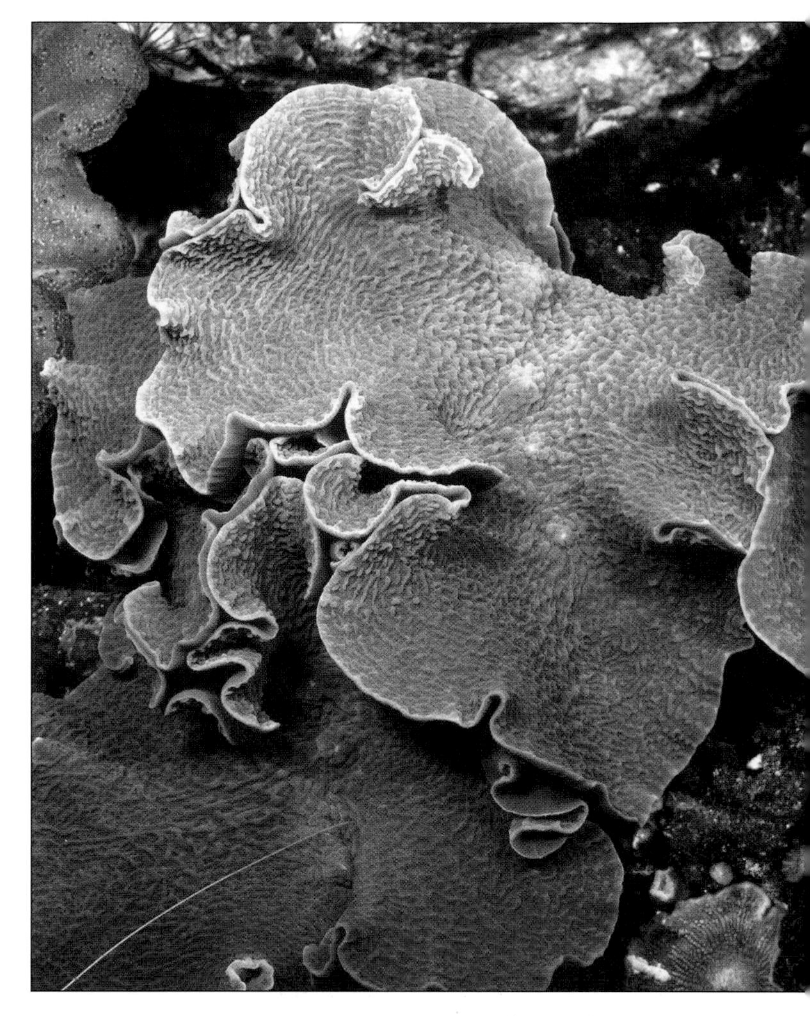

▲ *The flat disk of Giant Elephant Ears (Rhodactis species) can fold up to engulf small fish.*

MUSHROOM CORALS—TANK CONDITIONS AND CARE

HABITAT Rocky walls, outcrops, shallow reefs with very clear water. Best kept in reef aquaria.

Tank Size	More than 24 U. S. gal (20 Imp. gal/91 l)
pH	8.1–8.3
Temperature	75–79°F (24–26°C)
Ammonia	Zero
Nitrite	Zero
Nitrate	Less than 10 ppm (preferably zero)
Specific Gravity	1.022–1.025
Phosphates	Less than 0.5 ppm (preferably zero)
Redox Potential	350–450 mv

Filtration Trickle filtration is preferable. Efficient protein skimming and activated carbon filtration as standard.

Water Changes 15 to 25% change every two weeks using high-quality, filtered water.

Fish Stocking Level Absolute maximum of 1 in per 7.2 U. S. gal (1 in per 6 Imp. gal/2.5 cm per 27 l).

Water Circulation Moderate flow over the colonies.

Lighting Choose lighting to suit specific colonies and ensure correct placement of colonies. If polyps shrink or retract, they require less light. If they extend, they require more intense lighting. Rich colors of deepwater species can be maintained by subdued lighting.

FEEDING *Rhodactis* spp. should be offered pieces of mussel, squid, or lancefish. Smaller species may be fed on live or frozen rotifers, frozen fine zooplankton, and the juices from other frozen fish foods. Liquid fry food and artificial substitutes are not recommended, since they may cause pollution. Many colonies will survive successfully without additional feeding supplements, but if feeding is necessary, do so very sparingly and certainly no more than twice a week.

HEALTH The two main health risks faced by mushroom polyps are detachment and shrinking. Large polyps can shrink from healthy disks down to tiny buttons less than 10% of their normal size. The main causes are poor water quality or unsuitable lighting. Once these conditions have been improved, mushroom corals should recover slowly.

diameter should be investigated before they are housed with small fish, shrimps, and crabs.

Mushroom polyps are capable of defending themselves. Not only do they produce a toxin to keep other invading corals at a distance, but they are also very resistant to toxins produced by encroaching species. In the confines of the invertebrate aquarium such poisons will need to be removed on a continuous basis by the use of activated carbon filtration.

Although all mushrooms are disk shaped, the sheer variety of colors, spots, stripes, dimples, fringing tentacles, and textures is astounding. The vast majority of these fascinating creatures have yet to be classified properly, and most specimens are referred to simply as *Actinodiscus* species.

SEA ANEMONES

Like many corals, sea anemones often possess zooxanthellae in their tissues. This is particularly true of those anemones that tend to be brownish or green in color. In addition, their tentacles are equipped with special stinging cells, called nematocysts, which enable them to capture prey directly, effectively by harpooning it with these barbed stinging cells. They release a poison into the cells of any victim that swims within range.

The sea anemones kept in aquaria tend to be those that attach to rocks and other structures on the reef, rather than free-swimming examples of the group, such as jellyfish. Members of the genus *Heteractis* are the most popular group in saltwater circles, because they will be home to anemonefish (clownfish), which hide under their protective tentacles. Unfortunately, this behavior is less likely to be displayed by tank-bred anemonefish than their wild-caught relatives, suggesting that it is acquired rather than instinctive behavior.

Anemones need to be positioned so that their tentacles can move freely and are not partially buried. A good water flow is important for wafting edible particles toward the anemones, and it is possible to set up wavelike motions in the tank for this purpose. Anemones will occasionally withdraw their tentacles, resembling a small ball, but this should be temporary. If it lasts longer than a day it may be a sign that the anemone is ill, water conditions are poor, or the level of illumination is inadequate and is having a harmful effect on the zooxanthellae.

ANEMONES—TANK CONDITIONS AND CARE

HABITAT For tank and water conditions see Mushroom Corals, page 273.

FEEDING A healthy anemone can be fed small pieces of lancefish, squid, cockle, and mussel once each week to keep it in good condition. Small pieces should be pressed lightly into the tentacles and never forced into the mouth as this will cause serious damage. If food is rejected, remove it and do not try again for another week. Many anemones will remain perfectly healthy without such gross feeding, taking adequate nutrition from their zooxanthellae exclusively. Where fish are kept in the same tank, the juices from frozen fish food are often enough to keep anemones healthy. Nuisance anemones such as *Aiptasia* spp. must not be fed if they are to be kept under control.

HEALTH The most common ailment is when anemones turn white, shrink, and eventually die. This can be caused by several factors, including lack of light, poor water quality, or lighting in the wrong color spectrum. Anemones possess a symbiotic alga within their tissues, and as the alga dies the anemone loses its color and shrinks in response to a lack of nutrients and oxygen. Once the process of degeneration has begun, the anemone may lose its power to attach, and death usually follows soon after. This scenario is much more likely with the clownfish types (e.g., *Heteractis* spp.), rather than the hardier Caribbean species. Once an anemone starts to break up and disintegrate, it should be removed from the aquarium immediately to avoid massive pollution.

Among the easiest anemones to maintain in a reef tank are those belonging to the genus *Condylactis*. These anemones originate from the Caribbean region and are attractively colored, usually displaying pink tips to their tentacles. Yet, as with all anemones, you need to be careful of their stinging tentacles. Not only will they be painful if you brush against them, but anemones may also use them to "catch" other invertebrates and even small fish in the aquarium. Some types of anemone, such as the beautiful fireworks anemone (*Pachycerianthus mana*), a nocturnal species, have a particularly potent sting and are probably best avoided.

SPONGES

Sponges are grouped together in the phylum Porifera. These sessile members of the reef fauna are, unfortunately, very susceptible to drying out at any stage, which leaves them vulnerable during transportation. Individual species can be difficult to identify because their environmental conditions have a major impact on their individual shape.

Sponges are very primitive creatures with no organs. Their bodies are made up simply of masses of individual cells, and can grow in some cases to more than 6.5 feet (2.2 m) in height. They may also attain a similar width, but they are incapable of any movement. This means that nutrients must be wafted to them on the current, and they feed on tiny bacteria and other microbes present in the water. They reproduce sexually but also have remarkable powers of regeneration. If a piece breaks off, it may start to grow independently.

▲ *THE ATLANTIC ANEMONE (*CONDYLACTIS GIGANTEA*) IS ONE OF THE EASIEST ANEMONES TO KEEP IN A SALTWATER AQUARIUM.*

▶ *THE ORAL DISK OF RITTERI ANEMONE (*HETERACTIS MAGNIFICA*) IS DENSELY COVERED WITH FINGER-SHAPED TENTACLES.*

When buying anemones, try to pick the most brightly colored individuals, since this is a sign of good health as far as the symbiotic algae growing within their tissues are concerned. Anemones also need to be fed small pieces of shrimp or similar food once or twice a week to keep them in good health. Although they may appear sedentary, anemones can prove surprisingly mobile, as shown by various *Heteractis* species. The sand anemone (*H. aurora*) is one that prefers to be sited on the substrate rather than on rockwork. If disturbed, it will compact its body and withdraw itself down below the substrate.

▲ *RED TREE SPONGE* (HALICLONA COMPRESSA), *YELLOW TREE SPONGE* (AXINELLA VERICOSA), *AND RED BALL SPONGE* (CLIONA LAMPA).

Some sponges are exceedingly colorful and can become an attractive focal point in the reef tank. Examples include the red tree sponge (*Haliclona compressa*), although not all examples of this sponge appear fiery orange in color, and there are variations in shape as well. It pays to examine sponges carefully while they are still submerged, looking for any signs of damage resulting from dehydration in transit. Such damage is usually characterized by whitish patches forming on the sponge's projecting arms.

Positioning sponges in a reef tank needs careful consideration. When placing them in new surroundings, do not remove them from the water. Take them out only after submerging the bag in which they have been transported. Sponges should not be located in a brightly lit area—otherwise before long their bodies will become overgrown by algae, with fatal consequences.

Some sponges, such as the blue tubular sponges (*Adocia* species), can grow fast, thriving especially where there is no strong current. Cup-shaped sponges (axinellid species) are often available as well, so if you cannot incorporate a tall-growing variety, choose these instead. It is very important not to allow debris from the aquarium to accumulate in the cups, however, because it will choke the sponge.

SPONGES—TANK CONDITIONS AND CARE

HABITAT For tank and water conditions see Hard Corals, page 270.

Temperature Within the range 70–78°F (21–25.5°C). Higher temperatures will cause stress.

Lighting Subdued conditions preferred, as they dislike being smothered by the algae normally encouraged by intense lighting. If bright conditions are required by other invertebrates sharing the same tank, position the sponges in a shady position, perhaps in a cave or rock crevice.

FEEDING Direct feeding is not strictly necessary. Juices from frozen fish food will be enough to sustain most sponges. Extra feeding may cause pollution.

HEALTH If aquarium conditions are poor, sponges can deteriorate, leaving only spicules or a spongin skeleton.

SEGMENTED WORMS

Some invertebrates—such as the segmented worms in the phylum Annelida—can be introduced inadvertently to a reef tank on live rock. While some of these creatures are desirable, others may prove to be harmful to the other occupants.

Most worms favored by the aquarist are sedentary tubeworms belonging to two families: the family Sabellidae, commonly known as featherdusters or fanworms, and the family Serpulidae, often called Christmas tree worms. All these species live permanently within tubes and exhibit one or more feathery plumes of tentacles with which they feed and breathe. Frequently, the plume is most attractive and brightly colored in blue, red, yellow, white, black, orange, green, purple, mauve, beige, and brown.

These worms trap tiny particles of food in their feathery crowns. From there it is passed to a central rib and into a mucus stream that flows down to a central mouth. The crowns themselves are extremely sensitive to movement in the immediate vicinity and they often withdraw rapidly. Reactions are controlled by giant nerve fibers that run the whole length of the body within a central main nerve cord.

Featherdusters (fanworms) are usually found in the sand or mud of shallow intertidal zones. They mix tiny particles of mud with mucus to form a parchmentlike tube in which to live. Sizes vary from 1 inch (2.5 cm) in length to more than 4 inches (10 cm) depending on the species. Featherdusters are gregarious and they live in extremely large colonies, especially where the food supply is abundant and rich. As a general rule, colors are somewhat muted and are limited to beige, brown, black, dark red, mauve, and white. The species that are most commonly available to the marine aquarist are *Sabellastarte magnifica* and *S. sanctijosephi.*

▼ *FEATHERY CROWNS OF THE FEATHERDUSTER* SABELLASTARTE MAGNIFICA *EMERGE FROM HIDDEN TUBES.*

▲ *Crown of a Coco Worm (Protula species).*

◄ *Looking like a pair of Christmas trees, the crown of a Christmas Tree Worm (Spirobranchius species) has emerged.*

Christmas tree and coco worms differ from the sabellid worms in that they produce a hard calcareous tube. Species such as *Protula magnifica* are solitary animals. Their tube may reach 12 inches (30 cm) in length, with an opening of up to 1 inch (2.5 cm). The incumbent worm may display one or several colorful plumes and is much in demand by the invertebrate keeper. They are referred to as coco worms by exporters, wholesalers, and hobbyists, but the reason for their name is unclear.

The much smaller Christmas tree worms embed their tubes within living hard corals such as *Porites*, often establishing large communities. The spiraling double crowns are normally 0.4 inches (1 cm) in diameter and as many as 50 worms may occupy a single rock of 11 square feet (1 sq. m). Resembling tiny twin Christmas trees, even a small portion of rock can display radioles (crowns) in a stunning variety of colors. *Spirobranchius giganteus* is possibly the most commonly available species, but in the aquarium it rarely achieves its potential size of 6 inches (15 cm) across the radioles.

Some fish will attack the worms, and they can lose their projections (called cilia), especially after being moved or if there is a sudden deterioration in water quality. This does not necessarily mean that they are dead, and it may be possible to regenerate them by trimming back the head, which will encourage new cilia to develop. Another cause of cilia loss in established fanworms is as a prelude to breeding. They reproduce sexually, with dark jets of sperm and eggs being spurted out of the tube. After fertilization some of the offspring may establish themselves elsewhere in the aquarium, allowing the colony to expand. It is thought that the fanworms lose their cilia during this period so that they cannot consume their own offspring.

SEGMENTED WORMS—TANK CONDITIONS AND CARE

HABITAT For tank and water conditions see Mushroom Corals, page 273. Christmas Tree worms should be given optimum water conditions as for Hard Corals, page 272.

FEEDING Tubeworms benefit from occasional feedings with live rotifers and brine-shrimp nauplii. Squeezing the juices of a thawed mussel around the crowns can also be helpful.

HEALTH Occasionally the worm will leave its tube. It can survive for many weeks in this state, or it may die almost immediately. Owing to the lack of suitable materials, the worm cannot rebuild a new tube and is ultimately doomed. Individual Christmas tree worms may fail to appear one by one until the whole colony is no longer showing. The cause of this is usually deterioration in water quality and lack of nutrition.

CRUSTACEANS

Some of the most brightly colored and lively of all reef invertebrates belong to the phlyum known as Crustacea, which encompasses crabs, lobsters, prawns, and shrimps. They have a segmented body that is divided into a head, a thorax, and an abdomen. In some cases—in prawns, for example—the head and the thorax are fused.

Crustaceans are excellent scavengers. Their front legs are modified into pincers that can be used to grab pieces of food or used as weapons in combat. They tend to walk around the aquarium, often retreating into nooks and crannies, but larger species can often be lured from their favorite hiding places by morsels of thawed meaty foods.

CRABS

Although crustaceans are relatively well protected by their hard body casing, many fish will prey on them, so companions for these invertebrates need to be chosen carefully. Some crabs have evolved a way of protecting themselves that is similar to that of anemonefish: they retreat into the stinging tentacles of anemones if danger threatens. Anemone crabs (*Neopetrolisthes* species) are about 1 inch (2.5 cm) in diameter, which means they can be vulnerable to larger crustaceans. The anemone benefits from the association, because the crabs bring back food for them. Equally intriguing perhaps is the boxing crab (*Lybia* species), which is roughly the same size. This aptly named crab carries small anemones in its claws to defend itself, waving them at would-be predators. Crabs molt their hard body casing regularly as they grow, so during this period boxing crabs have to leave aside their anemones, but they pick them up again as soon as possible.

Many crabs arrive in the saltwater aquarium as passengers on living rock. They include the arrowhead crab (*Stenorhynchus seticornis*), which resembles an underwater spider. These crabs are best kept singly because they will fight with each other. They can be useful in the aquarium, however, because they eat bristleworms, keeping this pest under control.

Hermit crabs (*Dardanus* species) live in the empty shells of gastropods, so a supply of these is needed if the crabs are to grow. Hermit crabs come in a variery of colors and sizes, with leg color varying from red, yellow, and blue to gray. They are ideal scavengers in a fish-only tank but they can do considerable damage to invertebrates. The unusual anemone hermit crab (*D. pedunculatus*) places sea anemones on its shell for protection and takes its anemones with it when it moves shells. The decorator crab (*Camposcia retusa*) covers its shell and legs with algae and detritus to camouflage itself and avoid detection by predators.

SHRIMPS

Crabs are not the only crustaceans to use anemones in defense. There are also anemone shrimps (*Periclimenes brevicarpalis*) that will seek sanctuary among the

The Anemone Crab (Neopetrolisthes maculatus) lives in close association with sea anemones, which rely on the crab for food.

▲ *THE CLEANER SHRIMP (*LYSMATA AMBOINENSIS*) GROOMS FISH ON THE REEF AND MAY REPLICATE THIS BEHAVIOR IN THE AQUARIUM.*

tentacles of anemones. These shrimps may even share their refuge with anemonefish. Their semitransparent coloration helps them blend in, although they will be at risk from many predatory fish.

Other shrimps make a point of revealing their presence, notably the pistol shrimp (*Synalpheus* species). It has one greatly enlarged claw, which it uses to generate a shock wave, creating a sound like a pistol firing. This stuns smaller shrimps, which it preys upon in the wild. Unfortunately, however, these particular shrimps prove to be very shy in aquarium surroundings.

Much more conspicuous—partly because of their bright coloration—are the candy, or dancing, shrimps (*Rhynchocinetes* species). These shrimps will live together in groups, and the males are discernible by their larger claws. If you are offered a crustacean that has lost a claw and the injury has obviously healed, there is a strong possibility that it could regenerate once the crustacean molts.

The cleaner shrimp (*Lysmata amboinensis*) is also a species that can be kept in a group. At just over 3 inches (8 cm) in length, it is slightly larger than the candy shrimp. As its name suggests, it will clean and groom fish on the reef, removing parasites from their bodies. These shrimps are stunningly attractive in their red and white livery with their long white antennae. Equally attractive is the blood, or scarlet, shrimp (*L. debelius*), which is bright red to dusky crimson in color.

CRUSTACEANS—TANK CONDITIONS AND CARE

HABITAT For tank and water conditions see Mushroom Corals page 273.

FEEDING It is preferable to feed shrimps, crabs, and lobsters regularly on suitably sized meaty foods rather than let them scavenge. Shy feeders may be tempted out of hiding by offering a small piece of squid in a pair of aquarium tongs close to their regular retreat.

HEALTH Shrimps, crabs, and lobsters will normally remain healthy as long as water conditions are good.

LOBSTERS

Lobsters have a more squat appearance than shrimps and they can turn out to be more predatory in terms of their feeding habits, preying not just on fellow invertebrates but sometimes taking small fish as well. Among the smaller species is the red dwarf lobster (*Enoplometopus occidentalis*), which grows to nearly 5 inches (13 cm) long. Other attractive reef lobsters (*Enoplometopus* species) grow a little bigger. Another lobster sometimes found in aquaria is the purple spiny lobster (*Panulirus versicolor*), which can reach 8 inches (20 cm).

▼ *ALTHOUGH A COLORFUL MEMBER OF ANY REEF AQUARIUM, THE SCARLET SHRIMP (*LYSMATA DEBELIUS*) MAY PREFER TO HIDE DURING THE DAY.*

MOLLUSKS

The phylum Mollusca is another large and diverse group of invertebrates, embracing cephalopods such as cuttlefish (*Sepia* species) and octopus (*Octopus* species) as well as bivalves such as clams.

CUTTLEFISH

Cuttlefish are more active predators than their close cousins the octopuses. They are fast swimming, have excellent eyesight, and catch crabs, shrimps, and fish. They must therefore be kept in aquaria only with sessile invertebrates such as corals and anemones. One of the attractions of these animals is their amazing and rapid ability to change color in a split second. Some species produce kaleidoscopic shows of color change in waves down their bodies, particularly during courtship.

Coldwater species that are suitable for the saltwater aquarium are the diminutive Atlantic cuttlefish (*Sepia atlantica*), which reaches just 2 inches (5 cm), and the common cuttlefish (*S. officinalis*), which grows to about 12 inches (30 cm). Both these species can be housed in warmer aquaria if they are raised from eggs. Adults adjust poorly to being transferred to warmer conditions.

OCTOPUSES

It is possible to keep smaller octopus species successfully in aquarium surroundings, although they are not suitable companions for most fish or invertebrates, which form their natural prey. An octopus must be kept in a covered aquarium, because it will be able to use the suckers on its legs to climb, push off the tank cover, and escape, with disastrous consequences. Do not disturb these cephalopods unnecessarily either because, if stressed, an octopus is likely to release a cloud of "ink," which can prove toxic in the confines of an aquarium. The common tropical octopus (*Octopus cyanea*) is the most popularly available species in the aquarium trade. It measures about 12 inches (30 cm) across its full span and can become very tame. Other very similar species are available from time to time.

COWRIES

Univalves such as the tiger cowrie (*Cypraea tigris*) possibly represent more accurately most people's idea of a mollusk. This sea snail has a very attractive shell and it

▲ *The Tiger Cowrie (Cypraea tigris) is at its most active at night. The results of its activity can often be seen the next morning.*

CUTTLEFISH—TANK CONDITIONS AND CARE

HABITAT Cuttlefish occupy open water and require plenty of clear swimming space. It is not necessary to have a lot of rockwork.

Tank Size juveniles can be housed in tanks as small as 48 x 15x 18 in (122 x 38 x 46 cm) but mollusks will require an aquarium in excess of 72 x 18 x 24 in (183 x 46 x 61 cm) as fully grown adults.

pH	8.1–8.3
Temperature	70–75°F (21–24°C)
Ammonia	Zero
Nitrite	Zero
Nitrate	10 ppm total NO_3 or less
Specific Gravity	1.021–1.024
Dissolved Oxygen	6–8 ppm

Filtration Efficient protein skimming and activated carbon filtration as standard. Biological filtration must be efficient and capable of coping with the large amounts of waste produced.

Lighting Preferably subdued but will adapt to moderate or bright conditions.

Water Circulation Moderate.

Water Changes 15 to 25% every two weeks with high quality water.

FEEDING Cuttlefish will greedily accept live river shrimps, frozen prawns, cockles, mussels, and all meaty marine foods.

HEALTH Given good water conditions, cuttlefish suffer no particular health problems.

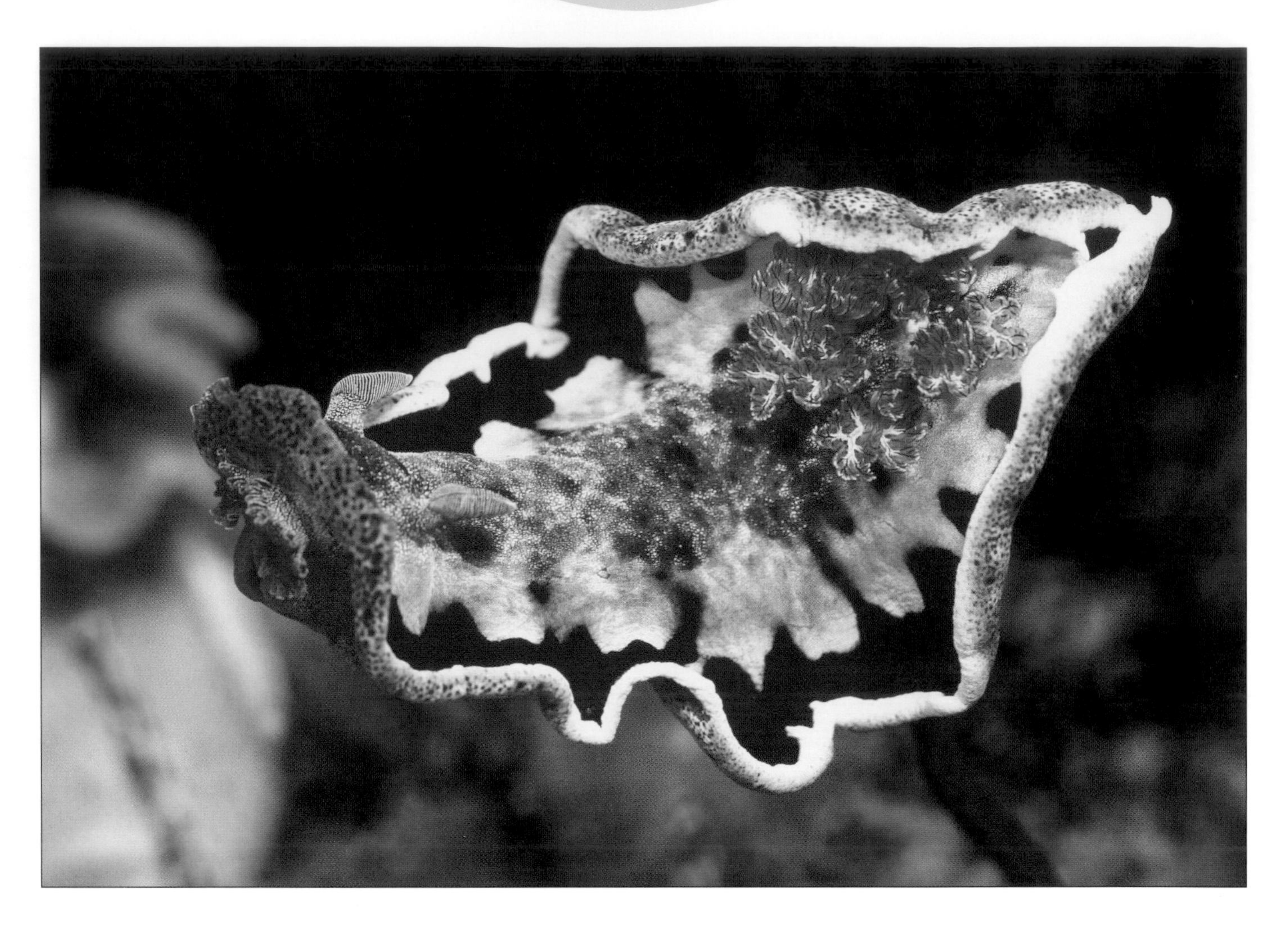

will wander over the rockwork of a reef tank. Because of their bulk, however, larger specimens are likely to dislodge or damage sessile invertebrates. Another drawback is that tiger cowries may browse on the more ornamental strains of *Caulerpa* algae you are trying to cultivate, although they will also feed on less decorative forms of algae. These cowries are at their most active after dark and may not be very conspicuous, tending instead to leave a trail of damage overnight.

SEA SLUGS—TANK CONDITIONS AND CARE

HABITAT For tank and water conditions see Mushroom Corals, page 273.

FEEDING Sea hares need a constant supply of green algae, both micro- and macro-. They consume vast quantities of nuisance algae but will not discriminate against decorative algae and can decimate carefully cultured growths. Most highly colored sea slugs starve to death in the aquarium.

HEALTH Given good water quality, sea slugs do not suffer from any particular diseases. However, insufficient food will cause shrinking and inactivity; death usually follows.

▲ *SPANISH DANCER (HEXABRANCHUS IMPERIALIS) SWIMMING AND SHOWING FEATHERY GILL TUFTS ON ITS BACK.*

SEA SLUGS

A variety of colorful sea slugs (also called nudibranchs) are also seen in saltwater aquaria. Some, such as the sea hare (*Aplysia* species) can swim well, freed from the constraint of having to carry a heavy shell. The Spanish dancer (*Hexabranchus imperialis)* is a graceful and agile swimmer. However, most creep over the seabed and have little or no ability to swim.

Many sea slugs are beautiful animals with brilliant coloration. In the wild such colors are intended as a warning to potential predators that they are poisonous and should be left alone. The feathery tufts that adorn their backs are external gills.

Aplysia species browse on algae and can do so at such a high rate that they are useful for keeping problem algae under control. Some sea slugs are much more omnivorous in their feeding habits, and some are even highly carnivorous. They can cause a problem because they will attack corals, sponges, anemones, and other invertebrates. Others can be very specific in their nutritional requirements and are much harder to maintain in aquarium surroundings as a result.

CLAMS AND OTHER BIVALVES

Clams are bivalves, a group that also includes oysters, mussels, and scallops. Bivalves are more sedentary members of the phylum Mollusca, and they have a shell that is in two parts. Several clam species are popular in the saltwater aquarium hobby, including members of the giant clam genus *Tridacna,* such as *T. crocea, T. maxima,* and *T. gigas*.

The inner part of the clam's body, known as the mantle, has zooxanthellae present. These symbiotic algae help meet the nutritional needs of the filter-feeding clams. Like corals, clams must therefore receive adequate illumination in order to remain in good health. Choose clams with well-colored mantles, since they are most likely to be healthy. Special food for filter feeders of this type (supplementing that supplied by their symbiotic algae) can be given using a pipette. Giant clams kept in aquarium surroundings will grow to about 6 to 12 inches (15–30 cm) overall.

▲ *BLUE CLAM (*TRIDACNA CROCEA*) WITH EXPOSED MANTLE THAT CONTAINS SYMBIOTIC ALGAE.*

Flame scallops (*Lima scabra*) are rather more mobile by nature than clams and, as their name suggests, their tentacles and interior are bright red in color. They can move some distance by closing the two halves of their shell together. If necessary, they can also shed their sticky tentacles, which are likely to stick to a fish's body and cause irritation. Provided they are not molested, these scallops, which originate from the Caribbean, are inoffensive and may even start to multiply in the aquarium. They too need feeding with a suitable filter food, and have healthy appetites.

◀ *FLAME SCALLOP (*LIMA SCABRA*), SHOWING ITS BRIGHT RED TENTACLES.*

CLAMS—TANK CONDITIONS AND CARE

HABITAT For tank and water conditions see Mushroom Corals, page 273.

Lighting Moderate to intense lighting is essential. This can be fluorescent, mercury vapor, or metal halide, but it is important that 10 to 12 hours of light falls directly onto the mantle every day.

FEEDING In the brightly lit aquarium, the symbiotic algae within the mantle provide all the food a clam requires. Liquid foods tend to pollute the tank because they are difficult to regulate.

HEALTH Clams suffer from very few diseases provided that the water quality is high and that water circulation is good. If the animal is mistreated, however, the muscle holding the two shells together sometimes detaches and the clam separates into a "gaping" position. Some specimens recover but others will deteriorate quickly and should be removed. Predatory bristleworms can be a problem because they can crawl into the siphons and devour the clam from the inside. Therefore, all tanks containing clams should be cleared of these worms at regular intervals.

ECHINODERMS

▲ *THE RED STARFISH (*FROMIA MONILIS*) IS A POPULAR STARFISH FOR SALTWATER AQUARIA.*

The phylum Echinodermata (spiny-skinned animals) is made up of starfish, brittlestars, sea urchins, and sea cucumbers. Echinoderms have no head, brain, or complex sense organs. Most members have defensive spines on the outside of the body.

STARFISH—TANK CONDITIONS AND CARE

HABITAT For tank and water conditions see Mushroom Corals, page 273.

FEEDING: Initially offer meaty foods such as cockles, mussels, lancefish, prawns, squid, or mysid shrimps to establish a preferred diet. Blanched spinach may also be offered. Feed every 1 to 3 days. If there are fish or crustaceans that would steal the food first, place a small piece on the floor of the aquarium and place the starfish directly over the top. Otherwise, position the food right next to the starfish and let it move over it.

HEALTH If water quality deteriorates and the correct diet is not provided, starfish often suffer bacterial infections leading to open wounds. Being sedentary, they may also attract the unwanted attentions of crustaceans and inquisitive fish, which may do some damage. Tankmates must therefore be chosen with care.

STARFISH AND RELATIVES

Starfish are among the most distinctive of all marine invertebrates, although individual species differ widely in coloration—from combinations of white and red through to blue. Their characteristic five arms enable them to move slowly over the reef. Under each arm are rows of tube feet with suckers at the tips, which the starfish uses to anchor itself to rocks and other surfaces as well as to pry apart prey such as bivalve mollusks.

The diameter of starfish can vary significantly, as does the width of their arms, with some being significantly larger than others—measuring up to 12 inches (30 cm) in the case of the large red-knobbed starfish (*Protoreaster lincki*). The loss of an arm may not be critical. As in crustaceans, it may regenerate over time. But avoid any starfish that have pale blotches on

SEA URCHINS—TANK CONDITIONS AND CARE

HABITAT For tank and water conditions see Mushroom Corals, page 273.

FEEDING Most species need large quantities of algae to graze on. This may be macro- or microalgae. They do not favor nuisance algae.

HEALTH The most common health problem is casting off spines. Little is known about why some urchins shed their spines all at once, or a few at a time. The shock of being moved can trigger it, as can poor water conditions such as low pH and low levels of calcium. Some "bald" urchins survive and regrow their spines very slowly if conditions improve, but most die.

their bodies, since this could be indicative of a parasitic infection to which these invertebrates are susceptible.

Starfish are predatory by nature and will feed on pieces of food such as prawns. Some may even attack crustaceans directly, as well as eating smaller members of their own kind. Other starfish frequently available to aquarists include the attractive blue starfish (*Linckia laevigata*), the bun starfish (*Culcita* species), red starfish and orange starfish (*Fromia* species), and the common knobbed starfish (*Pentaceraster mammillatus*).

▼ *THE ADULT BUN STARFISH (CULCITA NOVAEGUINEAE) LOOKS QUITE UNLIKE A TYPICAL STARFISH. YOUNG HAVE THE USUAL FIVE-POINTED STAR SHAPE BUT AS THEY MATURE THEIR ARMS THICKEN AND BROADEN.*

Brittlestars have longer, narrower arms and are far more agile than ordinary starfish. As the common name implies, brittlestars are very fragile and frequently lose arms, which are quickly regenerated.They prove to be valuable scavengers in a saltwater aquarium, often reaching particles of food in inaccessible areas before they can start to pollute the water. Brittlestars may also associate with sea urchins by entwining in their spines, which gives them extra protection.

Crinoids, or featherstars, are similar to brittlestars, and their bodies can also be easily damaged. They need careful acclimatization at first. If an arm is broken it may regenerate, as long as an infection does not develop. They eat fine particles of food and brine shrimps.

The arms of basketstars are subdivided many times to form a delicate tracery trap in which to capture minute food particles. When fully emerged, these arms may extend more than 20 inches (50 cm), so basketstars are not suitable for a small aquarium. Nor are they recommended for the novice aquarist.

SEA URCHINS

Sea urchins are typically covered in spines that vary from short and blunt to very long and sharp. Although it is possible to include sea urchins in a reef tank, there

▲ A SEA CUCUMBER KNOWN AS THE SEA APPLE (PSEUDOCOLOCHIRUS AXIOLOGUS). THE HEAD OF TENTACLES HAS BEEN WITHDRAWN.

is always a risk that you could catch a hand or an arm on their spines, which will inject a painful venom. Should this happen, medical advice is to run very hot water over the affected area. The long-spined sea urchin (*Diadema savignyi*), for example, has very sharp venomous spines. As sea urchins move over the reef, their spines can also harm other reef tank occupants, such as sea anemones.

Sea urchins graze on algae using five powerful jaws, so they will help control algae in the aquarium and may also browse on leftover meaty foods. Good species for the beginner are the pencil urchin (*Heterocentrotus mammillatus*) and the mine urchin (*Eucidaris tribuloides*) because both have very blunt spines and are incapable of stinging. They are also less sensitive to aquarium conditions than other species.

▼ HEAD OF A SEA CUCUMBER, SHOWING THE FEATHERY TENTACLES WITH WHICH THE ANIMAL TRAPS FOOD.

SEA CUCUMBERS

Bearing little resemblence to other echinoderms, sea cucumbers have a group of feathery tentacles at the end of their mostly elongated body. Like all echinoderms, however, they move by using rows of tube feet located under their bodies. Sea cucumbers will occasionally breed, especially in a well-established saltwater setup. They are frequently very colorful in appearance and grow to just a few inches long.

The sea apple (*Pseudocolochirus axiologus*) is a popular example. It uses its feathery tentacles to extract food from the water and must not be housed with fish that might nibble at these projections. Small livefoods, such as rotifers and brine shrimps, should form the basis of the diet of this and other sea cucumbers.

SEA CUCUMBERS—TANK CONDITIONS AND CARE

HABITAT For tank and water conditions see Mushroom Corals, page 273.

FEEDING When the tentacles are active, copious quantities of brine-shrimp nauplii and/or live rotifers should be introduced into the vicinity. Large specimens will need feeding at least twice a day. Switch off all circulatory pumps when feeding to keep food from being swept into the filters.

HEALTH Sea cucumbers will remain healthy if the correct food and water conditions are provided. Shrinking bodies and inactive tentacles are sure signs that these vital requirements have been neglected.

SEA SQUIRTS

Despite using the definition of invertebrates as animals without backbones, this strange group of creatures—making up the class Ascidiacea in the phylum Chordata—show what some taxonomists believe is the very beginning of the development of a spinal cord while they are free-swimming larvae. Adult sea squirts look rather like small bags with two tubelike siphons. Their bodies are described as tunics and have no bony support.

Sea squirts feed by drawing water in through one siphon, filtering it for food particles, and then expelling the water through another. The aquarist can make good use of this behavior to improve water quality. Tiny amounts of a filter food given by pipette will help keep them in good health.

They are often introduced accidentally to the reef on pieces of living rock. Adult sea squirts are sedentary by nature, but they make an interesting addition to the aquarium, since they occur in a wide range of colors and live in colonies. They do very well in the invertebrate aquarium and may spread quite freely.

SEA SQUIRTS—TANK CONDITIONS AND CARE

HABITAT For tank and water conditions see Mushroom Corals, page 273.

FEEDING Sea squirts are efficient filter feeders and can often survive without any special attention. However, live brine-shrimp nauplii, live rotifers, and the juices from a freshly thawed mollusk are all readily accepted and will assist with growth and the preservation of good health.

HEALTH Ascidians normally remain healthy even if water conditions are less than perfect.

▼ *A SEA SQUIRT (*RHOPALAEA CRASSA*), SHOWING CLEARLY ITS FLASK-SHAPED STRUCTURE.*

MARINE ALGAE

Algae are an integral part of the reef ecosystem, providing food for a variety of the inhabitants, and also helping maintain water quality. In the aquarium, the larger macroalgae are much more desirable than the microscopic microalgae. The growth of microalgae (including so-called slime algae, which are not true algae but are most closely related to bacteria) tends to coat the sides of the tank and the equipment, even spreading to sessile invertebrates such as sponges, potentially killing them. Under ideal conditions, algae of all types will grow rapidly, but it is much easier to control the growth of rampant macroalgae—the aim being to keep balanced growth.

GROWING CONDITIONS

Algae do not flower; in common with plants they do photosynthesize, however. The level of illumination in the aquarium is critical to their well-being, although different types of algae have different lighting requirements, reflecting in part the chlorophyll pigment present in their tissues. Green algae require high illumination, whereas brown or red algae will grow much more slowly in an area of the tank where the light level is lower. This is because they naturally tend to be found in deeper water, where there is less light.

Calcareous algae are distinctive, utilizing calcium from the aquarium water to support their growth. Unfortunately, in spite of often being very attractive, they are more difficult to cultivate than other macroalgae, but should grow well under the same conditions as hard coral. They will thrive only where both lighting and water conditions are optimal, extracting calcium from the water to sustain their growth.

INTRODUCING ALGAE

Microalgae are very easily introduced to an aquarium in droplets of water, while macroalgae are likely to be present on living rock, and can also be purchased separately. Members of the genus *Caulerpa* are among the

▼ *A SPECIES OF SAILOR'S EYEBALLS (*VALONIA AEGAGROPILA*), WHICH CAN BECOME A PEST IN THE AQUARIUM BUT IS EASILY REMOVED.*

▲ CACTUSLIKE GROWTHS OF THE CALCAREOUS MARINE ALGA HALIMEDA OPUNTIA.

easiest to cultivate, either in a fish-only aquarium, where their presence can soften the overall appearance, or in a reef tank. It is important not to subject macroalgae to drying out before introducing them to the tank, since their lack of an internal structure will make them shrivel rapidly. It is also very important to acclimatize them carefully to their new surroundings. Float the bag in which they were purchased on the surface of the aquarium for about 20 minutes. This allows the water temperature to equilibrate. Algae do not adapt well to sudden changes in specific gravity, either.

REPRODUCTION AND GROWTH

All algae produce spores, which can be carried through the water to new locations. Existing macroalgae will also increase in size by means of runners, which spread out away from the established plant over the substrate, anchoring by means of so-called holdfasts. They look like roots but they differ significantly in that they do not absorb any nutrients. In addition, pieces of macroalgae that break off from the main growth may take hold elsewhere in the aquarium, giving rise to another colony.

Microalgal growth will need to be curtailed on glass by using an algal scraper, and must be wiped off equipment. The growth of macroalgae may also need to be kept in check by what is sometimes described as "harvesting." This refers to thinning the plant—cutting out the old growth—to encourage more attractive fresh growth and to keep it in check in the case of some of the more rampant species.

CULTIVATING ALGAE AS FOOD

If you have the space available for a spare aquarium, you can plant algae on a number of the rocks in it and watch colonies develop. This should mean you will have access to a constant supply of fresh algal growth, which can be invaluable for acclimatizing and maintaining a range of vegetarian fish, such as tangs, in aquarium surroundings. In fact, it is better to have this food available before you acquire the fish, rather than trying to transfer them at once to artificial diets.

ALGAL DIE-BACK

Marine algae do not always respond well to medication added to the aquarium water, so avoid using copper-based medications. If they die back, they may release toxins into the water, threatening the health of other tank occupants. The most common reason for algal die-back is inadequate aquarium lighting, but it can also be caused by poor water conditions. It is obvious when it happens because the algal colonies become very pale (although calcareous algae tend to lose their color naturally at night). If algae die unexpectedly, try to identify the cause and cut them back to the holdfasts without delay, to minimize pollution in the water. In due course, assuming the plants are not entirely dead, they should show signs of regeneration from here.

POPULAR SPECIES OF ALGAE

Caulerpa prolifera—rampant and easy to grow.

Caulerpa brachypus—similar to *C. prolifera* but with smaller blades. It demands high water quality.

Caulerpa racemosa, C. sertularioides, C. taxifolia, C. mexicana, and C. cupressoides—fast growing and ideal for the beginner, but can overwhelm sessile invertebrates.

Acetabularia sp.—a delicate cuplike species demanding good lighting and high water quality.

Valonia ventricosa—commonly called sailor's eyeballs. Can become a nuisance by growing in among sessile invertebrates but is easily removed with a fingernail.

Rhodophyceae—a group of decorative red algae that are very slow growing.

Calcareous Algae

Codiacea spp.—delightful red algae that are slow growing

Halimeda spp.—attractive cactuslike algae that grow quickly in optimum conditions.

Penicillus capitatus—resembles a shaving brush but is very difficult to culture.

INVERTEBRATE PESTS

The marine aquarium keeper must be vigilant in order to prevent troublesome pests from ruining an otherwise impressive display. Some pests appear to be attractive additions to the aquarium and only later manifest themselves as harmful plagues or vicious hunters that are extremely difficult to eradicate.

The following examples are well known to the experienced hobbyist, but most aquarists will only become aware of their presence when the situation is already out of hand. Therefore, newcomers to the hobby are advised to familiarize themselves with the more common pests and avoid their introduction in the first place.

Bristleworms (Class Polychaeta)

There are several species of bristleworm that find their way into the home aquarium, and they can all multiply rapidly either sexually or asexually. They are segmented marine worms with fully functioning destructive mouthparts and rows of barbed stinging hairs flanking the body. The smaller species are only about 2 inches (5 cm) long and as wide as a small elastic band, but the larger species grow well in excess of 12 inches (30 cm) long and reach the width of an adult's little finger.

Bristleworms are scavengers and predators. They sift through detritus searching for algae and morsels of food. They also attack live prey such as clams, sessile invertebrates, tubeworms, and fish that "lock" themselves into crevices at night. Large specimens can bite humans, and even the smallest produce a painful rash. Beware of sliding a hand beneath rockwork in an effort to move it, as this is a likely way to get bitten or stung.

Feeding forays are nearly always at night, while the day is spent hiding among rocks or buried in the substrate. The best way to gauge how many occupy the hidden depths is to examine the tank with the beam from a dull flashlight after it has been in total darkness for several hours. If it is badly infested, dozens will be moving over rocks, sand, and corals. Almost every tank will possess one or two specimens, and this situation may be acceptable. Tanks that are badly infested with bristleworms, however, must be cleared to ensure the safety of other livestock as well as that of the hobbyist.

Bristleworms can be removed using a pair of plastic aquarium tongs during the hours of darkness. A dim light will help locate the smaller species, but you will probably need to set a trap to catch the larger, more destructive worms.

Commercial traps are available and work very well, but you can make a homemade trap using a PVC tube measuring 1 inch (2.5 cm) in diameter and 4 inches (10 cm) in length, capped at both ends. Each cap has a small drilled hole through which the worms squeeze to get to a bait of squid, mussel, or cockle. Unable to find the exit, the worms are trapped—to be disposed of at leisure. The trap is best baited and positioned between a rock and the substrate as the lights go out. It can be emptied in the morning and reset that night. Repeat every night until no more are captured. It would be wise to set the traps about every two months.

Some fish will eat bristleworms but can get badly stung in the process. There are reports of worms wrapping themselves around the face of the fish, blinding it permanently and causing a premature death.

Brown, or planaria, flatworms (*Convolutriloba retrogemma*)

Although this flatworm species is only a fraction of an inch long, it can reproduce both sexually and asexually, so it takes only one specimen to infect a whole tank. Given the right conditions, flatworms will spread over everything that does not move, including rocks, glass, equipment, algae, corals, and sand. Within months the sheer weight of numbers will smother sessile invertebrates to the point of destruction. If left unchecked, the end result could be a brown, lifeless mass.

Brown flatworms arrive on living rock and corals, generally from tanks that are already infected. They can be difficult to locate and have the ability to move out of sight into the safety of tiny cracks in the rocks relatively quickly. Not only do they feed on microorganisms, but they also have a symbiotic alga within their tissues providing sustenance, so they thrive in brightly lit aquaria, often occupying the most brilliant locations.

Fish-only aquaria can be treated with a copper-based medication, but alternatives must be found for the reef aquarium. If small colonies are spotted at an early stage, then siphoning the area several times each day should clear the problem quickly. Some species of fish have been known to eat flatworms—they include mandarins, wrasses, damsels, butterflyfish, and tangs.

Consistently reliable and specific chemical controls are unknown for the invertebrate aquarium. Once the tank has become badly infected, the unlucky aquarist can only limit the damage by frequent siphoning.

Smaller aquaria may eventually recover, but larger tanks may house controlled populations for the rest of their existence.

Triffid, glass, or rock anemones (*Aiptasia* spp.)
Triffid anemones are the proverbial wolves in sheep's clothing. They are beautiful creatures, so impressive and attractive to look at that they are often given priority treatment, only to multiply into a dangerous and unsightly plague. They have a prodigious sting capable of killing (as a prelude to eating) small fish. Even larger fish can sustain a nasty injury.

Reef tanks suffer particularly badly, since as the anemones spread among the polyps, corals, clams, and other anemones, they tend to sting everything they touch. As a result, if their neighbors are not killed, they will show the "burn" marks from having brushed against these fearsome predators.

One anemone can produce copious offspring in a few weeks, owing to an extremely rapid reproductive ability, especially if the tank is consistently overfed. Several months after that, a whole aquarium may be covered in a mass of these anemones if nothing is done.

Triffid anemones have an elongated body stem lodged far down in a rock crevice or among a polyp colony, making it virtually impossible to remove. At the slightest disturbance it will disappear into the safety of its inaccessible home. To make matters worse, if the whole animal is not destroyed, it can regenerate itself from a small piece of remaining tissue. They can live in total darkness quite successfully, so check all water routes, especially pipes and other areas of limited access, when trying to eradicate them from the system.

Fish-only aquarists are relatively fortunate and have a range of options. The easiest is to remove any pieces of rock where colonies are established and scrub them under hot running water. Alternatives are to introduce fish species that regard these anemones as food (larger angelfish and most butterflyfish, for example) or to kIll the anemones with copper-based medication.

The reef hobbyist is not so lucky. Fish and invertebrates that consume anemones tend not to discriminate where other livestock is concerned, and may feed on the very corals the aquarist is trying to save. The most effective option may be to inject the offending anemone with a lethal dose of calcium additive. This has the added advantage of being safe should any of the solution escape into the surrounding water. In fact, almost any additive could be used if the first proves ineffective. Another option is to fill a syringe with very hot water and squirt it at the base of the anemone. One drawback is that it could prove harmful to invertebrates in the immediate vicinity, and so it can only be used on isolated individuals. An effective (although seemingly brutal) strategy is to push a sharpened or red-hot screwdriver into the anemone's hiding hole and grind it out.

Amphipods (Order Amphipoda)
Amphipods are small crustaceans generally found feeding in the detritus at the bottom of an aquarium, within the filters, or even on the front glass. They are easily identified as laterally compressed, gray, shrimplike creatures, half-moon in shape.

Amphipods rarely do any direct damage but some species may carry disease that they can pass to other livestock. They are quite hardy and capable of multiplying into plague proportions when there is enough food, usually provided by an overenthusiastic hobbyist.

An immediate cure is to remove the creatures by siphoning them out of the tank. They are found under rocks during the daytime, and quick action will be needed because they move surprisingly fast. In the long term, a drastic reduction in the food supply will see a steady decline in numbers until they are barely noticeable. Many fish would enjoy amphipods as a food if they had access to their hiding places. Regular disturbance of the rockwork may reduce numbers as the fish learn to associate such activity with a potential meal.

Mantis shrimps (*Odontodactylus* spp.)
Mantis shrimps are highly developed predators. They can emerge from a cave at lightning speed to beat to death a likely meal with their clublike claws. Fish, other shrimps, crabs, lobsters, and aquarists' fingers are all potential targets. Their alternative common name of "thumb-splitter" is particularly apt. Mantis shrimps normally arrive in pieces of living rock. They travel very well and rarely die in transportation. Depending on the particular species, size may range from 2 to 12 inches (5–30 cm) in length.

Ridding a tank of mantis shrimps is not an easy matter. They are intelligent and can avoid traps set for them. If a shrimp settles in a favorite cave, you will have to remove the whole rock—taking care that any back entrance is covered and that the creature does not slip out as the rock is lifted from the water. If no other solution can be found then, as a last resort, you may have to push scissors, skewers, or a sharp stick into the cave to kill the shrimp by impaling or decapitating it.

Other potential pests include giant elephant ears (*Rhodactis* spp.), a mushroom coral (see page 273), and pistol shrimps (*Synalpheus* spp., see page 280).

GLOSSARY

Words in SMALL CAPITALS can be looked up elsewhere in this Glossary.

absorption The process of taking in and retaining a substance—the way a dry sponge takes in water. Liquid vitamins added to marine flake act in this manner.

acrylic A lightweight form of plastic used in the manufacture of molded aquaria.

activated carbon Material used to remove chemical pollutants from aquarium water by drawing it into its structure. Can inactivate medications, so it is not recommended for treatment tanks.

adsorption The process by which organic molecules are bonded to a medium, such as ACTIVATED CARBON.

aerobic Requiring oxygen.

airstone The piece of equipment that breaks up the air drawn into the aquarium via the air pump, releasing it as a constant stream of small bubbles. Can become blocked by debris, affecting its efficiency, and will therefore need to be cleaned or even replaced regularly.

algae Primitive organisms, which may be microscopic or large (e.g., kelp). They have plant characteristics, are almost exclusively aquatic, but do not flower.

alkalinity Water conditions that exceed a reading of 7.0 on the PH scale. (*See also* page 44.)

ammonia (NH_3) A gas that is the first by-product of decaying organic material; common aquarium sources include fish's waste and uneaten food. Highly toxic to fish and invertebrates. (*See also* page 44.)

anaerobic Not requiring oxygen.

anal fin A single fin positioned vertically below the fish, near the VENT.

asexual reproduction Reproduction without the fertilization of eggs with sperm, as in corals that "bud off," with new pieces forming a new colony.

barbel A whiskerlike growth around the mouth, which helps detect food. Often seen in fish living in dark surroundings, in which it acts as a feeler.

Berlin system A filtration technique using living rock and a powerful PROTEIN SKIMMER only.

biological filtration A means of filtration using bacteria, *NITROSOMONAS* and *NITROBACTER*, to change otherwise toxic AMMONIA into a safer substance such as NITRATE.

bivalve A mollusk or shell-dwelling animal with two respiratory VALVES.

black worm A type of LIVEFOOD, known scientifically as *Lumbriculus variegatus*.

bleaching The process by which corals or anemones lose or expel their ZOOXANTHELLAE to shock or pollution. They turn a pale color or white.

brackish water Water that has about 10 percent seawater; found where freshwater rivers enter the sea.

brine shrimp A saltwater crustacean, *Artemia salina*, whose dry-stored eggs can be hatched to provide LIVEFOOD for fish or invertebrates.

buffer A chemical that helps maintain the water at a relatively constant PH.

buffering action The ability of a liquid to maintain its desired PH value. (*See also* CALCAREOUS.)

byssus gland A gland found in BIVALVE mollusks. It produces sticky attachment threads, helping the animal stay in place and not be swept around by the current.

calcareous Formed of, or containing, calcium carbonate, a substance that can help aquarium water maintain a high PH.

calcium An important element found in seawater, the metallic basis of lime. (*See also* page 45.)

caudal fin The fin at the rear of the fish's body, also called the tail or tail fin.

caudal peduncle The part of a fish joining the CAUDAL FIN to the body.

cephalopod A group of predatory marine mollusks without shells, which often attain a large size. Includes octopuses, squid, cuttlefish, and nautiluses.

cirrus The bristlelike jointed appendage seen in some fish, such as hawkfish, present on the DORSAL FIN in this case. The plural is cirri.

communal tank An aquarium housing either a number of different, unrelated species or fish, or fish and invertebrates.

copepod A crustacean belonging to the subclass Copepoda. Can be an important source of food for fish and even other invertebrates.

copper A metal used in copper sulfate form as the basis for marine aquarium remedies. It is poisonous to fish in excess, and to invertebrates at trace levels. (*See also* page 45.)

coral sand Coral exoskeletons that have been crushed to the consistency of sand.

corallivore A fish that feeds exclusively on live corals. These fish, including various butterflyfish, are difficult to maintain in the aquarium as a result.

counter-current skimmer An efficient PROTEIN SKIMMER in which the water flows against a current of air, thereby giving a longer exposure time for collection of waste, or sterilization if OZONE is used.

crepuscular A creature that is most active naturally at dawn or dusk.

cyanobacteria A primitive life form, having some characteristics of ALGAE and bacteria but regarded as different from both. Often called slime algae or blue-green algae.

deionizer A mains water filter that uses several purifying ion-exchange resins. It removes impurities from ordinary household water.

demersal Meaning "close to the sea floor." Demersal eggs are heavier than water and are laid in prepared spawning sites on the seabed. The fertilized eggs are then guarded by one or both adults until hatching occurs.

denitrification The process in which NITRATE is changed by ANAEROBIC bacteria into nitrous oxide and then into free nitrogen gas.

detritus Unwanted evident waste matter that may build up on the SUBSTRATE.

diffuser Another name for an AIRSTONE.

diurnal A creature that is naturally active during the hours of daylight.

dorsal fin A single vertical fin on a fish's back; some fish have two dorsal fins, one behind the other. Many marine species have venomous rays in the dorsal fin.

feeding stick A device that enables pieces of food to be held in the aquarium for individual fish, encouraging them to take pieces. Useful for potentially dangerous species.

filter feeder An animal (fish or invertebrate) that sifts water for microscopic food, e.g., pipefish, tubeworms.

filter medium Used in filtration systems to remove dissolved or suspended organic substances from water.

fluidized bed A biological filtration method. Unclean water is pumped through a cylinder containing millions of tiny "beads" on which nitrifying bacteria have become established.

foam fractionation A method of separating proteinous substances from water by a foaming action. Also called protein skimming.

fry Very young fish (*See also* LARVAE).

gill flukes Trematode parasites, such as *Dactylogyrus*.

gills Membranes through which fish absorb dissolved oxygen from the water during respiration.

grass shrimps Small crustaceans used as LIVEFOOD for some saltwater fish.

gravel tidy Plastic mesh fitted between layers of gravel to protect biological filtration systems from being exposed (and thus rendered ineffective) by burrowing fish.

greenstuff Plant matter provided as food. May refer to vegetables such as peas or lettuce.

heaterstat A combined unit for regulating water temperature in the aquarium, consisting of a heater and thermostat, sealed within a glass tube.

hydrometer A device for measuring the SPECIFIC GRAVITY (SG) of water, especially vital when making up synthetic saltwater mixes. May be either a free-floating or swing-needle type.

hydrophilic Attracted to water.

hydrophobic Water-hating.

ick/ich A relatively common parasitic problem seen in marine fish, also known as "white spot" because of the tiny white spots that develop over the body and fins.

impeller An electrically driven propeller that produces water flow through filters.

invertebrate A creature without a vertebral column, which forms the backbone.

invertebrate-only setup An aquarium that houses invertebrates, but no fish.

juvenile A young individual that often may display markedly different coloration than an adult fish of the same species.

KH A means of measuring the hardness of water, in degrees on the KH scale.

krill Marine crustaceans with a shrimplike appearance.

larvae (1) The first stage of fish development after hatching; underdeveloped fish fry. (2) The first reproductive stage of many invertebrates.

lateral line A line of perforated scales along the flanks of fish, connected to a pressure-sensitive nervous system, used to detect vibrations in the water.

livefoods Invertebrates used specificially as fish food, although they may not be of marine origin, and might have been preserved, e.g., frozen.

macroalgae Seaweedlike ALGAE, which are clearly visible to the naked eye.

marine fungus An organism that causes cottonwool-like growths on the body. May be linked with a bacterial infection or a traumatic injury.

meaty foods General term for food that is of animal rather than plant origins. It rarely describes meat as such in the case of marine fish diets.

Melanesia A group of islands that are in the Oceania group, lying to the south of MICRONESIA. Includes New Guinea, the Solomon Islands, Vanuatu, and Fiji, as well as New Caledonia.

microalgae Microscopic ALGAE that lack the obvious form of MACROALGAE to the naked eye.

Micronesia The northwestern group of more than 2,000 islands which form part of the Oceania group lying in the Pacific Ocean. Includes the Caroline, Mariana, and Marshall Islands.

mouthbrooder A fish that incubates fertilized eggs in its mouth.

mulm A very fine particulate muddy deposit.

mysid shrimp Commercially available marine shrimp that is used as live and frozen food.

nauplii Newly hatched BRINE SHRIMP.

nighttime tubes Used to illuminate aquaria, enabling the interior and its occupants to be seen. Creates conditions that replicate moonlight rather than sunlight, for viewing nocturnal activity.

nitrate (NO_3) A compound derived from (and less toxic than) NITRITE by *NITROBACTER*. (*See also* page 44.)

nitrate-nitrogen (NO_3-N) A measurement of NITRATE levels in the water.

nitrification The process by which toxic nitrogenous compounds are converted by AEROBIC bacteria into less harmful substances, e.g., AMMONIA to NITRITE to NITRATE.

nitrite (NO_2) A toxic compound derived from AMMONIA by *NITROSOMONAS*. (*See also* page 44.)

Nitrobacter AEROBIC bacteria used in the biological filter to convert NITRITE into less harmful NITRATE.

Nitrosomonas AEROBIC bacteria used in the biological filter to convert AMMONIA into NITRITE.

nuisance algae Hair (filamentous) or slime ALGAE that can overrun a tank. (*See also* CYANOBACTERIA.)

Oceania An area comprising some 25,000 islands through the Pacific Ocean, from New Guinea to Easter Island in the east and Hawaii to New Zealand in the south.

ocellus A spot that resembles an eye in appearance on the fish's body, frequently near the rear of the DORSAL FIN. It often serves to confuse predators about which end of the fish carries the head.

opercular A protective spine that lies on the OPERCULUM. This makes it difficult for a predator to swallow a fish protected in this way, because the spines may stick in the throat, causing the fish to spit out its prey.

operculum The covering over the gills on the sides of the fish's head behind the eyes. It moves as water flows out of the body over the gills.

ORP *See* REDOX POTENTIAL.

osmolator Equipment for replacing evaporated water so as to maintain the desired SPECIFIC GRAVITY.

osmosis The passage of a liquid through a semipermeable membrane to dilute a more concentrated solution.

osmotic stress An adverse reaction caused in livestock when the SALINITY of its environment changes significantly. Also called osmotic shock.

Oxygen Reduction Potential (ORP, or REDOX POTENTIAL). A measurement of the water's ability to cleanse itself.

ozone (O_3) A three-atom, unstable form of oxygen used as a disinfectant.

ozonizer A device that uses high-voltage electricity to produce OZONE.

pectoral fins Paired fins, one on each side of the body, immediately behind the gill cover in most cases.

pelagic Meaning "of the open sea." Pelagic eggs are lighter than water and are scattered after an ascending spawning action between a pair of fish in open water, compared with DEMERSAL eggs.

pelvic fins Paired fins, one on each side of the body, immediately below the gill cover. (Not all marine fish have them.)

pH A measure of acidity or ALKALINITY; the scale ranges from 1 (very acid) through 7 (neutral) to 14 (very alkaline). Aquarium water should be kept in the range pH 7.9–8.3. (*See also* page 44.)

phosphates (PO_4) Compounds that are waste products generated by livestock, also present in unfiltered mains water. (*See also* page 44.)

photoperiod The length of time that aquarium lights remain on.

phytoplankton Very small plants (e.g., unicellular ALGAE) that drift in water.

plankton The encompassing term for PHYTOPLANKTON and ZOOPLANKTON.

polygamous Does not maintain a pair bond, but mates randomly.

power filters Also called canister filters, they suck water into the filter media with the unit.

power head An electric IMPELLER system fitted to biological filter return tubes to create water flow.

ppm An abbrievation for "parts per million," a measurement of the concentration of a substance.

protein skimmer A device that removes proteinous substances from aquarium water. Also used in conjunction with ozonized air for sterilizing water. (*See also* OZONIZER.)

rays Bony supports in the fins of fish.

reactor An isolated container, usually in or near a SUMP, used for performing a particular task, e.g., increasing calcium or oxygen in the water, before it is reintroduced to the main system.

redox potential The oxidation and reduction process, when electrons are lost from one atom or molecule and gained by another. (*See also* OXYGEN-REDUCTION POTENTIAL.)

reef aquarium An aquarium that incorporates INVERTEBRATES and is similar to a reef in appearance, compared with a fish-only setup.

refugium An area that has remained unaffected by changes in climate over a period of time that have affected neighboring areas, and thus has a more ancient flora and fauna.

reverse-flow filtration A biological filtration system in which water flows up into the tank through the base covering instead of having the more usual downward flow.

reverse osmosis A technique whereby water is forced through a membrane in the reverse way to OSMOSIS, so that the mineral salts are left behind, resulting in pure water.

rotifers PLANKTONIC organisms, used as a first food for the fry of many marine fish.

salinity The measure of saltiness of aquarium water. The hobbyist can measure salinity and SPECIFIC GRAVITY as one and the same. (*See also* page 44.)

seagrass An important retreat for the young of many marine fish, close inshore. Also known as eel grass (*Vallisneria* species).

sexual reproduction Reproduction in which the eggs of the female are fertilized by the sperm of the male.

silicone sealant A slightly flexible adhesive used to bond glass or stop leaks in aquaria, or to create rock and coral formations in the aquarium itself.

siphon (1) A length of tube used to move water from one vessel to another. (2) The inhalant and exhalant organs of a mollusk.

spawning The part of the reproductive process involving fertilization of eggs.

species-only setup An aquarium that is restricted to displaying just one species.

specific gravity (SG) The ratio of the density of a measured liquid to the density of pure water. Natural seawater has an SG of around 1.025, but marine aquarium fish are normally kept at (1.020–1.023). (*See also* SALINITY.)

Spirulina A type of marine ALGA highly valued as fish food, available in various formulated diets created especially for herbivorous fish.

stripping The process by which the crust of a gorgonian (the part housing the polyps) falls away.

substrate Aquarium base covering.

sump An undertank reservoir that usually holds a TRICKLE FILTER, mechanical and chemical filters, as well as probes and REACTORS.

sushi noir A type of food used for some aquarium fish, which is more commonly sold through human food outlets.

sweeper tentacles Long tentacles used by aggressive hard corals to sting other nearby corals in order to establish a territory.

swimbladder The hydrostatic organ enabling fish to maintain their chosen depth and position in water.

total system An aquarium with built-in sophisticated filtration and other management systems, providing full water treatment.

trace elements Elements found in seawater in very small quantities, often much less than 1ppm. (*See also* PPM.)

trickle filter A biological filter, filled with inert media .

tunicate Sea squirts and related organisms belonging to the subphylum Tunicata.

turnover The water flow rate through a filter. For marine aquaria a high turnover is recommended.

ultraviolet sterilizer An ultraviolet light tube, enclosed in a water jacket, through which aquarium water is passed to sterilize it.

undergravel filter The SUBSTRATE of an aquarium used as a biological filter.

valve In BIVALVES, one half of a two-valved shell.

vent In fish, the uro-genital opening found between the anus and the ANAL FIN.

ventral Situated at, or related to, the lower bottom side or surface.

ventral fins *See* PELVIC FINS.

water change The replacement of a proportion (usually 20 to 25 percent) of aquarium water with fresh seawater mix.

zooxanthellae Symbiotic ALGAE found within the tissue of many corals, anemones, and clams.

zooplankton Extremely small animals (and their larval stages) that drift in the water. (*See also* LARVAE.)

Further Reading

Alderton, D. *Encyclopedia of Aquarium and Pond Fish* (Dorling Kindersley, London and New York, 2005).

Blasiola, G. C. *The New Saltwater Aquarium Handbook* (Barron's Educational Series Inc., New York, 1991).

Borneman, E. H., and S. W. Michael. *Aquarium Corals: Selection, Husbandry, and Natural History* (TFH Publications & Microcosm Ltd., Neptune City, NJ, and Charlotte, NC, 2001).

Burgess, W E., Axelrod, H. R., and R. E. Huniziker. *Dr. Burgess's Atlas of Marine Aquarium Fish* (TFH Publications Inc., Neptune City, NJ, 1988).

Campbell, A., and J. Dawes (eds.). *The New Encyclopedia of Aquatic Life* (Facts On File Inc., New York, 2004).

Dakin, N. *Complete Encyclopedia of the Saltwater Aquarium* (Firefly Books Ltd., Buffalo, NY, 2003).

Dakin, N. *The Marine Aquarium Problem Solver* (Tetra Press, Blacksburg, VA, 1996).

Debelius, H., and H. A. Baensch. *Marine Atlas 1–2* (Mergus, Melle, Germany, 1994, 2005).

Erhardt, H., and H. Moosleitner. *Marine Atlas 3: Invertebrates* (Mergus, Melle, Germany, 1998).

Fenner, R. M. *The Conscientious Marine Aquarist* (TFH Publications & Microcosm Ltd., Neptune City, NJ, and Charlotte, NC, 1998, 1999, 2001).

Garratt, D., and T. Hayes, T. Lougher, D. Mills. *500 Ways to Be a Better Saltwater Fishkeeper* (Firefly Books Ltd., Buffalo, NY, 2005).

Halls, S. *Understanding Marine Fish* (Interpet Publishing Ltd., Dorking, U.K., 2001).

Hargreaves, V. B. *The Complete Book of the Marine Aquarium* (Thunder Bay Press, San Diego, CA, 2002).

Holliday, L., and G. Rogers. *Coral Reefs* (Tetra Press, Blacksburg, VA, 1990).

Maître-Allain, T., and C. Piednoir. *Aquariums* (Firefly Books Ltd., Buffalo, NY, 2005).

Michael, S. W. *Angelfishes and Butterflyfishes* (TFH Publications & Microcosm Ltd., Neptune City, NJ, and Charlotte, NC, 2004).

Michael, S. W. *Basslets, Dottybacks, and Hawkfishes* (TFH Publications & Microcosm Ltd., Neptune City, NJ, and Charlotte, NC, 2004).

Michael, S. W. *Marine Fishes* (TFH Publications & Microcosm Ltd., Neptune City, NJ, and Charlotte, NC, 2001).

Michael, S. W. *Reef Fishes* (Vol. 1) (TFH Publications & Microcosm Ltd., Neptune City, NJ, and Charlotte, NC, 2001).

Mills, D. *A Practical Guide to Setting Up Your Marine Aquarium* (Barron's Educational Series Inc., New York, 2001).

Mills, D. *The Practical Encyclopedia of the Marine Aquarium* (Salamander Books Ltd., London, 1987).

Shimek, R. L. *Marine Invertebrates* (TFH Publications & Microcosm Ltd., Neptune City, NJ, and Charlotte, NC, 2004).

Tullock, J. H. *Natural Reef Aquariums: Simplified Approaches to Creating Living Saltwater Microcosms* (TFH Publications & Microcosm Ltd., Neptune City, NJ, and Charlotte, NC, 1997, 1999, 2001).

Tullock, J. H. *Water Chemistry for the Marine Aquarium* (Barron's Educational Series Inc., New York, 2002).

Useful Websites

General sites
www.aquariumhobbyist.com
www.fishlinkcentral.com
www.aquarticles.com

Lists of clubs and organizations
www.caoac.on.ca (Canadian Association of Aquarium Clubs)
www.faas.info (Federation of American Aquarium Societies)
www.masna.org (Marine Aquarium Societies of North America)

Useful sites for those interested in reef tanks
www.reefindex.com
www.reefs.org

Magazines
www.aquariumfish.com
www.famamagazine.com
www.practicalfishkeeping.co.uk

Classification, queries about common names, etc.
www.fishbase.net
www.itis.usda.gov

Public aquaria listings
www.touchthesea.org/aquariums.htm

INDEX

Numbers in *italics*, e.g., *158*, refer to illustrations.

A

B

C

H

I

J

K

L

M

N

O

P

R

S

T

U

V

W

X

Z